THE GENERA OF THE MESEMBRYANTHEMACEAE

THE GENERA OF THE MESEMBRYANTHEMACEAE

Including a full set of botanical drawings by the artists of the
Bolus Herbarium of the University of Cape Town, and others, and also
distribution maps for each genus, identification keys, a scientific
system, a contribution concerning the poisonous genus, and also a history
of the introduction, with notes on and portraits of the various scientific
workers from the beginning to the present day by

H. HERRE, DIPL.G.I., MEMBER I.O.S.
Ex-Curator of the Botanical Garden of the University of Stellenbosch,
Cape Province, South Africa

Illustrations by Harry Bolus, Beatrice Carter, Mary Page and Maisie Walgate
Distribution maps by H. Herre and Dr H. Friedrich

A. A. BALKEMA / ROTTERDAM / 1973

ISBN 90 6191 002 1

First Edition 1971
Tafelberg-Uitgewers Beperk
Cape Town and Johannesburg, and
Nasionale Boekhandel (Publishers) Ltd, London

Additional botanical drawings: Robert Darroll, Pieter de Weerdt,
Rudolf Luithardt and Mary Maytham-Kidd

Printed in South Africa

In remembrance of our work together over a period of forty years,
and in honour of an indefatigable spirit, it is my pleasure to dedicate this
book to the late

DR H.M.LOUISA BOLUS

More than sixty years ago this Nestor of the scientific workers in the field
of the Mesembryanthemaceae began her investigations, and this
group of plants held her attention until she was ninety years old. Without
her work this book could not have been written.

Acknowledgements

It is with sincere gratitude that I record the names of the following persons and institutes; without their valuable help this book would not have been possible.

Prof. P. G. Jordaan, of the University of Stellenbosch, who at all times, and gladly, gave unstinting help and advice;

Prof. A. E. Schelpe, Curator of the Bolus Herbarium, University of Cape Town, who very kindly made available the priceless drawings used in this book, and who helped in other ways; also the staff of the Herbarium who rendered valuable assistance in connection with the drawings and herbarium sheets in instances where no drawings were available;

The Council for Scientific and Industrial Research, whose grant enabled me to complete the research necessary for this book;

The Director, Botanical Research Institute, Pretoria, who kindly permitted me to make use of the text of *The Genera of South African Flowering Plants*, by E. P. Phillips, in the genera descriptions;

Dr H. Friedrich, of the University Botanic Gardens at Nymphenburg, Munich, who re-drew my distribution maps for each genus;

Prof. Dr H.-D. Ihlenfeldt, of the University of Hamburg, who kindly allowed me to use his lecture entitled "Some Aspects of the Biology of Dissemination of the Mesembryanthemaceae", given at Stellenbosch;

Prof. Dr H. Straka, of the University of Kiel, and Prof. Dr H.-D. Ihlenfeldt who kindly allowed me to use their article "On the Delimitation and the Taxonomic Rank of the Mesembryanthemaceae";

Dr H. Jacobsen, of Kiel, whose advice and photographs of some of the scientific workers were most valuable;

Dr S. Dupont, of Nantes, who kindly allowed me to use the article "On the Interest of the Seedlings and the Epidermis of the Mesembryanthemaceae";

Dr D. A. Roller, of Vienna, who helped to obtain pictures and details of several of the scientists;

Mrs M. O'Connor-Fenton, of the Bolus Herbarium, University of Cape Town, for her valuable assistance in obtaining plant material and herbarium sheets, in helping with the proofreading of this manuscript and the compiling of the Glossary of botanical terms used in this book;

Miss Alice Mertens, Head of the Photographic Division of the Department of Fine Arts of the University of Stellenbosch, whose advice and help greatly facilitated this work;

Mr W. Wisura, of the National Botanical Gardens of South Africa (Kirstenbosch), who helped to read the proofs of this book and who supplied some plant specimens;

Mr F. J. Stayner, Curator of the Karroo Botanic Gardens, Worcester, Cape Province, who helped to obtain some plant specimens;

Miss H. P. Gunn, Head of the Library of the Botanical Research Station in Pretoria, who unstintingly gave advice on and assistance with the historical aspects of this book;

My late wife, Ada Herre, deserves special recognition for the many years of affectionate support which was her contribution in the realization of this book.

Contents

Preface

Lovers and students of the Mesembryanthema or "vygies" will be delighted that Mr Herre has produced this book. It has been compiled in such a way that it will be of interest to the amateur and of great value to the student and prospective researcher.

There was a time when everyone interested in the South African succulents was acquainted with the name of Mr Herre. Today there may be many who do not know him. To them, in particular, I should like to address a few words of introduction.

Hans Herre was born on 7 April 1895 in Dessau, Germany. In September 1914 he joined the army and in July 1916, he was seriously wounded in the left leg in the battle of the Somme. He was a brave soldier and officer and was awarded the Iron Cross (first class).

After his discharge from hospital in March 1919, he continued with his studies of horticulture, this time in Dahlem, Berlin, where he came to know Prof. A. Engler and Prof. L. Diels, and where he received the diploma of Horticultural Inspector. During 1924 and 1925 the University of Stellenbosch was looking for a horticulturist to manage and develop its Botanical Garden, which had been created on a new site in the centre of the town. On the recommendation of Prof. L. Diels, Mr Herre was appointed, and he began his task on 1 August 1925. He stayed on as curator of the Botanical Garden until 31 December 1960, when he retired.

In developing the Botanical Garden, Mr Herre had the whole-hearted co-operation of Prof. G. C. Nel, the professor of Botany and the Head of the Garden, and the advice of the distinguished Prof. R. Marloth, who had a very sound knowledge of South African plants and the botanical problems they present.

As the Botanical Garden in Stellenbosch was fairly small, it was decided to concentrate on building up a collection of succulents. Mr Herre made by far the largest contribution to the growth of the collection. He collected mainly in Namaqualand, to which he made four expeditions, in 1929, 1930, 1933 and 1939, before he was interned during the Second World War. Under his guidance the collection of succulents was, in the late forties, one of the largest and one of the most representative in the world.

Many of the plants cultivated in the Garden belonged to undescribed species. As it was impossible for Professor Nel, the only botanist attached to the Garden, to describe all these new species, the plants had to be sent to specialists: to Dr L. Bolus and her colleagues, and to Prof. R. Marloth in Cape Town, Prof. S. Schönland in Grahamstown, Dr R. Dyer in Pretoria, Dr N. E. Brown in Kew, Prof. G. Schwantes, Dr K. von Pöllnitz, Dr A. Tischer, Prof. H. Straka, Prof. H.-D. Ihlenfeldt and Dr H. Friedrich in Germany, A. White and B. L. Sloane in California, and to many others. Mr Herre wrote often to these botanists. Through his letters which were, in fact, botanical articles, and by supplying plant material, he formed a link between students of succulents all over the world. These students held him in high esteem for his remarkable enthusiasm and unselfish assistance.

Mr Herre gained expert knowledge of the conditions under which these plants grow in nature and applied this knowledge to his methods of cultivating them. The result was that he cultivated succulents and xeromorphic plants with very great success. His greatest achievement in this respect was the success he had in cultivating *Welwitschia mirabilis* from seed to seed. The success with which he cultivated *Heliamphora nutans* and several other species which are difficult to grow, shows how talented a gardener he was.

When he was interned, from 1940 to 1946, Mr Herre's undaunted spirit tried to make the best of the situation, and he studied the capsules and the descriptions of the mesems very thoroughly. This intensive study paved the way for most of his subsequent publications. While still in the internment camp, he and Prof. O. H. Volk of Wurzburg drew up a key for the genera of the mesems. Later Mr Herre described the results of his pre-war expeditions to Namaqualand and the geographic distribution of the mesems. In collaboration with Dr H. Friedrich of München, he described and mapped the distribution of several of the genera.

As Mr Herre was very well acquainted with the bibliography and descriptions of the mesembryanthema, and as he had a sound knowledge of these plants in nature and in cultivation, he was urged to apply to the Council for Scientific and Industrial Research for a grant to compile a bibliography on the genera and to amend and supplement where necessary the published descriptions from his own experience. This work, which is unfortunately not yet duplicated, has to a great extent been incorporated in the present book and ought to be of great value to researchers.

For his work in the cultivation and study of the succulents, he was made an honorary member of the *Deutsche Kakteen-Gesellschaft* in 1959. In recognition of his contribution to botany, he was specially honoured in 1963 by the Republic of West Germany, and as a token of its esteem a valuable book was presented to him on 18 February 1963.

Although Mr Herre cultivated about 300 new species in the Botanical Garden, he himself has described only about six species and the two genera *Arenifera* and *Jensenobotrya*. About thirty new species, amongst them *Lithops herrei, Conophytum herrei, Anacampseros herreana, Crassula herrei, Sarcocaulon herrei, Stapelia herrei, Othonna herrei, Haworthia herrei,*

Gethyllis herrei, *Homeria herrei*, and two genera, *Herrea* and *Herreanthus*, were named after him.

Mr Herre was in the midst of the explosive growth in the knowledge of the mesems. He was a spectator of the race to describe new species and genera and the resulting increase in their numbers. Today there are so many species and genera that a survey such as Mr Herre's present one is a necessity. It is hoped that the illustrations of typical species in each genus will be a very great help in the reviewing of the variation in the group.

Although so many species and genera have been described, very little is known about the morphology, anatomy and cytogenetics of the different species. Only when the species and genera are better known, can it be decided how they are related, how they can best be classified, and whether and how their number can be reduced. Very little is known about the ecology and reproduction of the species under natural conditions, and it is a noteworthy recommendation that the essay of Prof. Ihlenfeldt on the Biology of Dissemination has been included in this book.

Botanically and biologically, the mesembryanthema is a very remarkable group and the elucidation of the problems they present will widen the perspective in botany and biology. Mr Herre's book will undoubtedly be a contribution to a renewed interest in the study of the group.

P.G.JORDAAN, Professor of Botany, University of Stellenbosch, Stellenbosch, Republic of South Africa. May, 1971

THE GENERA OF THE MESEMBRYANTHEMACEAE

The Classification of the Mesembryanthemaceae *by Prof. Dr G. Schwantes*

Revised and completed by Prof. Dr H. Straka and Prof. Dr H.-D. Ihlenfeldt in accordance with Prof. Dr G. Schwantes

Subfamilies:
I. MESEMBRYANTHEMOIDEAE IHL., SCHW. et STR. (syn. *Aptenioideae* Schwant).
II. HYMENOGYNOIDEAE SCHWANT.
III. CARYOTOPHOROIDEAE IHL., SCHW. et STR.
IV. RUSCHIOIDEAE SCHWANT.

A. CENTRAL PLACENTATION

I. Subfamily: MESEMBRYANTHEMOIDEAE IHL., SCHW. et STR. Gynaecium isomeric, does more or less project beyond the axillary calyx (flower perigyn), nectaries koilomorphic. Fruit: loculicidal, hygrochastical capsule, numerous seeds in each cell.

Tribe 1. Apteniinae SCHWANT.
Petals soft, not rigid.

Subtribe 1. Apteniae SCHWANT.
Perennial herbs. Capsule 4-celled.
Aptenia N. E. BR.
Platythyra N. E. BR.

Subtribe 2. Mesembryanthemeinae IHL. SCHW. et STR.
Annual or biennial herbs with flat to cylindrical leaves. Capsule 5-celled.
Hydrodea N. E. BR.
Mesembryanthemum L. emnd. L. BOL.
Eurystigma L. BOL.
Halenbergia DINT.
Opophytum N. E. BR.
Callistigma DINT. et SCHWANT.
Synaptophyllum N. E. BR.

Subtribe 3. Preniinae SCHWANT.
Perennial. Main stem with shortened internodes, flowering shoots elongated. Capsule 4-5-celled.
Prenia N. E. BR.
Sceletium N. E. BR.

Subtribe 4. Aridariinae SCHWANT.
Erect or creeping shrubs, often with tuberous or turnip-shaped roots. Capsule 4-5-celled.
Aridaria N. E. BR.

Subtribe 5. Psilocaulinae HERRE.
Shrubs with leaves deciduous or persisting as spines. Capsule 5-celled.

Amoebophyllum N. E. BR.
Psilocaulon N. E. BR.

Tribe 2. Dactylopsideae
Flower in *Dactylopsis* with rigid petals. Dwarf shrubs or stemless. Leaves partly alternate, with very long leaf sheaths, digitate, very succulent. Capsule 5-celled.
Dactylopsis
Aspazoma

II. Subfamily: HYMENOGYNOIDEAE SCHWANT.
Annual herbs with flat leaves with idioblasts in the epidermis. Gynaecium pleiomeric, flower epigyn. Carpels far separated into two halves by false septums, the halved segments carry only a single seed. Nectaries: an intermediate stage between koilomorphic and lophomorphic type. Fruit: loculicidal-septicidal, many-celled fractionising fruit, the seeds remain embedded in a pocket of the partial fruitlets (clausae).
Hymenogyne Haw.

III. Subfamily: CARYOTOPHOROIDEAE IHLF., SCHWANT. et STR. Perennial herbs with a perennial rootstock. Gynaecium oligomeric (3-4 carpels, this occurs but once within the Mesembryanthemaceae), flower epigyn. Carpels divided into two halves by false septums with the exception of a minute opening near the mouth of the style channel; each half carries one (rarely two) seeds. No further seeds in the carpels. The nectaries represent an intermediate stage between the koilomorph and lophomorph type. Fruit: loculicid fractionising fruit; the narrow communication channel between the two halves of a cell becomes recognisable as small lateral openings in the partial fruitlets (double clausae) when the fruit is splitting, the segments of fruit are woody and remain connected at first by the axillary calyx. The seeds remain enclosed in the segments.
Caryotophora LEISTN.

B. BASAL TO PARIETAL PLACENTATION

IV. Subfamily: RUSCHIOIDEAE SCHWANT.
Placentation basal to parietal with all transitional stages. Gynaecium isomeric or pleiomeric, flower epigyn. Nectaries lophomorphic.

Tribe 1. Ruschieae SCHWANT.
Gynaecium isomeric or pleiomeric, numerous seeds in the cells. Fruit: loculicidal hygrochastical capsule.

Subtribe 1. Ruschiinae SCHWANT.
Shrubs or stemless. Capsule 5- or many-celled, with placental tubercles. Valve wings wanting or rudimentary. Cell lids often with a closing mechanism.

Ruschia SCHWANT.
Eberlanzia SCHWANT.
Acrodon N. E. BR.
Astridia DINT. et SCHWANT.
Bergeranthus SCHWANT.
Ottosonderia L. BOL.
Carruanthus SCHWANT.
Tischleria SCHWANT.
Hereroa DINT. et SCHWANT.
Rhombophyllum SCHWANT.
Bijlia N. E. BR.
Machairophyllum SCHWANT.

Subtribe 2. Leipoldtiinae SCHWANT.
Shrubs or stemless. Capsule many-celled, with cell lids and (except *Vanheerdea*) placental tubercles. Cell lids often with a closing mechanism.

Leipoldtia L. BOL.
Cephalophyllum N. E. Br.
Fenestraria N. E. BR.
Cheiridopsis N. E. Br.
Vanheerdea L. BOL.
Cylindrophyllum SCHWANT.
Calamophyllum SCHWANT.
Schlechteranthus SCHWANT.
Odontophorus N. E. BR.
Polymita N. E. BR.
Octopoma N. E. BR.
Vanzijlia L. BOL.

Subtribe 3. Lampranthinae SCHWANT.
Shrubs or stemless. Capsule 5-celled with cell lids and sometimes placental tubercles.

Lampranthus N. E. BR.
Oscularia SCHWANT.
Ebracteola SCHWANT.
Braunsia SCHWANT. (syn. *Echinus* L. BOL.)
Cerochlamys N. E. BR.
Dicrocaulon N. E. BR.
Disphyma N. E. BR.

Subtribe 4. Jacobseniinae SCHWANT.
Shrub with cylindrical leaves, juvenile stage of the papillose leaves corpuscular. Capsule with cell lids and valve wings, 5-celled (*Lampranthus*-type).

Jacobsenia L. BOL. et SCHWANT.

Subtribe 5. Delospermatinae SCHWANT.
Stemless or shrubby with a perennial rootstock. Capsule without cell lids or with rudimentary cell lids, 5-6-celled.

Delosperma N. E. BR.
Drosanthemum SCHWANT.
Trichodiadema SCHWANT.
Mestoklema N. E. BR.
Ectotropis N. E. BR. (*Delosperma*)

Subtribe 6. Psammophorinae SCHWANT.
Stemless or shrubby, leaves with sand adhering. Capsule without cell lids, with valve wings. (*Delosperma*-type) 5-8-celled.

Psammophora SCHWANT.
Arenifera HERRE.

Subtribe 7. Erepsiinae SCHWANT.
Shrubs or stemless and of tufted habit. The axillary calyx projects beyond the ovary in shape of a tube. The stamens and staminodes in part or wholly bent into the corolla tube. Capsule 5-celled, with cell lids, with or without placental tubercles.

Erepsia N. E. BR.
Semnanthe N. E. BR.
Smicrostigma N. E. BR.
Kensitia FEDDE.
Argyroderma N. E. BR.
Nelia SCHWANT.

Subtribe 8. Nananthinae SCHWANT.
Plants forming rosettes or clumps. Leaves generally tuberculate. Capsule many-celled, with cell lids, without or with small placental tubercles.

Nananthus N. E. BR.
Aloinopsis SCHWANT.
Titanopsis SCHWANT.
Khadia N. E. BR.
Rabiea N. E. BR.

Subtribe 9. Pleiospilinae SCHWANT.
Stemless. Leaves marked with numerous dark dots. Capsule many-celled, with cell lids and placental tubercles.

Pleiospilos N. E. BR.

Subtribe 10. Stomatiinae SCHWANT.
Creeping shrubs or stemless. Leaves tuberculate-papillose. Capsule 5-celled, with cell lids, without placental tubercles.

Stomatium SCHWANT.
Chasmatophyllum DINT. et SCHWANT.
Rhinephyllum N. E. BR.
Neorhine SCHWANT.
Neohenricia L. BOL.

Subtribe 11. Jensenobotryinae SCHWANT.
Plant depressed with long woody stems as thick as an arm, creeping over rocks. Leaves nearly globose. Capsule 5-celled.

Jensenobotrya HERRE.

Subtribe 12. Dracophilinae SCHWANT.
Compact, prostrate shrubs with thick leaves. Capsules many-celled, with developed or small rudimentary cell lids or without these.

Dracophilus DINT. et SCHWANT.
Juttadinteria SCHWANT.
Namibia DINT. et SCHWANT.

Subtribe 13. Lithopinae SCHWANT.
Stemless. Leaves of the seedlings forming globose or cone-shaped bodies, later more separated from each other; in *Lithops*

only does the juvenile form persist. Flowers without bracts
(not sepals!). Capsule 5-celled, with valve wings, only in
Lapidaria with cell lids.
>   *Lithops* N. E. BR.
>   *Schwantesia* DINT.
>   *Lapidaria* SCHWANT.
>   *Dinteranthus* SCHWANT.

Subtribe 14. Frithiinae SCHWANT.
Stemless; leaves cylindrical, the end windowed. Petals at
the base united into a tube. Capsule 5-celled, without cell lids
and valve wings.
>   *Frithia* N. E. BR.

Subtribe 15. Gibbaeinae SCHWANT.
Stemless or nearly stemless. Capsule 6-many-celled, usually
with cell lids, without placental tubercles.
>   *Gibbaeum* HAW.
>   *Antegibbaeum* SCHWANT.
>   *Imitaria* N. E. BR.
>   *Didymaotus* N. E. BR.
>   *Muiria* N. E. BR.
>   × *Muirio-Gibbaeum* JACOBS.

Subtribe 16. Conophytinae SCHWANT.
Stemless or nearly stemless. Leaves short and thick, or united
to cordate or conical bodies. Petals united into a tube. Capsule
with valve wings, without cell lids (*Delosperma*-type).
>   *Conophytum* N. E. BR.
>   *Berrisfordia* L. BOL.
>   *Herreanthus* SCHWANT.
>   *Oophytum* N. E. BR.
>   *Ophthalmophyllum* DINT. et SCHWANT.

Subtribe 17. Faucariinae SCHWANT.
Stemless, usually leaves with dentate margins. Capsule 5-celled.
The cell walls gape in their upper portion and bend over
the cells, whereby the cells are closed up to narrow openings
the same as by true cell lids (*Faucaria*-type).
>   *Faucaria* SCHWANT.
>   *Orthopterum* L. BOL.

Subtribe 18. Malephorinae SCHWANT.
Shrubs, stemless or nearly stemless. Capsule many-celled,
with cell lids and valve wings, in *Malephora* with adaxial seed
pockets.
>   *Malephora* N. E. BR.
>   *Glottiphyllum* HAW.

Subtribe 19. Dorotheanthinae SCHWANT.
Annual herbs with flat or more or less cylindrical leaves.
Capsule 5-celled, with or without cell lids.
>   *Dorotheanthus* SCHWANT.
>   *Aethephyllum* N. E. BR.
>   *Pherolobus* N. E. BR.
>   *Micropterum* SCHWANT.

Subtribe 20. Mitrophyllinae SCHWANT.
Shrubs or stemless. Pronounced heterophylly. Capsule 5-celled

with cell lids (except *Mitrophyllum*), with or without poorly
developed placental tubercles.
>   *Mitrophyllum* SCHWANT.
>   *Monilaria* SCHWANT.
>   *Conophyllum* SCHWANT.
>   *Mimetophytum* L. BOL.
>   *Meyerophytum* SCHWANT.
>   *Maughaniella* L. BOL.
>   *Diplosoma* SCHWANT.

Subtribe 21. Carpantheinae SCHWANT.
Annual herbs with flat leaves. Capsule many-celled. Lamellae
of the cell walls diverging, pressed over the cell by domed connecting arch leaving narrow fissure only through which
the seeds must pass.
>   *Carpanthea* N. E. BR.

Subtribe 22. Stoeberiinae FRIEDR.
Shrubs, intricately branched. Leaves smooth, somewhat
clavate, dotted, horizontally spreading. Flowers terminal,
dichotomously divided. Capsule turbinate-globose, 5-(6)-10-
locular, with cell lids and with tubercle; capsules break easily
into parts so that with Ruschianthemum it becomes a
schizocarp.
>   *Stoeberia* DINT. et SCHWANT.
>   *Ruschianthemum* FRIEDR.

Tribe 2. Apatesieae SCHWANT.
Annuals, biennials or perennials. Without idioblasts in the
epidermis. Gynaecium pleiomeric, in each cell there are formed
two seed pockets at the upper (outer) end of the placenta by
an imperfect false septum. In each seed pocket one seed is
developed (abaxile seed pocket). Fruit: transitional form from
hygrochastic schizocarp (with tendencies to a fractionizing)
fruit (*Apatesia*) to dry opening loculicidal (with additional tendencies to septicidal splitting) fractionizing fruit (*Herrea*).
Partial fruitlets: double clausae. Seed pockets always 1-seeded
and a few seeds are still produced in the ovary-chambers.
>   *Apatesia* N. E. BR.
>   *Conicosia* N. E. BR.
>   *Herrea* SCHWANT.

Tribe 3. Skiatophyteae (STRAKA) IHLENF.
Biennials or perennials with flat leaves. Gynaecium isomeric,
(4)5-7-carpels. An imperfect false septum and a horizontal placental wall form, at the upper end of the placenta in each
seed cell, two seed pockets (abaxile seed pockets), in which one
seed develops always. Fruit: hygroscopical (opens one time
and remains open) loculicidal schizocarpic capsule. Beside the
two seeds in the pockets a few seeds are still produced in the
cells.
>   *Skiatophytum* L. BOL.

Tribe 4. Saphesieae (SCHWANT.) IHLENF.
Shrubs. Gynaecium isomeric, only first indications of seed
pockets in the cells. Fruit: dry opening, 5-celled, loculicidal
capsule; numerous seeds in each cell.
>   *Saphesia* N. E. BR.

Tribe 5. Carpobroteae SCHWANT.
Creeping shrubs. Gynaecium pleiomeric. Fruit: many-celled
indehiscent fruit, numerous seeds in each cell. On ripening
the axial calyx and especially the surrounding bases of the sepals
become fleshy. The edible fruit may be called a berry.
Carpobrotus N. E. BR.

GENERA OF UNCERTAIN CLASSIFICATION:

Amphibolia L. BOL.   *Ruschianthus* L. BOL.
Anisocalyx L. BOL.   *Scopelogena* L. BOL.
Enarganthe N. E. BR.   *Stayneria* L. BOL.
Mossia N. E. BR.   *Wooleya* L. BOL.
Namaquanthus L. BOL.   *Zeuktophyllum* N. E. BR.
Subtribe I. Ruschiinae

Explanation of some Botanical Terms

by Prof. Dr H.-D. Ihlenfeldt

The aggregation of carpels (pistil) is termed *gynaecium*.
This is *isomeric*, if its members are equal in number with those
of the other floral whorls, viz. the calyx and the whorl of
primordias of stamens; it is *oligomeric* if its members are less in
number and it is *pleiomeric*, if they are more numerous. In
Mesembryanthemums with a pleiomeric gynaecium the number of carpels is increased appreciably (about 2- to 3-times)
over the number of the members of each floral whorl.

Relating to the situation of the *placentas* in the centre, at the
base or on the side-walls of the ovaries, we distinguish a
placentation which is central (axile), or basal, or parietal.

Staminal-primordias are the early stage of development of
the *androecium* (i.e. the complex of the male organs) in a
flower-bud. With the Mesembryanthemums their number is
always equal to the number of the sepals (segments of calyx);
they are isomeric. During the development of the flower
(ontogenesis) these primordias will be divided into numerous
petals and stamens, a division which is typical of Mesembryanthemums. Therefore the petals of the Mesembryanthemums ought to be called *staminodes* (transformed stamens), to be
more exact.

Capsules are fruits possessing the following characters:
(i) they are dry; (ii) they are composed of more than one carpel;
(iii) they are dehiscent (in contrast to *indehiscent fruits*), with
the Mesembryanthemums by valves; (iv) they do not separate
into single *partial fruitlets* (as with *schizocarpic* and with
fractionizing fruits). With the schizocarps each partial fruitlet
corresponds to a single *complete carpel*, whereas with
fractionizing fruits each partial fruitlet consists of parts other
than a complete carpel (e.g. half of a carpel). *Dehiscence* of
capsules may happen *loculicidally* (splitting along the midrib
of the carpel), or *septicidally* (dividing the true septa), or
more rarely in both ways at once.

Clausae and *double clausae* are fruitlets of fractionizing
fruits. A "clausa" is a partial fruitlet which originates from
a partial compartment of an ovary-chamber (loculus), which is
produced by a false septum. A "double clausa" consists of
two such clausae of adjoining carpels, united by a true septum.

By an unequal shrinkage of different tissues of the fruit-
wall ripe capsules can burst open by *desiccation* (*dry-opening*).
In other cases they dehisce on wetting and remain open
permanently as in the preceding case (*hygroscopical opening*)
or they close again when becoming dry and reopen when
becoming wet (*hygrochasy*).

Nectaries which are sunk like a pit are called *koilomorphic*,
and when raised like a crest *lophomorphic*.

A *seed pocket* is a separate part of an ovary-chamber, in which
one or more seeds are located, and in which the seeds will be
retained *totally*, while the other seeds of the main chamber
will be dispersed – supposing that seeds are developed in
the main chamber at all. With the *Ruschioideae* seed pockets
can be formed either by the upper (outside) end of the
placenta (an *axile seed-pocket*), or by its lower (inner) end
(*adaxile seed pockets*).

Idioblasts: Special cells which differ from their neighbouring
cells by their shape, their contents or the thickening of their
walls.

Key to the Genera of the Mesembryanthemaceae *by H. M. L. Bolus*

The key is constructed according to the following basic principle:

A. Placentation axile.
A. Placentation parietal.
  B. Fruit indehiscent, juicy and edible.
  B. Fruit capsular.
    C. Loculi of the capsule without a covering membrane.
    C. Loculi of the capsule with a covering membrane.
      D. Capsule with a covering membrane, without a placental tubercle.
      D. Capsule with a covering membrane, with a placental tubercle.

A. Placentation axile.
  1. Leaves flat.
    2. Fruit a schizocarp.
      3. Perennial; petals white; fruit consisting of 3-4 nut-like, bilocular mericarps which contain a single seed in each locule .. .. **Caryotophora**
      3. Annual; petals yellow; fruit consisting of 18-21 membranous 1-seeded sections .. .. **Hymenogyne**
    2. Fruit a capsule.
      4. Leaves persisting as membranous skeletons with margins and veins intact, sometimes until the next season's growth .. .. **Sceletium**
      4. Leaves not as above.
        5. Stigmas red .. .. **Callistigma**
        5. Stigmas green or pallid.
          6. Leaves opposite.
            7. Annual; leaves conspicuously united .. .. **Synaptophyllum**
            7. Perennials; leaves not conspicuously united.
              8. Petals lemon yellow; valves of the expanded capsule winged .. **Platythyra**
              8. Petals magenta; valves wingless .. .. **Aptenia**
          6. Leaves alternate on flowering branches.
            9. Papillae conspicuous; leaves of the central tuft (when they occur) usually forming a rosette, margins undulate .. .. **Mesembryanthemum**
            9. Papillae not, or scarcely, visible; leaves of the central tuft erect, margins flat .. .. **Prenia**
  1. Leaves terete, subterete, or semiterete.
    10. Perennials.
      11. Leaves and sepals spinescent from their early stages .. .. .. **Amoebophyllum**
      11. Leaves and sepals not spinescent from their early stages.
        12. Internodes entirely enclosed in embracing leaf tissue, or with age becoming evident.
          13. Leaves semiterete, up to 3,5 cm long and 8 mm diam., sepals very unequal; corolla fugitive, up to 2,5 cm long (Shrubby) .. .. **Aspazoma**
          13. Leaves terete and finger-like, up to 7 cm long and 2 cm diam.; sepals nearly equal; corolla long-lived, up to 1,5 cm long, segments stiff. (Forming clumps) .. .. **Dactylopsis**
        12. Internodes not thus enclosed as in 12.
          14. Branches usually constricted at the nodes; leaf-bases sheathing, persistent; staminodes usually petaloid.

15. Leaf-sheaths fringed with deflexed hairs at the base, conspicuous (or in *B. marlothii* at the tip of the leaf-sheaths) .. **Brownanthus**
15. Leaf-sheaths nude at the base, inconspicuous .. .. .. **Psilocaulon**
14. Branches very rarely constricted at the nodes; leaf-bases not conspicuously persistent; staminodes filamentous, usually absent in *Aridaria* and *Sphalmanthus*.
16. Papillae scarcely visible, branches and peduncles woody or wiry; cymes usually strong dichasial; flowers nocturnal .. **Aridaria**
16. Papillae usually evident; young branches and peduncles externally herbaceous; flowers 1-3-nate, usually diurnal.. .. **Sphalmanthus**
10. Annual or biennial.
17. Stigmas broad and somewhat flattened .. .. **Eurystigma**
17. Stigmas subulate or filiform.
18. Expanded capsule with inflexed valve-wings .. .. **Mesembryanthemum**
18. Expanded capsule with spreading, or erect valve-wings.
19. Corolla not, or slightly, exceeding the sepals, violet red or citron yellow; valves of the expanded capsule erect, bifid .. .. **Hydrodea**
19. Corolla well exceeding the sepals, pure white or pale straw; valves spreading or recurved, not bifid (wings bifid at the apex).
20. Valve-wings erect .. .. **Opophytum**
20. Valve-wings spreading .. .. **Halenbergia**
A. Placentation parietal.
B. Fruit indehiscent, juicy and edible .. .. **Carpobrotus**
B. Fruit capsular.
C. Loculi without a covering membrane.
1. Nectary composed of 5 pits or hollows, sometimes surmounted by rudimentary glands.
2. Annual; herbaceous parts papillose, leaves flat, entire or pinnatifid .. **Micropterum**
2. Perennial, herbaceous parts not papillose; leaves thick, usually acutely keeled and dentate .. .. **Stomatium**
1. Nectary not composed of pits or hollows.
3. Capsule dehiscing in dry conditions; valves without expanding keels, remaining more or less erect (except in *Herrea* where the capsule breaks up into separate segments).
4. Leaves flat; stigmas 5; capsule closing again if placed in water .. **Saphesia**
4. Leaves triquetrous or terete or subterete; stigmas more than 10.
5. Annual growths deciduous; dissepiments reaching to apex of valve.. **Herrea**
5. Annual growths not deciduous; dissepiments reaching about halfway up the valve .. .. **Conicosia**
3. Capsule dehiscing in wet conditions; valves with expanding keels, widely spreading or recurved when fully expanded.
6. Nectary composed of separate glands, which in *Mossia* are so close as to form an almost annular nectary.
7. Petals united at the base; staminodes and stamens joined at the corolla tube .. .. **Ruschianthus**
7. Petals not united at the base.
8. Leaves entirely united to form an ovoid or subglobose body, pubescent with deflexed hairs .. .. **Muiria**
8. Leaves not as above.
9. Leaves fenestrate at the apex .. .. **Frithia**
9. Leaves not fenestrate.
10. Capsule very small with thin walls, sometimes breaking up into 5 parts .. .. **Ectotropis**
10. Capsule not like above.
11. Compact plants with internodes enclosed.
12. Leaves usually rough (in *R. macradenium* smooth), not papillose; petals yellow .. .. **Rhinephyllum**
12. Leaves very smooth; bracts present .. .. **Carruanthus**

11. Plants more or less loosely branched with internodes visible.
  13. Plants creeping; leaves dentate   ..   ..   ..   **Mossia**
  13. Plants occasionally creeping; leaves entire   ..   ..   **Delosperma**
6. Nectary annular.
 14. Petals stiff and rather thick; staying fresh longer than is known
  in any other genus except *Dactylopsis*   ..   ..   **Nelia**
14. Petals not as above more or less fugitive.
 15. Sepals and petals free to the base.
  16. Leaves expanded and flat.
   17. Flowers opening in the afternoon; petals yellow;
   plants annual, sun-loving   ..   ..   **Apatesia**
   17. Flowers opening in the morning; petals white, plants
   not annual, shade-loving   ..   ..   **Skiatophytum**
  16. Leaves not expanded and flat.
   18. Plants resembling clusters of grapes on account of the
   rounded grape-like or club-shaped leaves   ..   ..   **Jensenobotrya**
   18. Plants not as above.
    19. Peduncle with bracts.
     20. Herbaceous parts with sand adhering to the
     viscid epidermis ..   ..   **Psammophora**
     20. Herbaceous parts not as above.
      21. Leaves acutely keeled upwards; sepals 6 ..   **Herreanthus**
      21. Leaves not acutely keeled; sepals 5.
       22. Shrubby; petals whitish; staminodes
       present   ..   ..   **Zeuktophyllum**
       22. Dwarf compact plants; petals yellow,
       with or without a central red stripe;
       staminodes absent   ..   ..   ..   **Nananthus**
    19. Peduncle bractless.
     23. Compact plants with internodes enclosed in the
     leaf-sheaths.
      24. Sepals 4   ..   ..   **Juttadinteria**
      24. Sepals 5-10.
       25. Stigmas 9-25   ..   ..   **Namibia**
       25. Stigmas 4-10.
        26. Leaf-pairs forming bodies.
         27. Petals in the lower half white, in
         the upper half purplish   ..   **Oophytum**
         27. Petals yellow or white.
          28. Apex of leaves more or less
          flattened, sepals 4-7   ..   **Lithops**
          28. Apex of leaves not flattened, if
          flattened (as in *D. pole-evansii*
          and *D. vanzijlii*) sepals 8-10   **Dinteranthus**
        26. Leaf-pairs not forming bodies   ..   **Schwantesia**
    23. Plants with elongated branches; some of the
    internodes visible.
     29. Creeping; leaf-pairs of one form; flowers
     nocturnal   ..   ..   **Neohenricia**
     29. Erect; leaf-pairs dimorphic, or at flowering-
     time trimorphic; flowers diurnal ..   ..   **Mitrophyllum**
 15. Sepals and petals united to form a calyx-tube and corolla-tube.
  30. Receptacle usually produced beyond the ovary to form a
  tube partly adnate to calyx-tube   ..   ..   **Ophthalmophyllum**
  30. Receptacle not produced beyond the ovary; corolla-tube
  free from calyx-tube.
   31. Annual growth consists of one pair of leaves   ..   **Conophytum**
   31. Annual growth consists of two pairs of leaves   ..   **Berrisfordia**

C. Loculi of the capsule with a covering membrane, which is sometimes
incomplete.

    D. Loculi without a placental tubercle. (Tubercle occasionally present in
Drosanthemum and *Trichodiadema*.)

      1. Annuals.

        2. Ovary with 5 processes (lobes) alternating with the stigmas..   .. **Pherelobus**

        2. Ovary without processes.

          3. Leaves pinnatifid   ..   .. **Aethephyllum**

          3. Leaves entire.

            4. Stigmas 5, hardening and accrescent with age and persisting
throughout the fruiting-stage   ..   .. **Dorotheanthus**

            4. Stigmas 12-20, withering with the rest of the flower before the
fruit matures   ..   .. **Carpanthea**

     1. Perennials.

      5. Leaf-pairs dimorphic or trimorphic.

        6. Internodes invisible; lower pair of leaves not always evident.

          7. Upper leaf-pair spreading or prostrate on the ground, much
more united on one side than the other; flowers sessile   .. **Diplosoma**

          7. Upper leaf-pair erect or suberect or ascending, symmetrically
united below; flowers usually pedunculate.

            8. Roots fibrous; stems and branches not moniliform;
staminodes present   ..   .. **Maughaniella**

            8. Roots more or less woody, stems and branches usually
moniliform; staminodes absent   ..   .. **Monilaria**

        6. Some internodes visible; lower pair of leaves always evident.

          9. Leaf-pairs trimorphic at flowering-time   ..   ..   .. **Mitrophyllum** (*Conophyllum*)

          9. Leaf-pairs dimorphic throughout.

            10. Sepals 4; staminodes present   ..   ..   .. **Dicrocaulon**

            10. Sepals 5-6; staminodes absent.

              11. Lower leaf-pairs oviform in earlier stages; petals
rather lax, rose or rose-purple ..   .. **Meyerophytum**

              11. Lower leaf-pairs bead-like; petals dense, white, or
white and rose in the same species   ..   ..   .. **Monilaria**

     5. Leaf-pairs of one form (or in *Drosanthemum diversifolium* dimorphic).

      12. Nectary composed of separate glands, which rarely may be con-
tiguous and form an almost annular nectary.

        13. Sepals and petals connate below, style present   ..   .. **Imitaria**

        13. Sepals and petals not connate; style absent.

          14. Herbaceous parts papillose.

            15. Leaves usually crowned with spreading or erect
bristles, or their vestiges   ..   .. **Trichodiadema**

            15. Leaves without terminal bristles (tubercle in loculi of
capsule very rarely present)   ..   .. **Drosanthemum**

         14. Herbaceous parts not papillose.

          16. Plants creeping.

            17. Petals yellow   ..   .. **Chasmatophyllum**

            17. Petals white ..   .. **Mossia**

          16. Plants not creeping.

           18. Flowers cymose   ..   .. **Oscularia**

           18. Flowers solitary.

            19. Leaves dentate, teeth ending in a bristle   .. **Faucaria**

            19. Leaves entire.

             20. Peduncle bracteate.

              21. Petals yellow   ..   .. **Rabiea**

              21. Petals purple-rose   ..   ..   .. **Cerochlamys**

             20. Peduncle bractless ..   .. **Gibbaeum**

      12. Nectary annular or indistinct.

        22. Petals yellow (red-orange in *Titanopsis hugo-schlechteri*).

23. Stigmas 10 .. .. **Vanheerdea**
23. Stigmas 5-7.
    24. Compact plants with internodes enclosed in the leaf-sheaths.
        25. Glabrous; leaves warted or pustulate in the upper part; sepals and stigmas 6 .. .. **Titanopsis**
        25. Receptacle and sepals pubescent; leaves without warts or pustules (microscopically granulated); sepals 7; stigmas 6-7 .. .. **Lapidaria**
    24. Erect, decumbent or creeping plants with internodes visible.
        26. Leaf-keel serrate .. .. **Erepsia**
        26. Leaf-keel entire
            27. Valves of the expanded capsule winged .. **Lampranthus**
            27. Valves of the expanded capsule not winged .. **Scopologena**
22. Petals rose-purple, pink, red, pallid or white.
    28. Stigmas 8-14.
        29. Petals clawed, the blade obovate or elliptic, equal to or shorter than the claw .. .. **Kensitia**
        29. Petals not as above.
            30. Stigmas minute, completely hidden by staminodes and stamens.
                31. Leaf-keels entire .. .. **Smicrostigma**
                31. Leaf-keels lacerate .. .. **Semnanthe**
            30. Stigmas not as above.
                32. Peduncle bracteate.
                    33. Low compact plants; sepals 5 .. .. **Dracophilus**
                    33. Shrubby plants, 15-30 cm high, sepals 4.
                        34. Herbaceous parts smooth; stamens not papillose, seeds echinate .. .. **Namaquanthus**
                        34. Herbaceous parts "puberulus"; stamens papillose; seeds smooth .. .. **Wooleya**
                32. Peduncle bractless.
                    35. Staminodes present; valves of the expanded capsule not winged .. .. **Khadia**
                    35. Staminodes absent, valves of the expanded capsule winged .. .. **Anisocalyx**
    28. Stigmas 5-6.
        36. Herbaceous parts more or less papillose.
             37. Sepals and stigmas 5 .. .. **Jacobsenia**
            37. Sepals and stigmas 6 .. .. **Drosanthemum**
        36. Herbaceous parts not papillose.
             38. Flowers terminating the previous year's vegetative growth, otherwise developing axillary shoots .. .. **Didymaotus**
            38. Flowers not as above.
                39. Flowers in conspicuously bradched pedunculate cymes; rootstock remarkably large and sometimes tuberous .. .. **Mestoklema**
                39. Flowers and rootstock not as above.
                  40. Herbaceous parts smooth.
                    41. Seeds echinate .. .. **Braunsia**
                    41. Seeds not as above.
                        42. Valves of the expanded capsule winged .. .. **Lampranthus**
                        42. Valves of the expanded capsule not winged .. .. **Khadia**
                40. Herbaceous parts hairy .. .. .. **Braunsia**

D. Loculi of the capsule with a placental tubercle (sometimes obscure or
absent in *Aloinopsis*, very small in *Antegibbaeum* and uncertain in *Calamophyllum*).
   1. Stigmas 5-6
      2. Petals purple-red, pink or white.
         3. Nectary composed of separate glands   ..   ..   **Cerochlamys**
         3. Nectary annular.
            4. Placental tubercle 2-lobed   ..   ..   **Disphyma**
            4. Placental tubercle not 2-lobed.
               5. Peduncles not spinescent.
                  6. Herbaceous parts with sand adhering to the viscid epi-
                      dermis   ..   ..   **Arenifera**
                  6. Herbaceous parts not as above.
                     7. Capsule in the lower part isolated into mericarps .. **Ruschianthemum**
                     7. Capsule not as above.
                        8. Valves of the expanded capsule winged.
                           9. Shrubby; internodes visible; cymes often conspi-
                             cuously branched.
                              10. Capsule after first opening, not closing again .. **Stoeberia**
                              10. Capsule not as above   ..   ..   ..   **Amphibolia**
                         9. Dwarf compact plants; internodes enclosed,
                          flowers solitary or 1-3-nate.
                            11. Leaves cylindrical or subcylindrical; flowers
                               solitary.
                                 12. Bracts large, sometimes exceeding the
                                    flowers; sepals 5   ..   ..   ..   **Cylindrophyllum**
                               12. Bracts very small; sepals 6 ..   ..   **Antegibbaeum**
                           11. Leaves laterally compressed in the upper
                             part; flowers finally 2-3-nate   ..   ..   **Ebracteola**
                      8. Valves of the expanded capsule wingless.
                         13. Seeds echinate or rough   ..   ..   ..   **Astridia**
                       13. Seeds not as above.
                           14. Stigmas plumosely papillate ..   ..   **Acrodon**
                           14. Stigmas not plumosely papillate, or if so, the
                             leaves not triquetrous ..   ..   ..   **Ruschia**
               5. Some, or all, of the peduncles spinescent   ..   ..   **Eberlanzia**
      2. Petals yellow, golden or orange-red.
         15. Leaf-pairs dimorphic at flowering-time; nectary annular   ..   **Mimetophytum**
         15. Leaf-pairs of one form; nectary composed of separate glands.
            16. Wings of the loculi of the expanded capsule erect ..   ..   **Orthopterum**
            16. Wings not erect.
               17. Leaves short and thick, polymorphic; valves erect in
                   expanded capsule ..   ..   **Bijlia**
              17. Leaves not as above; valves widely spreading or
                   recurved.
                  18. Leaves usually triquetrous.
                        19. Filaments papillate; stigmas 5; keels of valves in
                           expanded capsule diverging from the base;
                           placental tubercle large ..   ..   ..   **Bergeranthus**
                     19. Filaments epapillate; stigmas 5-6; keels parallel
                          below; placental tubercle small ..   ..   **Machairophyllum**
              18. Leaves not triquetrous, one of a pair often slightly
                   differing in form from the other.
                  20. Placental tubercle with a dividing line in the
                      middle   ..   ..   **Rhombophyllum**
                    20. Placental tubercle not as above ..   ..   **Hereroa**
  1. Stigmas more than 6 (or in *Khadia beswickii* 6).
     21. Leaves fenestrate at the apex   ..   ..   **Fenestraria**

21. Leaves not as above.
   22. Petals filiform-linear, white   ..   ..   **Polymita**
   22. Petals not as above.
      23. Plants dwarf, compact.
         24. Peduncle bractless.
            25. Petals pink or rose-purple; staminodes present   .. **Khadia**
            25. Petals yellow or rarely white; staminodes absent .. **Glottiphyllum**
         24. Peduncle with bracts.
            26. Leaf-pairs dimorphic   ..   .. **Cheiridopsis**
            26. Leaf-pairs of one form.
               27. Receptacle produced above the ovary into a tube; stamens deflexed; stigmas obscurely defined, forming a viscid circular pulvinus   .. **Argyroderma**
               27. Receptacles, stamens and stigmas not as above.
                  28. Herbaceous parts glabrous.
                     29. Rootstock (where known) tuberous   .. **Aloinopsis**
                     29. Rootstock not tuberous.
                        30. Nectary composed of separate glands **Machairophyllum**
                        30. Nectary annular (unknown in *Calamophyllum*).
                           31. Petals pallid, rose, rose-purple or deep-red.
                             32. Flowers solitary.
                                33. Sepals 5   ..   ..   .. **Cylindrophyllum**
                              33. Sepals 4   ..   ..   .. **Calamophyllum**
                           32. Flowers not solitary ..   .. **Cephalophyllum**
                        31. Petals yellow, golden or orange-red ..   .. **Pleiospilos**
                  28. Herbaceous parts pilose or pubescent
                     34. Leaves obtusely keeled, dentate or denticulate   ..   .. **Odontophorus**
                     34. Leaves entire or rarely with the keels only dentate, in which case they are acutely keeled upwards   ..   ..   .. **Cheiridopsis**
      23. Plants with more or less elongated branches with some, at least, of the internodes visible, except in the shrubby *Ottosonderia.*
         35. Inflorescence persisting for several years and eventually becoming an obconic cyme, the axils of the previous year's bracts normally producing a bracteated flower which represents the annual growth   ..   ..   .. **Ottosonderia**
         35. Inflorescence not as above.
             36. Leaf-pairs dimorphic   ..   .. **Vanzijlia**
             36. Leaf-pairs of one form.
                37. Peduncle with bracts.
                  38. Sepals 4.
                     39. Leaves entire, 4-6 cm long.
                        40. Filaments papillose; stigmas 8; seeds smooth   ..   .. **Enarganthe**
                        40. Filaments not papillose; stigmas 8-12, seeds echinate   ..   .. **Namaquanthus**
                     39. Leaves denticulate, less than 2 cm long.. **Octopoma**
                  38. Sepals 5-6.
                     41. Sepals 5.
                       42. Leaves less than 2 cm long: flowers sessile or subsessile   ..   .. **Octopoma**
                       42. Leaves more than 2 cm long; flowers pedunculate.

43. Herbaceous parts pubescent.
    44. Leaves dentate   ..   ..   **Odontophorus**
    44. Leaves entire ..   ..   ..   **Cheiridopsis**
43. Herbaceous parts glabrous.
    45. Leaves rough, sometimes dentate   ..   ..   ..   **Cheiridopsis**
    45. Leaves not as above.
        46. Usually decumbent and often creeping; stigmas 10 or more   ..   ..   **Cephalophyllum**
        46. Usually erect or if decumbent not creeping; stigmas 10 or fewer.
            47. Leaves obtusely keeled; loculi of the capsule winged ..   ..   ..   **Leipoldtia**
            47. Leaves acutely keeled; loculi of the capsule not winged ..   ..   ..   **Stayneria**
  41. Sepals 6   ..   ..   **Schlechteranthus**
37. Peduncles bractless.
    48. Leaves of a pair dissimilar and unequal in length, the larger more than 5 cm long, sometimes with a marginal tooth or hump, or with a conspicuous white margin ..   ..   ..   **Glottiphyllum**
    48. Leaves of a pair similar and about equal in length, less than 5 cm long, entire and without a white margin   ..   ..   ..   **Malephora**

Key to the Genera of the Mesembryanthemaceae *by H. Herre and Prof. Dr O. H. Volk*

Directions for use. In this key under each cipher opposed qualities or characters are given for choice; one is marked by the numeral, the other by a dash (–). Beginning with number one we have to see which of the two paragraphs is applicable to the plant given and whether all other items mentioned prove correct. At the end of the applicable section we find again a cipher, which tells us where to continue testing. This has to be repeated until we find at the end of the paragraph a name instead of a number. This is the name wanted. As far as possible, such characters or qualities were used as are more or less easy to detect.

1 Placentation AXILE; plants often papillate, mostly annual; leaves often flat, ovaries and capsules with 4-6 chambers or loculi.
   Subfamily **MESEMBRYANTHEMOIDEAE** Ihl., Schwant., Str.
   (syn. *Aptenioideae* Schwant.) 2
– Placentation PARIETAL, rarely axile but then the fruit separating into 1- or 2-seeded fruitlets (*Hymenogyne*), plants sometimes papillate, mostly perennial; leaves rarely flat,* ovaries and capsules with 4 to many chambers.
   Subfamilies **RUSCHIOIDEAE** Schwant. and **HYMENO-**
   GYNOIDEAE Schwant. 15
2 Leaves ALTERNATE or opposite, but not at the same time flat and connate at their whole base 3
– Leaves opposite, flat and CONNATE at their whole base, resembling an AMPLEXI-CAUL more or less orbiculate leaf; a prostrate, soft, branched succulent annual; flowers whitish, small, about 1 cm diameter .. .. **Synaptophyllum**
3 Old leaves not persistent and spinescent· tips of calyx soft 4
– Old leaves persistent, SPINESCENT or HARDENING; calyx-lobes spinescent; dwarf perennial shrubs with erect branches, leaves of the upper parts alternate .. **Amoebophyllum**
4 Leaves with amplexicaul LONG SHEATHS, mostly alternate, more or less terete 5
– Leaves with VERY SHORT or NEARLY MISSING sheaths 6
5 A dwarf, highly succulent, perennial plant; leaves alternate, very succulent, terete; sheaths hiding the stem; flowers white; several weeks unchanged open; calyx-lobes 5, EQUAL .. .. **Dactylopsis**
– A bushy, branched succulent perennial; flowers similar to those of *Aridaria*, soon marcescent, yellowish-white; calyx-lobes 4-5, VERY UNEQUAL; stigmas 4-5, subulate; capsules like those of *Aridaria* .. .. **Aspazoma**
6 Leaves at least in the floral region ALTERNATE 7
– Leaves all OPPOSITE 8
7 PERENNIALS with a THICK fleshy or woody rootstock; staminodes almost ever present .. .. **Aridaria**
– Short-lived ANNUAL or biennial herbs with THIN, fibrous roots; leaves flat or more or less teretely narrowed towards the base, or more or less terete .. .. **Mesembryanthemum** s.l.
   The following genera belong here:
   A   Leaves or branches or both with papillae .. .. B
   –   Plants without papillae C
   B   Capsule-valves ·BIFID; branched, prostrate, stout annuals with sessile, stout, terete, obtuse leaves .. .. **Hydrodea**
   –   Capsule-valves ENTIRE; annual or biennial herbs with distinct internodes and flat or terete fleshy leaves, covered with often glittering papillae.. .. **Mesembryanthemum**

* It seems that *Caryotophora* Leistn. belongs here (perennial plant).

C Flowers whitish to violet-red, rarely yellow, but then stigmas not conspicuously purplish-red .. .. **D**
– Flowers bright yellow with purplish-red stigmas; calyx and corolla-lobes united in a short tube; annual herb with lanceolate leaves .. .. .. **Callistigma**
D Stigmas thick and broad, yellow .. .. **Eurystigma**
– Stigmas filiform; halophyte .. .. **Halenbergia**
(See also No. 13: *Opophytum* N. E. Br.)
8 Leaves THIN, usually flat and broad or narrow and canaliculate 9
– Leaves THICK, semiterete to terete or broadly triquetrous.. 12
9 Leaves PETIOLATE or petiolate-contracted; calyx-lobes 4, exceeding the petals; stigmas 4 10
– Leaves SESSILE; calyx 4-5-partite; stigmas 4-5 11
10 Leaves often cordate; calyx-lobes FREE to the ovary; petals united in a tube; perennial succulent herb with elongated branches.. .. .. **Aptenia**
– Leaves usually lanceolate; calyx produced in a short TUBE above the ovary; petals united in a tube; perennial succulent herb, rootstock with long fleshy roots, flower yellowish to white .. .. **Platythyra**
11 Leaf-nerves conspicuous, particularly on the lower side, old leaves withering to a persisting skeleton; perennial succulent herbs with prostrate branches .. **Sceletium**
– Leaf-nerves not hardening and persistent; succulent perennials with crowded or distant opposite leaves on long branches .. .. **Prenia**
12 Sheaths of leaves without a fringe of stiff hairs 13
– Sheaths of the leaves small, terete, fugitive and deciduous leaves.. .. .. **Brownanthus**
13 Perennials, the plant remains alive underground when the aerial parts die off and thrives again later on 14
– ANNUALS with thick, often articulated branches; leaves small, usually papillate; flowers on short stalks, small, whitish, terminal or lateral; calyx-lobes more or less unequal. (Some annual species of *Psilocaulon* described by L. Bolus as e.g. *P. annuum*, etc., should be classified here) .. .. **Opophytum**
14 Branches thick and fleshy, often constricted-articulate, sometimes beaded; leaves vanish soon, small terete; flowers small (not exceeding 2 cm) white to pink .. .. **Psilocaulon**
– Branches woody, neither fleshy nor constricted-articulate, if fleshy then rootstock tuberous or turnip-shaped; leaves persistent for some time, sometimes spinescent, terete or 3-angled, sometimes papillate, all or only the lower ones opposite, distant or crowded .. .. **Aridaria**
The following genera belong here:
A Leaves crowded, internodes very short; small, shortly branched plants, growing in clumps .. .. **B**
– Leaves distant, internodes distinct; large erect or prostrate undershrubs **C**
B Rootstock TUBEROUS, fleshy; leaves all or in the floral region only alternate; flowers subsessile or shortly pedicellate, dirty reddish or yellowish, small to large .. .. **Phyllobolus**
– Rootstock woody; leaves broadly triquetrous, thick, with large papillae, in the floral region alternate. (Flowers imperfectly known, position uncertain) .. .. **Derenbergiella***
C Prostrate or decumbent creeping succulents, often rooting at the nodes; rootstock thick and fleshy; leaves subterete, papillose, sometimes spinescent, opposite, in the floral region alternate; flowers several to 1, greenish to yellowish or reddish .. .. **Sphalmanthus**
– Erect, shrubby plants, if prostrate then without a tuberous rootstock; leaves all opposite, rarely papillate; flowers singular to several, of moderate size or large, with some species opening at night and scented, white or yellowish or reddish .. .. **Neo-Aridaria**
15 Erect or prostrate shrubby or herbaceous succulents with clearly VISIBLE INTERNODES between the distant pairs of leaves 16

* This genus is united with *Mesembryanthemum* L.emend. L. Bolus by G. Schwantes.

– INTERNODES INCONSPICUOUS between the leaves, very short, stemless, highly succulent plants, often growing in clumps with densely crowded leaves or single pairs of leaves; sometimes the whole plant is represented by only one pair of fused leaves (e.g. *Lithops*)..   73

16 ANNUAL herbs with FLAT and THIN, entire or pinnatifid leaves, often with glittering papillae; roots thin   17

– PERENNIAL shrublets or herbs, when dying off above the ground with a perennial rootstock; leaves THICK, semiterete to terete or more or less triangular, glabrous, rarely papillose or with non-glittering warts, rarely flat, but roots always woody or thick   24

17 Ovary with numerous stigmas or chambers (7-25)   18

– Ovary with few (4-)5(-7) stigmas or chambers   20

18 Leaves glabrous or very finely hairy   19

– Plant covered with scattered, VILLOUS WHITE HAIRS; leaves petiolated, broad-spathulate; flowering-stalks with one pair of large bracts ..   ..   ..   **Carpanthea**

19 STYLE PROMINENT, funnel-shaped with 7-12 small stigmas; glabrous herbs with prostrate twigs   ..   ..   **Hymenogyne**

– STYLE MISSING; ovary with numerous many-seeded chambers without wings at the expanding keels and without roofs in the capsule, opening when wetted   ..   **Apatesia**

20 Leaves with GLITTERING PAPILLAE; valves of the capsules with expanding keels, and repeatedly opening and closing when wetted and drying up..   ..   21

– Leaves SHINING, without papillae, lanceolate, opposite or alternate, central nerves thick, lateral nerves conspicuous   ..   ..   **Skiatophytum**

21 Leaves ENTIRE, but when pinnatifid then with pink flowers   22

– Leaves LYRATO-PINNATIFID; stigmas deciduous ..   ..   **Aethephyllum**

22 Anthers not carmine-red   23

– Anthers CARMINE-RED; top of the ovary with alternating grooves and conspicuous ridges; with loculi-roofs, expanding keels on the valves ending with awns, without wings ..   ..   **Pherolobus**

23 Chambers of the capsules OPEN, roofs reduced to a minor rim; valves with very contiguous expanding keels and broad wings; top of the ovary or capsule not ridged ..   ..   **Micropterum**

– Chambers of the capsules with well developed ROOFS; valves with diverging expanding keels and with narrow wings ending in awns; stigmas persistent, forming an appendix to the valves ..   ..   **Dorotheanthus**

24 (16) Ovary (capsule) with 4-7 chambers and stigmas; when more then placentation axile or at the apex of leaves a fringe of bristles or leaves covered by adhering sand   25

– Stigmas 8-25; placentation never axile; leaves not as above   54

25 Shape and form of the succeeding leaves SIMILAR to the preceding ones ..   26

– Shape and form of the succeeding leaves very DIFFERENT from the preceding ones   49

26 Leaves glandular-VISCOUS and therefore covered by dust and sand   ..   27

– Leaves not glandular-viscous   28

27 Stigmas 5-7; chambers of capsule with reduced or no roofs, without tubercles; calyx-lobes 4-6, viscous   ..   ..   **Psammophora**

– Stigmas 8; chambers of capsule closed by well developed roofs with a thick rim by tubercles, calyx-lobes 4, almost equal ..   ..   **Arenifera**

28 Stamens ERECT, not bent over the stigmas; sepals usually free to the ovary   29

– Stamens BENT over the stigmas, which are concealed by numerous staminodes; sepals inserted on in a tube above the ovary; branched shrublets with 2-edged twigs ..   ..   **Erepsia**

29 Leaves not warty   30

– Leaves warty ..   ..   **Chasmatophyllum**

30 Margins of the leaves and their keels toothed or finely toothed   31

– Margins and keels of the leaves not toothed   33

31 Flowers white to violet; leaves short   32

– Flowers yellow. *Circandra* N. E. Br. = *Erepsia* N. E. Br.

32 Leaves with a bluish-grey WAXY coat, shortly united, not decurrent at the internodes; flowers with 5 separate glands; chambers of the capsule roofed, without tubercles .. .. **Oscularia**
 — Leaves long JOINTED, without a waxy coat, long-decurrent on the internodes, chambers of the capsule with roofs and with tubercles; nectaries forming a ring.. **Ruschia**
33 Erect, rarely climbing, shrublets or dwarf plants 34
 — PROSTRATE or climbing shrubs; leaves with white CARTILAGINOUS margins, thick, smooth .. .. **Braunsia** (Echinus)
34 Stamens in many series, very numerous 35
 — Stamens in 1 series, about 25; dwarf perennial with very minute papillae; capsules small and thin and splitting up into 5 parts which fall off .. .. .. **Ectotropis** (Delosperma)
35 Leaves of many forms, but never flat and broad; capsules repeatedly opening when wetted and closing when drying 36
 — Leaves broad and flat; capsules with 5 chambers, but not closing when dry; expanding keels are missing; a herbaceous succulent with a long woody main root .. .. **Saphesia**
36 Nectary-glands in the flowers distant (usually 5) or in a rim with 5 deep notches (*Mossia*), rarely annular, but then the chambers of the capsule without roofs (*Delosperma*); expanding keels on the valves of the capsule contiguous, rarely diverging 37
 — Nectary-glands in the flowers forming an annular or cupular, sometimes crenulated rim; expanding keels of the valves of the capsule divergent, with or without wings, chambers roofed, with or without tubercles (*Ruschia* group).. 44
37 Flowers DEEP-YELLOW, with distant glands; capsule-valves with widely diverging expanding keels and membranous, pointed wings, chambers roofed and tubercled 38
 — Flowers white to violet or pale-yellow; capsule-valves with parallel or contiguous expanding keels, chambers with or without roofs 39
38 Chambers of capsules with large or plane tubercle; rootstock often fleshy .. **Rhombophyllum**
 — Chambers of capsules with small tubercles; rootstock woody .. .. .. **Hereroa**
39 Stamens usually collected in a cone; chambers of the capsule WITH ROOFS; all green parts of the plants with more or less GLITTERING papillae or with a tuft of bristles at the apex of the leaves 40
 — Stamens usually erect; chambers of the capsules WITHOUT roofs; papillae on the young branches or flowers only or completely absent 43
40 Flowering stalks NOT PERSISTING after the capsules have gone 41
 — Flowering stalks PERSISTING and growing spinescent after the capsules have gone; inner stamens bearded; often intricately branched shrub; leaves opposite but united at the base .. .. **Mestoklema**
41 Apex of the leaves and of the sepals WITHOUT a tuft of bristles 42
 — Apex of leaves and of sepals with a TUFT OF BRISTLES .. .. **Trichodiadema**
42 Nectary-glands annular (united in a ring); capsules hard and fairly large like those of *Lampranthus* .. .. **Jacobsenia**
 — Nectary-glands DISTANT, 5; capsules small; glittering papillae only on young branches, later hispid, setaceous or roughly dotted .. .. **Drosanthemum**
43 Valves of the capsules WITHOUT wings; a prostrate perennial herb, flowering at night .. .. **Mossia**
 — Valves of the capsules WITH WINGS.. .. .. **Delosperma**
44 (36) Stigmas subulate or filiform 45
 — Stigmas stout, broad and short, rarely with filiform projections 48
45 Valves of capsules WITH wings; leaves not decurrent 46
 — Valves of capsules WITHOUT wings; chambers with roofs and a hardened rim (*Verschlussleiste*), with tubercles; leaves more or less united, often decurrent, often toothed, thick, large-dotted, rarely with glittering papillae; flowers stalked, single to several, white to violet; bracts present, more or less similar to the leaves .. .. **Ruschia**
46 Erect shrublets; expanding keels of the valves of the capsules contiguous at their base; chambers with roofs and tubercles 47

– Procumbent plant, rooting at the nodes; expanding keels of the valves of the capsules distant at their bases .. .. **Disphyma**

47 Leaves large and thick (3-4 to 2 and 3 cm), falcate, smooth, finely VELVETY hyaline dotted .. .. **Astridia**

– Leaves smaller, oblong-terete to ovate, frequently mucronate, with dark hyaline dots .. .. **Eberlanzia**

48 Chambers of capsules with perfectly developed roofs and without tubercles; leaves smooth .. .. **Lampranthus**

– Chambers of capsules with roofs which are reduced to a rim, tubercles very large; leaves with dark dots .. .. **Stoeberia**

49 (25) First pair of leaves free, spreading, united only at the very base; second pair united for at least one third, forming a thick body-like sheath .. 50

– First pair of leaves forming a globose or ovoid body, more or less included in the marcescent leaf-sheath; second pair more or less oblong-terete, united for about half their length or less, often inflated 52

50 Chambers of the capsules with roofs 51

– Chambers of the capsules without roofs .. .. **Mitrophyllum**

51 Chambers of capsules without tubercles .. .. **Conophyllum**

– Chambers of capsules with tubercles .. .. **Mimetophytum**

52 Branches not conspicuously articulated, less than half a centimetre thick.. 53

– Branches very conspicuously ARTICULATED, joints beaded .. .. .. **Monilaria**

53 Flowers LARGE, bright RED; calyx-lobes free to the ovary; valves of the capsules with diverging expanding keels; chambers with or without tubercles; first pair of leaves shortly cleft, second pair spreading .. .. **Meyerophytum**

– Flowers SMALL, WHITE; calyx-lobes united in a fleshy ring above the ovary; valves of the capsules with parallel or contiguous expanding keels; chambers without tubercles; otherwise like preceding, but plants smaller .. .. .. **Dicrocaulon**

54 (24) Leaves without a prominent waxy coat 55

– Leaves with a WAXY coat; easily removable, shortly united, oblong-terete, thick and soft-fleshy .. .. **Malephora** (Hymenocyclus)

55 Stamens HIDDEN by the staminodes, which form a cupola above them .. 56

– Staminodes ERECT or absent, not hiding the stamens 59

56 Stigmas VERY SMALL; chambers of capsules without tubercles; expanding keels ending in awns 57

– STIGMAS FILIFORM-SUBULATE, conspicuous; chambers of capsules with tubercles, roofed; expanding keels with broad terminal wings; abruptly diverging .. **Polymita**

57 Margins of leaves not CARTILAGINOUS, more or less smooth; capsules repeatedly opening and closing 58

– Margins of leaves CARTILAGINOUS and more or less finely TOOTHED; capsules remaining open .. .. **Semnanthe**

58 Calyx-lobes FREE to the ovary, top of ovary nearly flat, dwarf succulent shrub .. **Smicrostigma**

– Calyx-lobes inserted on a cup above the ovary; top of ovary depressed .. .. **Kensitia**

59 (55) Calyx-lobes 4; erect shrubs 60

– Calyx-lobes 5 or more, rarely 4, but then leaves decurrent on the stems (*Octopoma*) 61

60 Shrub up to 60 cm high; leaves clavate up to 3 cm long; capsules with 8 chambers, these with roofs and with tubercles .. .. **Enarganthe**

– Undershrub up to 20 cm high; leaves falcate, up to 6 cm long; capsules with 8-12 chambers, tubercle none .. .. **Namaquanthus**

61 Leaves not warty and hairy 62

– Leaves warty and hairy, see *Odontophorus marlothii* N. E. Br. 96

62 Leaves united at the very base only or free.. 63

– Leaves united for half of their length or more, or succeeding pairs of leaves very dissimilarly shaped 72

63 Leaves longer than 5 cm; when shorter then flowers yellow 64

– Leaves shorter than 5 cm 69

64 Leaves CROWDED, distant on the floral branches only; fruits dry 65

– Pairs of leaves distant; fruits FLESHY, without valves and expanding keels .. **Carpobrotus**

65 Rootstock FLESHY, turnip-shaped; flowers yellow; capsule thin; branches not
   rooting at the nodes   68
  – Rootstock not fleshy, flowers rarely yellow; capsules firm; branches rooting
   at the nodes or forming clumps   66
66 Plants forming clumps   67
  – Plants producing branches, rooting at the nodes and ending at the short main
   stem with a tuft or crowded, subterete or 3-angled finely dotted, firm leaves ..   **Cephalophyllum**
67 Flowering stalks without bracts; calyx-lobes 5; stigmas (5-)6-8(-10), rather
   short, subulate, bent in a cone   ..   ..   **Cylindrophyllum**
  – Flowering stalks with 2 bracts; calyx-lobes 4; stigmas 10-20, diverging..   ..   **Calamophyllum**
68 Capsules SPLITTING up into many segments, without a firm central column;
   flowering stalks with 2 narrow bracts at the base; stigmas about 20   ..   **Herrea**
  – Capsules not separating into segments, remain open when once opened; flowering stalks without bracts; 10-20 filiform stigmas ..   ..   **Conicosia**
69 Pairs of leaves almost free at their bases, not decurrent along the stems ..   70
  – Pairs of leaves distinctly united at the base and decurrent along the stems   ..   **Octopoma**
70 Flowers violet to white, never bright yellow   71
  – Flowers yellow or red, erect shrub; leaves slightly united, not dotted; flowers
   remaining open, on short stalks   ..   ..   **Malephora**
71 Leaves with large PROMINENT DOTS; flowers in branched cymes; robust compact
   shrublets   ..   ..   **Leipoldtia**
  – Leaves without dots; flowers SOLITARY, yellowish-pink; small shrublets, 7-10 cm
   high   ..   ..   **Zeuktophyllum**
72 (62) Succeeding pairs of leaves like the preceding ones; chambers of the capsules 11, with recurved roofs with a hardened rim, with tubercles; expanding
   keels diverging in the upper part, without wings ..   ..   **Schlechteranthus**
  – Succeeding pairs of leaves different 72a
72a Lateral branches NOT rooting at their nodes; chambers of the capsules 10, roofed,
   without tubercles; expanding keels diverging, with wings; flowers solitary;
   stamens united in a cone; stigmas 10, erect, subulate; dwarf shrublets with
   woody branches, producing two kinds of pairs of leaves; first (resting period)
   completely united, except the small points; second with long terete points,
   united at the base only   ..   ..   **Vanzijlia**
  – Lateral branches rooting at their nodes; chambers of capsules with roofs and
   with tubercles   ..   ..   **Cephalophyllum**

HIGHLY SUCCULENT MESEMBRYANTHEMUMS
73 (15) Branches or shoots with SEVERAL leaves or pairs of leaves, rarely only one
   pair but then the two leaves not united for a long distance or the free segments
   of the leaves considerably longer than thick or acute   74
  – Branches or shoots with ONE pair of leaves only (except when the new growth
   appears), united for the most part, free part at least slightly longer than thick,
   truncate, never acute (those leaves will be called "bodies") (extremely succulent
   Mesembryanthema)   110
74 Stigmas more or less free   75
  – Stigmas united in a disc or knob   ..   ..   **Argyroderma**
75 Stigmas or chambers 4-6, rarely 7, but then leaves viscous (covered by sand),
   the two leaves of one pair very unequal and dissimilar (*Gibbaeum*)   ..   75a
  – Stigmas or chambers 7 or more, rarely 6, but then plants very small (5 cm high)
   and with few, very short, semi-ovoid leaves yolk-yellow flowers (*Dinteranthus*),
   or leaves with hyaline white dots (*Khadia*), never sticky   96
75a Base of the leaves somewhat constricted, hard; stigmas 5, chambers of capsules
   without roofs ..   ..   **Jensenobotrya**
  – Lamina of leaves not constricted   76
76 Leaves not sticky   77
  – Leaves GLANDULAR-VISCOUS, always covered by sand or dust   ..   ..   **Psammophora**
77 Leaves with warts or papillae or white spots or white cartilaginous margins   78
  – Leaves smooth and not as above   86

78 Leaves with warts or white spots or white cartilaginous margins 79
– Leaves with more or less distinct papillae .. .. **Delosperma**
79 Flowers red or pale violet; when white, then leaves long prismatic, keeled
without white dots and not toothed 80
– Flowers white to yellow, leaves not as above 81
80 Leaves SHORT, erect, very obtuse, warty; rootstock not fleshy .. .. .. **Neohenricia**
– Leaves LONG, prismatic, distinctly keeled towards the apex, much dotted;
rootstock very thick .. .. **Ebracteola**
81 Leaves with small, regularly distributed warts, often toothed 82
– Leaves with large irregularly tesselate warts, not toothed, short, broadest towards the apex, obtuse .. .. **Titanopsis**
82 Stigmas filiform, sometimes plumose 83
– Stigmas short, dilated or clavate to capitate; chambers of capsules with well
developed or reduced roofs and without tubercles (*Agnirictus* included differs
from *Stomatium* only by absence of the small tubiform union of sepals and
petals) .. .. **Stomatium**
83 Stigmas not plumose; rootstock fleshy, thick 84
– Stigmas plumose, as long or longer than the stamens; chambers of capsules
with roofs expanding keels diverging .. .. **Chasmatophyllum**
84 Flowers sessile or shortly stalked, solitary; bracts similar to the leaves .. 85
– Flowers LONG-STALKED, bracts different from the leaves (*R. rhomboideum*) .. **Rhombophyllum**
85 Stamens collected into a dense CONE; margins of the leaves usually with distant
irregular teeth; chambers or capsules without roofs, expanding keels of the
valves contiguous nearly to the top (*Neorhine* included, differs by a short tubiform union of the sepals and petals) .. .. **Rhinephyllum**
– Stamens DIFFUSE; margins of leaves regularly toothed; capsules deep, barrel-
like with roofed chambers; expanding keels on the flanks of the valves, touching
those of the neighbouring valves, ending in a vertical spreading winged awn .. **Faucaria**
86 Leaves without windows, but sometimes with dots 87
– Leaves with large HYALINE WINDOWS at the truncate apex, terete and usually
alternate on underground stems, fascicled, about 2 cm high .. .. .. **Frithia**
87 Leaves without a waxy coat 88
– Leaves with a conspicuous WAXY COAT, brown, clavate to oblique trigonous,
very firm, without dots .. .. **Cerochlamys**
88 Flowers white or violet 89
– Flowers yellow 94
89 Pairs of leaves very obliquely united or unequal in length.. 90
– Pairs more or less equal 91
90 Leaves nearly flat, strikingly very OBLIQUELY united in pairs, soft and pulpy,
faintly papillate, with hyaline dots; flowers solitary, sessile between the bases
of leaves; chambers of capsules usually 7, with stiff roofs, without tubercles;
expanding keels very broad, without wings, with awns .. .. **Diplosoma**
– Leaves subcylindrical, EQUALLY united for about one quarter of their length,
with large papillae, sometimes reddish; flowers with a few stamens only; capsule
similar to above .. .. **Maughaniella**
91 Stigmas elongate, subulate or filiform 92
– Stigmas short 93
92 Sepals and petals united in a short TUBE above the ovary; chambers of capsules
without roofs and without tubercles; expanding keels contiguous, forming a
ridge of the valves, wings large .. .. **Herreanthus**
– Sepals and petals FREE to the ovary; chambers of capsules with roofs and with
tubercles; expanding keels diverging, valves without wings .. .. .. **Ruschia**
93 Margins of the leaves SERRATO-DENTATE; flowers open for a short time only;
sepals and petals not united in a tube; stigmas 5; chambers of capsules with
roofs and large tubercles; expanding keels diverging, ending in awns .. .. **Acrodon**
– Margins of the leaves ENTIRE; flowers open for several weeks; sepals and petals
united in a short TUBE; nectary glands absent; chambers of capsules open, roofs
reduced to a rim; valves with contiguous keels and broad wings .. .. .. **Nelia**

94 (88) Bracts below the flowers SIMILAR to the other leaves; flowers usually solitary and short-stalked 95
  − Bracts DIFFERENT from the other leaves; inflorescences usually with several flowers, long-stalked .. .. **Bergeranthus** s.l.
    Here belong described as genera:
    A Flowers on long stalks; lobes of calyx almost equal .. .. .. B
    − Flowers almost SESSILE; lobes of calyx very unequal, with keeled and terete appendices .. .. **Bijlia**
    B Leaves, at least in transmitted light, DOTTED; chambers of capsules with tubercles C
    − Leaves not dotted; chambers of capsules without tubercles .. .. E
    C Dots prominent; tubercles small. See No. 38 .. .. **Hereroa**
    − Dots inconspicuous; tubercles big, large or plane .. .. .. D
    D Tubercles large, forming a more or less 3-sided knob; expanding keels very broad .. .. **Bergeranthus**
    − Tubercles plane .. .. **Rhombophyllum**
    E Margins of the leaves with several and prominent teeth; chambers of capsules open, roofs reduced to a rim; expanding keels without wings .. .. **Tischleria**
    − Margins of leaves with few or inconspicuous teeth; chambers of capsules roofed; expanding keels with broad wings .. .. **Carruanthus**
95 Leaves more than 1 cm broad; sepals almost EQUAL; ovary with 5 points, nectaries united in a crenulated ring; chambers of capsules open, without tubercles; expanding keels connivent in the middle of the valves, with wing-like processes .. .. **Schwantesia**
  − Leaves less than 1 cm broad; sepals very UNEQUAL; ovary truncate at the top or obconical, with 5 separate nectaries; capsules similar to Faucaria, barrel-like; expanding keels at the margins of the valves .. .. **Orthopterum**
96 (75) Leaves GLABROUS, not toothed, or very minutely hairy 97
  − Leaves covered by SOFT HAIRS and with distant teeth, laterally compressed, soft-fleshy, grey-green and with dispersed warts; capsules soft-hairy .. .. **Odontophorus**
97 Leaves with prominent, usually white, but not shining, warts; flowers yellow 98
  − Leaves smooth; flowers yellow, white to violet 101
98 Leaves more or less flat; stamens usually collected in a cone or more rarely in a column; chambers of capsules without tubercles (*Nananthus* group) .. 99
  − Leaves thick, more or less keeled; stamens diffuse; chambers of capsules with tubercles (see No. 104, 114) .. .. **Cheiridopsis**
99 Chambers of capsules with well developed ROOFS, sometimes recurved .. 100
  − Chambers of capsules OPEN, roofs absent or reduced; expanding keels ending in wings or awns .. .. **Nananthus**
100 Leaves SPATULATE; lobes of the calyx free to the ovary; chambers of capsules roofed, 6-14; expanding keels ending in membranous points .. .. .. **Aloinopsis**
  − Leaves ELONGATE-TRIANGULATE, broadest near the base, keeled near the apex; calyx on a short tube above the ovary; petals spreading in several planes; expanding keels of the valves of the capsules with acute tips and broad wings .. **Rabiea**
101 Leaves not club-shaped and without a window at the apex 102
  − Leaves CLUB-SHAPED with a large, hyaline window at the apex, opposite, fascicled, usually more or less embedded in the ground, fresh-green .. .. .. **Fenestraria**
102 Sepals free to the ovary, or if united then leaves with translucent dots; flowers never deep-yellow (*Khadia*) 103
  − Calyx-lobes 5-6, produced in a short tube above the ovary; 8-12-15 filiform stigmas (see No. 114) .. .. **Pleiospilos**

110 (73) Stigmas 4-24, FREE to the base or on a short style 111
  − Stigmas in a simple, sessile knob; leaves whitish, semi-ovoid (see No. 74) .. **Argyroderma**
111 Stigmas or chambers of ovary or capsules 8-20, rarely less, but then bodies without windows or the stigmas are plumose 112

  − Stigmas or chambers 4–7 115
112 Margins of the leaves SMOOTH 113
  − Margins of the leaves and bracts TOOTHED; leaves greenish-brown, usually minutely hairy (pocket-lens!) .. .. **Vanheerdea**
113 Leaves more or less densely dark-DOTTED; flowering stalks with bracts .. 114
  − Leaf-surface whitish, roughly granulated (pocket-lens!), plants resembling a *Lithops* but without windows (see No. 109) .. .. **Dinteranthus**
114 Leaves with numerous dark dots, BROWNISH-GREEN; stone-like, very thick; sepals united in a short tube (see No. 102) .. .. **Pleiospilos**
  - Leaves with scattered dark spots, whitish to grey-green, rarely red-brown; calyx-lobes free to the ovary (see Nos. 98 and 104) .. .. .. **Cheiridopsis**
115 Flowering-stalks with bracts (sometimes hidden within the bodies), rarely without but then sepals united into a tube or flowers on lateral stalks .. 116
  − Flowering stalks without bracts, produced in the centre of the bodies .. 118
116 Sepals free to the ovary; chambers of capsules with roofs; expanding keels diverging, without wings (except *Didymaotus*) 117
  − Sepals UNITED into a tube above the ovary; chambers of capsules not roofed, without tubercles; expanding keels contiguous, with wings .. .. .. **Conophytum** s.l.
    Allied to *Conophytum*, described as genera:
    A Also the PETALS united into a tube .. .. B
    − Petals FREE or very shortly united, white with reddish points; leaves fused, with a narrow orifice in the bodies, papillate or hairy (pocket-lens!), small, ovoid; bracts exserted from the orifice .. .. **Oophytum**
    B Bodies with small to MINUTE windows; orifice small and central or a diverging cleft on top of lobes rounded or flattened, rarely hairy, including *Berrisfordia* with hunch-backed dentate leaves .. .. **Conophytum**
    − Bodies with LARGE windows, with a cleft right across the whole upper side, free tops of the leaves (lobes) short and rounded; flowers never yellow with short styles .. .. **Ophthalmophyllum**
117 Flowering stalks lateral on both sides of the bodies, with small, thick bracts; leaves short, thick, broad-ovoid, keeled; flowers white to violet, filaments violet; chambers of the capsules with roofs and with tubercles .. .. .. **Didymaotus**
  − Flowers in the CENTRE of the bodies, violet to white; leaves on a branchlet only two, light grey-green; chambers of capsules with roofs and conspicuous tubercles (*Ruschia dualis, R.elevata*, etc.) .. .. **Ruschia**
118 Top of the bodies rounded to semiglobose, rarely compressed, but then without windows, often split into two unequal parts; flowers never yellow; chambers of capsules with thin roofs, rarely without tubercles.. .. .. **Gibbaeum** s.l.
    Allied to *Gibbaeum*, described as genera:
    A Chambers of capsules with roofs .. .. B
    − Chambers of capsules without roofs; expanding keels contiguous with broad wings; bodies silky-hairy with a small lateral slit-like orifice .. .. **Muiria**
    B Leaves nearly equal in length, smooth; flowering-stalks with two fleshy bracts with membranous margins; nectaries united in a ring; capsule large, firm; seeds large .. .. **Antegibbaeum**
    − Leaves unequal in length, mostly hairy; flowering stalks without bracts; nectaries separate; capsules small to medium-sized, more or less soft, light coloured, seeds minute .. .. **Gibbaeum**
  − Tops of the bodies truncate, with large hyaline windows or with foveolate or spotted windows; bodies turbinate to terete, more or less bedded in the ground; chambers of the capsules open, without tubercles; flowers yellow or white, opening in the afternoon .. .. **Lithops**

The Key shown on pages 15–23, those genera that are not divided by numbers but by the letters A–D and so on, are very closely related to one another. If the small differences mentioned are later found to be inconsistent, these genera marked by letters can be united with the one to which they are most closely related.

Supplement. Not included in the key are the last genera published by Dr L. Bolus in the *Journal of South African Botany* since 1959.

On the Delimitation and the Taxonomic Rank of the Mesembryanthemaceae

by Prof. Dr H.-D. Ihlenfeldt and Prof. Dr H. Straka[1]

Since the twenties the large genus **Mesembryanthemum** L. has been divided into new genera by L. Bolus, N. E. Brown, G. Schwantes and others by means of a new taxonomic item: the morphology of fruits. Jacobsen (1960) listed 122 genera with about 2 500 species in his *Handbook of Succulent Plants*. Splitting up the old Linnean genus in that way surely involves exaggeration, and a comprehensive revision; reduction will be necessary and has in fact been started by several authors. We do not intend to deal with these problems here; but we wish to discuss the following two questions:

1. In what manner can the Mesembryanthemums (i.e. all the derivates of the Linnean genus Mesembryanthemum) be delimitated from their nearest relatives?
2. Which taxonomic rank should be attributed to this group?

1. Delimitation of the Mesembryanthemums

Concerning the order Centrospermae, to which the Mesembryanthemums belong, it is well known that there are more or less gliding transitions between most of their taxa (see Pax and Hoffmann, 1934). It is striking to see how groups of this order are interconnected with one another even down to the species level. This is especially true for those taxa that Pax and Hoffmann classified under the family Aizoaceae "A. Br."[2] which, according to these authors, include the Mesembryanthemums.

It is therefore logical to state that the Mesembryanthemums can be defined and delimitated only by combining certain characteristics which may also be found in neighbouring groups:

1. The multitude of generally showy petaloid staminodes. Staminodes also occur in a lesser number with some genera of the Molluginaceae Hutch. (e.g. **Limeum** L., **Orygia** Forsk.) which also belong to the Aizoaceae s.l. (see p. 26).
2. The inferior ovary. With the Mesembryanthemums the ovary is included in the receptaculum for at least half its height. In other relatives, an inferior ovary occurs only in **Tetragonia** L. and **Tribulocarpus** S. Moore (i.e. Tetragoniaceae Nakai, see p. 26).
3. The characteristic hygrochastic capsules or fruit types – often connected by transitions with the former – that may be derived from these (Schwantes, 1952, Straka, 1955, Ihlenfeldt, 1959a, 1959b, 1960). In the whole vegetable kingdom, hygrochastically splitting capsules of similar

construction seem to occur in a second instance only with **Aizoanthemum** Dinter ex Friedr., also a representative of the Aizoaceae s.l.

There are other minor characteristics, which are of importance since they are more widespread among the relatives: the **morphology of pollen,** which is the same in the most aberrant genera of the Mesembryanthemums (Straka, 1955), as well as with the rest of the Aizoaceae s.l. (Erdtman, 1952); the **number of chromosomes:** as far as it is known to-day, all Mesembryanthemums have the basic number of $x = 9$, while the Tetragoniaceae Nakai have the number of $x = 8$; for the remaining Aizoaceae s.l. in some instances the numbers of $x = 8$ and $x = 9$ are given for one and the same species; a nearly **continuous distribution area:** at least 99 per cent of all Mesembryanthemums are indigenous to South Africa (only a few occur in other parts of Africa, in Australia, South America and the Mediterranean, and in some islands of the Atlantic and the Indian Ocean . . . one may ask whether these representatives have been introduced there by man); the distribution area, and in one case the centre of dispersal as well, is at least to a certain degree identical to that of related groups; the **nectary glands,** which are homologous in spite of their sometimes quite different outer appearance (Ihlenfeldt, 1960); on the form of the nectaries with the most closely related groups there is not yet exact information.

The two characteristics given first – petaloid staminodes and the inferior ovary – are found among the relatives only here and there; although these characteristics are rather simple, the probability of a very close homology is therefore very high (see Remane, 1952). The feature of the hygrochastic capsule even has an extraordinary high grade of homology, since capsules of this distinct construction are evidently unique in the whole vegetable kingdom. Thus one can assert with highest probability that the Mesembryanthemums, as they are delimitated here, are of phylogenetically uniform origin, i.e. they are a natural group. Up to now, this has been doubted only by Friedrich (1955), who did not give any reasons.

2. The taxonomic rank of the Mesembryanthemums

The taxonomic classification of the relatives of the Mesembryanthemums (Aizoaceae s.l.) has been discussed quite recently with different results (e.g. Hutchinson, 1926, 1959, Pax and Hoffmann, 1934, Herre and Volk, 1948, Friedrich, 1955, 1956, 1959, Schwantes, 1957, 1960, Herre and Friedrich, 1959, Volk, 1960). To begin our considerations, we choose the grouping of Pax and Hoffmann; we shall give an alternative proposal of our own in the following review. (With regard to the family names

---

[1] Extract from an article published in Ber. Dtsch. Bot. Ges. 74 (1961), 485.492 and revised by H.-D. Ihlenfeldt according to recent publications. The authors dedicated this paper to Mr Herre on occasion of his 75th birthday.

[2] In the following quoted as "Aizoaceae s.l."

and their authors, see Bullock, 1958a, 1958b, 1959, Buchheim, 1963 and ICBN, 1966.)

Proposal of Pax and Hoffmann, 1934	Proposal of the authors of this paper
Aizoaceae Rudolphi (= Aizoaceae s.l.) = Ficoidaceae Juss. nom. illeg.	
Tribes	
1. Gisekieae	Molluginaceae Hutch.[1]
2. Orygieae	(nom. conserv.)
3. Limeeae	(exl. **Acrosanthes** Eckl. et Zeyh.)
5. Mollugineae	
4. Mesembryanthemeae	
(*a*) Mesembryantheminae	Mesembryanthemaceae Fenzl
(*b*) Aizoinae	Aizoaceae Rudolphi (= Aizoaceae s.str.) (excl. **Glinus** L. and **Glischrothamnus** Pilger)
6. Tetragonieae	Tetragoniaceae Nakai

Following the general tendency of contemporary taxonomy to diminish the range of families, Hutchinson (1926) separated the tribes 1, 2, 3 and 5 as the distinct family Molluginaceae Hutch.[1] from the rest, which he called Ficoidaceae Juss. With regard to the diagnoses given by Hutchinson and later by Friedrich (1955), the differences between these two families are slight: the Molluginaceae show little (even very little) succulence in contrast with the expressed succulence of the Ficoidaceae (*sensu* Hutchinson); furthermore, the Molluginaceae are characterized by a herbaceous dryskinned perianth that is free whereas the succulent tepala form a tube with the Ficoidaceae (*sensu* Hutchinson). In both the families, the fruit types exhibit a great diversity.

Nakai (1942) was the first to realize that the remainder of the Aizoaceae s.l. (i.e. the Ficoidaceae of Hutchinson) was by no means a homogeneous group. He therefore separated the genera **Tetragonia** L. and **Tribulocarpus** S. Moore, which have three characteristics in common (inferior ovary, indehiscent fruits and apically inserted anatrope ovules), as the distinct family Tetragoniaceae. In this he was independently confirmed by Friedrich (1955, Tetragoniaceae "Link em. Friedr."). The logical consequence of also regarding the Mesembryanthemums, in the limits given above, as a distinct family, was decidedly rejected by Friedrich. This family had already been established by Herre and Volk (Mesembryanthemaceae "Herre et Volk") in 1948 and also by a number of the earlier authors. When declining the Mesembryanthemums as a distinct family, Friedrich gave two reasons: first, according to his conviction, the Mesembryanthemums are no natural group and secondly, there is a connecting link between the Aizoaceae (s. str.) and the Mesembryanthemums, the genus **Aizoanthemum** Dinter ex Friedr.

In our opinion, the arguments of Friedrich are of no consequence whatsoever. As has already been demonstrated, the Mesembryanthemums are most probably a natural group and the existence of transitions seems to be one of the characteristics of the whole order Centrospermae. Generally, one can state that transitions between closely related taxa must have existed at a time; the finding of fossilized remains or even living descendants of these connecting links must be fortuitous.

In these pages we shall briefly consider how far **Aizoanthemum** is really a connecting link between the Aizoaceae (s.str.) and the Mesembryanthemums, particularly the more primitive representatives classified in the subfamily Mesembryanthemoideae. Schwantes (1952) was the first to describe the remarkable fruit of this genus; Friedrich (1957) dealt with the genus in a monograph. The fruit of **Aizoanthemum,** like that of the typical Mesembryanthemums, is a hygrochastically splitting capsule, the opening mechanism of which is formed by expanding keels similar to those of the Mesembryanthemums. But the capsule is devoid of false septa, a decidedly characteristic feature of the Mesembryanthemums. (They may, however, be missing in some advanced members.) The ovary is completely free, whereas with the Mesembryanthemums the ovary is included in the receptaculum up to at least half its height. Like many advanced Mesembryanthemums, **Aizoanthemum** shows a tendency to increase the number of carpels (pleiomeric gynaecium) and stamina. The last-mentioned feature is widespread among the relatives. But whereas in **Aizoanthemum** there are no staminodes at all, they are always present in the Mesembryanthemums. The tepala of **Aizoanthemum** are coloured and equal in form and size, while those of the Mesembryanthemums are green and always unequal, in some cases to an extreme degree. The seeds of **Aizoanthemum** are covered by characteristic channels which give them a completely different appearance from those of the Mesembryanthemums, which show papillae, or which are smooth in the more advanced representatives. Summing up, one may state that without doubt **Aizoanthemum** does not belong to the Mesembryanthemums, and that it does by no means connect the Aizoaceae s.str. more perfectly with the Mesembryanthemums than some other genera do.

Moreover, the separation of the Mesembryanthemums as a family of its own seems to be desirable for practical purposes. The great number of **Mesembryanthemum** species described – it is, for example, higher than that of the Caryophyllaceae, a member of the same order – and the strange fact that small groups of isolated forms exist beside groups that comprise numerous species, require for survey a large number of gradated taxa, which is the practical task of taxonomy. It is then not advisable to assume for a taxon rich in species (as with Mesembryanthemums) a rank lower than circumstances demand.

The separation of the Mesembryanthemums as a distinct family adopted, the correct name of this family has to be Mesembryanthemaceae Fenzl (1836), as has been pointed out by Buchheim, 1963, notwithstanding the fact that the Mesembryanthemaceae of Fenzl comprised only what we now call the sub-family Ruschioideae. According to the ICBN, the name Ficoidaceae Juss. has been illegitimate since 1961, as it was based on the illegitimate genus name Ficoides Miller. The earlier names Mesembryaceae Dumortier, Anal. Fam. Pl. 37, 41 (1829) ("Mesembryneae") and Mesembryaceae Link, Handb. 2: 12 (1831) ("Mesembrinae") are also illegitimate, as they were founded on the illegitimate Mesembryum Adanson. Moreover, the name Mesembryanthemaceae Fenzl has been listed by the ICBN since 1966 under the conserved family names (*nomina familiarum conservanda*).

---

[1] *Nomen conservandum* against Glinaceae Link (see Buchheim, 1963 and ICBN, 1966.)

Literature Cited

Buchheim, G., 1963: Die Benennung der Familien sukkulenter Pflanzen. Kakt. u.a. Sukk. 14 (1963): 73–75 and 92–94.

Bullock, A. A., 1958a: Indicis Nominum Familiarum Angiospermarum Prodromus, Taxon 7 (1958): 1–35.

–, 1958b: Indicis Nominum Familiarum Angiospermarum Prodromus. Additamenta et Corrigenda I. Taxon 7 (1958): 158–163.

–, 1959: Nomina Familiarum Conservanda Proposita. Taxon 8 (1959): 154–181.

Erdtman, G., 1952: Pollen Morphology and Plant Taxonomy. Angiosperms. (An Introduction to Palynology. I.). Stockholm and Waltham, Mass., U.S.A.

Fenzl, E., 1836: Monographie der Mollugineen und Steudelieen, zweier Unterabteilungen der Familie Portulacaceen. Ann. Wiener Mus. Naturgesch. 1 (1836): 337–384.

Friedrich, H. C., 1955: Beiträge zur Kenntnis einiger Familien der Centrospermae. Mitt. bot. München 11 (1954–58): 56–60.

–, 1956: Studien über die natürliche Verwandtschaft der Plumbaginales und Centrospermae. Phyton 6 (1956): 220–263.

–, 1957: Aizoanthemum Dinter ex Friedr., eine wenig beachtete Gattung der Ficoidaceae aus Südwestafrika. Mitt. bot. München 2 (1954–58): 339–349.

–, 1959: *Reihe Centrospermae*. In HEGI, Illustrierte Flora von Mitteleuropa, II/2, 1. Lfg.: 453–457. München.

Herre, H. and H. C. Friedrich, 1959: *Die geographische Verbreitung der Ficoidaceae-Mesembryanthemoideae in Südafrika. 1. Teil: Aptenieae.* Mitt. bot. München 3 (1959–60): 44–70.

–, and O. H. Volk, 1948: Mesembryanthemaceae Herre et Volk, familia nova. Sukkulentenkunde, Jb. schweiz. Kakteen-Ges. 2 (1948): 38.

Hutchinson, J., 1926: *The Families of Flowering Plants*. Vol. 1. London.

–, 1959: *The Families of Flowering Plants*. 2nd Ed., Oxford.

Ihlenfeldt, H.-D., 1959a: Über die Entwicklung der Blüte und den Bau der Frucht von Caryotophora skiatophytoides Leistn. (Ficoidaceae Juss. em Hutchinson – Formenkreis Mesembryanthemen). Z. Bot. 47 (1959): 490–504.

–, 1959b: Über die Samentaschen in den Früchten der Mesembryanthemen. Ber. Dtsch. Bot. Ges. 72 (1959): 33–342.

–, 1960: Entwicklungsgeschichtliche, morphologische und systematische Untersuchungen an Mesembryanthemen. Fedde's Repert. 63 (1960): 1–104.

Jacobsen, H., 1960: *A Handbook of Succulent Plants*. Vol. III. Mesembryanthemums. Jena.

Nakai, T., 1942: Journ. Jap. Bot. 18 (1942): 103 (not seen in original).

Pax, F. and Hoffmann, K., 1934: *Aizoaceae*. In ENGLER-PRANTL, *Die natürlichen Pflanzenfamilien* 16c, 2. Aufl. Leipzig.

Remane, A., 1952: *Die Grundlagen des natürlichen Systems, der vergleichenden Anatomie und der Phylogenetik*. Leipzig.

Schwantes, G., 1952: Die Früchte der Mesembryanthemaceen. Vierteljschr. naturforsch. Ges. Zürich 97 (1952), Beih. 2.

–, G., 1957: *Ficoidaceae (Juss.) em. Hutchinson. Kakt. u.a. Sukk.* 8 (1957): 156–158 and 167–169.

–, 1960: The Mesembryanthemums. In Jacobsen, H., 1960: *A Handbook of Succulent Plants*. Jena.

Straka, H., 1955: Anatomische und entwicklungsgeschichtliche Untersuchungen an Früchten paraspermer Mesembryanthemen. Nova Acta Leopoldina N.F. 17 (1955). Leipzig.

Volk, O. H., 1960: Flowers and Fruits of the Mesembryanthemums. In Jacobsen, H., *A Handbook of Succulent Plants*. Jena.

Some Aspects of the Biology of Dissemination of the Mesembryanthemaceae
by Prof. Dr H.-D. Ihlenfeldt

The epoch-making hypotheses of Charles Darwin on the origin of the species have not only put modern systematics on a new basis but have also decisively influenced many other branches of biology. One of the branches that was influenced by Darwin's teachings, is the study of the dispersal mechanisms of plants, a branch of botany which in recent years has become known as the biology of dissemination.

How could this branch of botany be stimulated by Darwin's writings? The answer is simple. Plants were examined for characteristics which enable them to survive in the struggle for existence, or which may even give them an advantage over competing species. It soon became clear that "dissemination units" are of fundamental importance in this connection. "Dissemination unit" is the term commonly used today for parts of the plant which bring about its propagation, whether they are vegetative parts of the plant which become detached, or whether they are fruits, seeds or parts of fruits.

On closer inspection it became apparent that, in order to survive, it is not enough for a plant to produce a sufficient number of dissemination units; it is of much greater importance that these units are, in fact, dispersed. It is obvious that in the case of a perennial plant the locality in which the dissemination units are produced, is already occupied, and no new individual can develop there. In most cases it is therefore essential that the dissemination units are transported over a certain distance. If we continue this line of thought, the following emerges:

If new space is in one way or another created for colonization, it will be the species with the most effective dissemination units which stand the best chance to include this new *Lebensraum* in their area of distribution. When we speak of the creation of new space, we do not necessarily think of new islands or areas rising out of the sea, although such occurrences were most certainly important in ages past; it suffices to think of an area that has been denuded of its vegetation by floods, fires or glaciers.

The botanists of the last century thought along similar lines, and they began to search the plant kingdom for adaptations which bring about an effective dispersal of the dissemination units. The number of dispersal mechanisms that were found is staggering, and we have today massive volumes in which they are enumerated and described. It is impossible to give even a survey of this multitude of mechanisms in a short paper. I shall merely point out that two main groups of dispersal mechanisms are generally distinguished; the first is classified as autochorism, and the second as allochorism. In autochorism, the dissemination units are actively transported by the mother plant without the aid of foreign agents. In this category we have the numerous explosion mechanisms, the best known of which is probably the squirting cucumber, as it is commonly known. In the second group – dispersal mechanisms classed as allochorism – the dissemination units are transported from the plant by foreign agents such as water, wind, animals or man. Two examples in the indigenous flora are Pterodiscus, which has winged fruits which are windborne, and Harpagophytum, of which the hooked fruit is transported by animals. The adaptations for the promotion of dispersal were studied so intently, that it was completely overlooked that some plants possess adaptations which, paradoxical though it may sound, prevent or inhibit the dispersal of dissemination units. The complex of phenomena which comprises the adaptations inhibiting dispersal is today covered by the term antitelechorism. Some cases of inhibited dispersal have been known for a long time. Examples are the ground-nut, which buries its fruit in the vicinity of the mother plant, or *Linaria cymbalaria*, which hides its seed capsules in rock crevices, a fact which caught the attention of physiologists, who studied it in detail. Generally these examples were simply passed over, as they did not fit in or did not seem to fit in with preconceived ideas of the selective value of an effective dispersal of the dissemination units.

Let us ask ourselves the following questions:

1. How did it happen that the examples of antitelechorism, which are by no means rare, were so long overlooked?
2. Is inhibited dispersal always to be regarded as a disadvantage, or are there cases in which a certain inhibition of dispersal is advantageous?

It must be remembered that, in most cases, inhibition is not complete. The very fact that inhibited dispersal is so common proves that it cannot always be a disadvantage.

The answers to our two questions are closely linked, and I shall deal with them at the same time. The fact that the antitelechoric phenomena were so long overlooked can be easily explained when we consider that the botanists who studied telechoric mechanisms, those that encourage dispersal, worked in Europe, where antitelechoric adaptations are extremely rare. This emphasizes how important it is for a botanist to visit other countries and to investigate other habitat conditions.

The first botanist who intensively studied antitelechoric phenomena, was Zohary (1937), who worked in Palestine, a country in which many plants possess antitelechoric adaptations. Basing his conclusions on extensive statistical investigations, Zohary found that antitelechoric adaptations are commonly met only in arid and semi-arid regions. This conclusion is the key to the understanding of the fact that inhibited dispersal need not be disadvantageous, but that it can in certain conditions be a distinct advantage.

In areas such as Central Europe, where the climate is moist or moderate, we find an almost closed vegetation cover. Furthermore, the majority of plants are perennials. Only a small fraction of the dissemination units is therefore likely to find suitable living conditions on or near the site of the mother plant. Let us, for example, picture a closed forest community. No dissemination unit (or nearly none) which fall near the mother plant is likely to find favourable conditions. Only at some distance from the mother plant, namely outside the forest, does it have a chance to develop.

In arid regions we find completely different conditions. The flora is characterized by a very high percentage of xerophytes, plants which complete their life-cycle in a short time and die off rapidly. In many cases, therefore, the space occupied by the mother plant becomes vacant after every vegetation period. Moreover, the vegetation cover in arid regions is generally sparse. The total available surface is much greater than the area capable of sustaining plants. In this case, partially inhibited dispersal can be a distinct advantage, as it will prevent the dissemination unit from being transported far away from the mother plant. We need only picture the mother plant growing in a slight depression in which water collects during the rainy season. The moisture in that locality may be just enough to support that particular plant species, while in the surrounding area it would be insufficient. It is easy to see that in such a case, all the dissemination units that are transported some distance away from this mother plant, are doomed. On the other hand, units which settle near the mother plant in the depression, do have a chance of growing in these suitable conditions. What we considered so far, was inhibition of dispersal in space. There is, however, also an inhibition of dispersal in time. For this phenomenon, Zohary chose the somewhat unsuitable term aestatiphorism, which means that dissemination units are held back during summer. This term is apt enough for conditions in Palestine, which is a winter rainfall area. For areas in which the rain falls during summer and where dissemination units are held back during winter, the term is obviously inaccurate.

Generally speaking, we can say that aestatiphorism, the inhibition of dispersal in time, is restricted to arid and semi-arid

Fig. 1: Longitudinal section of a young fruit with parietal placentation. r = receptacle, sz = stalk zone of the carpel, fs = false septum arching down into the locule from the midrib of the carpel, ps = false septum raising the placenta, p = placenta.

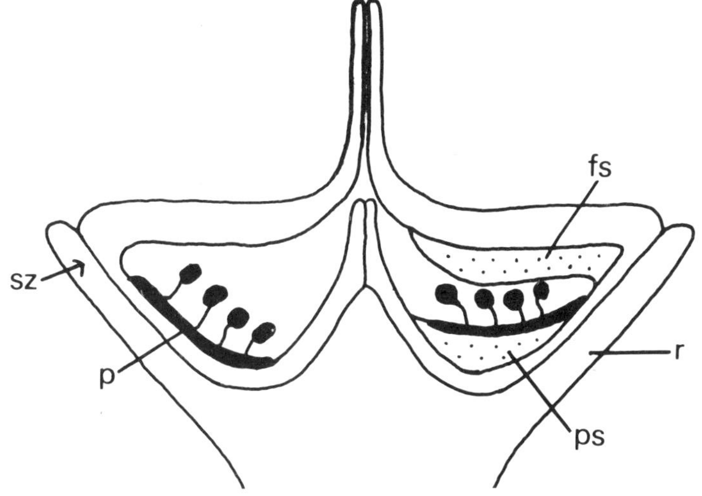

regions with periodic rainy seasons. During the dry season, the dissemination units – usually seeds – are held back, and they are released only during the rainy season which favours growth. It is clear that a complete inhibition of dispersal in time is at the same time an inhibition of dispersal in space. We need only compare a light seed which is released at the beginning of the dry period and is then blown about by the wind for several months, with a seed which is released at the beginning of the rainy season which shortly afterwards anchors itself by means of its growing radicle.

But enough of theory. Taking the Mesembryanthema as an example, I would like to show how complicated matters really are; I would like to demonstrate how closely inhibition and encouragement of dispersal may be interlocked, and in what unique fashion they are instrumental in adapting the Mesembryanthema to their surroundings.

In order to understand the construction of the fruits of the Mesembryanthema, we must first look at the structure of the ovary. The gynaecium of the Mesembryanthema consists of three to many carpels, which are sunk more or less deeply into the receptacle (fig. 1). By studying the course of the vascular bundles, it can be shown that the stalk zone of the carpels of the Mesembryanthema is situated either on the columella of the obconical receptacle (central placentation) or along the upper edge of the receptacle (parietal placentation), as we find it in the Ruschioideae. A shifting of the stalk zone from a central into a parietal position can be observed during the ontogenesis of the Ruschioideae. The tubular carpels are fused laterally, and form thus the true septa of the ovary. The tip of each carpel is drawn out into a styloid organ which also bears the stigma. During the course of the ontogenesis each carpel, and thus each locule, may be more or less completely divided lengthwise by two false septa. The septum most commonly formed arches down into the locule from the mid-rib of the carpel. When the capsule bursts open in a loculicidal fashion, this false septum breaks up into two lamellae and forms membranous appendages on the valves. The second type of false septum is quite rare. It is formed in the region of the placenta, opposite the false septum mentioned. During its development it raises the placenta.

There are a few exceptions, which I shall deal with later in some detail, but at least 98 per cent of all known Mesembryanthemums have a loculicidal, hygrochastic capsule. The term loculicidal means that the carpels split open along the dorsal suture; hygrochastic means that the capsule opens when it becomes moist, and closes when it dries out again. The opening mechanism of Mesembryanthema capsules was first described in detail by Steinbrinck (1883), while the anatomy of the structures involved was studied by Garside and Lockyer in 1930. The special structure which brings about the opening of the capsule when it becomes wet, originates (in most of the investigated cases) in the epidermal lining of the locule. The only exception known is Dorotheanthus, in which deeper layers also take part in the forming of the expanding tissue. Straka (1955) found that the cell walls of the expanding tissue show a pronounced secondary thickening. The cells of the expanding tissue – which, incidentally, are dead – have a rhomboidal shape, with one diameter of the cell being much less than the other. On swelling, the shorter axis of the rhombus elongates con-

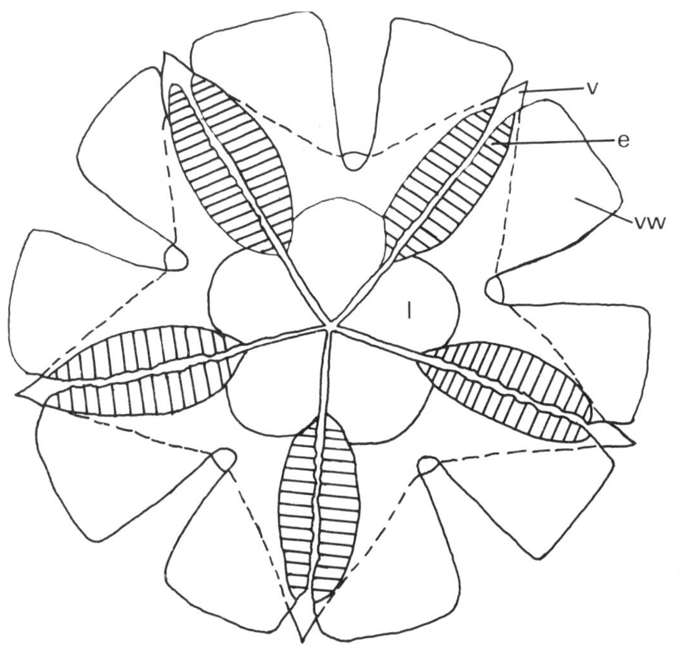

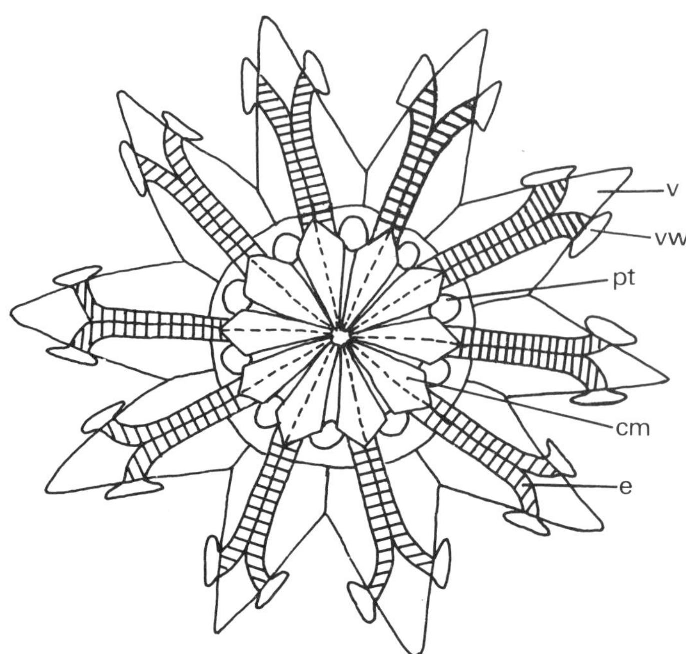

Fig. 2a: Open capsule of a primitive type (*Delosperma*) without covering membranes, seen from above. v = valve, vw = valve-wings, 1 = locule, e = expanding keel.

Fig. 2b: Open capsule of a complicated type (*Cephalophyllum*) with covering membranes or cell lids (cm) and placental tubercles (pt).

siderably, while the longer axis of the expanding tissue remains almost unchanged. This results in a pronounced elongation of the expanding tissue in one direction. The other tissues of the capsule do not change substantially when they become wet. They form the abutment for the expanding keels, so that the valves fold outwards when they are wet.

Seeds are, indirectly, ejected from the open capsule by raindrops. This process was first correctly described by Berger (1908), later by Garside and Lockyer (1930), and in recent years by Volk (1961). The authors mentioned experimented with artificial rain and found that in the simple types of capsules in which the seeds are exposed when the capsule is open (fig. 2a), the seeds were ejected to a distance of 30 cm. Unfortunately we have no exact observations which were made in the natural habitat. The size of the raindrops probably has an important influence on the distance which the ejected seeds cover, but we have no exact information about this. It is possible that, in natural conditions, seeds are ejected much farther.

Let us now consider the ejection mechanism in relation to conditions in the natural habitat. As the seed coats of the Mesembryanthema are not very hard nor very robust, it is probably a great advantage that the seeds remain enclosed in the capsule during the dry season, subjected to total inhibition of dispersal in space as well as in time. The capsule, on the other hand, is in most cases very robust and resistant. It is known that in the natural habitat the capsule of many species resists decomposition for several years. As the seeds are usually light and covered only by a soft testa, they would be dispersed for long distances by the wind and might even be damaged by sharp sand grains if the capsules were to open at the beginning of the rainy season. When the capsules open after the first rains, the seeds are transported by the process described for the desired short distances only. As most seeds germinate rapidly

(a phenomenon which I shall deal with later) the seeds are quickly anchored in the soil. A secondary dispersal by agents such as water and wind would therefore not play an important role in the case of the Mesembryanthemums.

As we gather from an extensive study by Stopp (1958), the phenomenon of aestatiphorism characteristic of the Mesembryanthemums, is frequently met in South African flora. Up to this point, the dissemination ecology of the Mesembryanthemums is of no particular interest. Most Mesembryanthemum capsules however have a more complicated structure than those I mentioned up to this point. The morphology of the capsules plays an important role in the systematics of the Mesembryanthemums. The numerous complications in capsule structure have unique results, which we shall now investigate in some detail.

The most common additional structures of the Mesembryanthemum capsule are the so-called covering membranes of the locules (fig. 2b). When these are present, the seeds are not exposed when the capsules open, but the locules are more or less covered by membranes which emanate to the left and right of the true septa of the capsule. The covering membranes are elastic and are never fused in the middle of the locule, although they differentiate from the endocarp lining the roof of the locules. The division in the centre line of the locule is always brought about by the valve-wings which are formed by the splitting of the false septum which projects into the locule from above. The valve-wings always remain firmly attached to the valves. When the capsule opens for the first time, they act like the rip-cords of a parachute, thus bringing about the exact divisions of the two parts of the covering membrane. Numerous intermediate forms between capsules with and without covering membranes are known. It is interesting to note that the more perfectly developed the covering membranes are, the shorter the expanding keels become. Moreover, they tend to shift more and more from the margins of the valves to the base of the valves. This relationship between covering membranes and expanding keel was first noticed by Schwantes (1929), who could offer no explanation for it. But the explanation is quite simple. If one considers that both the expanding mechanism and the covering membranes originate in the endocarp of the roof of the locule, it is easy to see that no expanding keels can be formed in the areas where the endocarp differentiates into covering membranes. A further consequence important for our considerations, is that the covering membranes can never cover the locules completely, as a certain portion of the endocarp on the roof of the locule is required for the formation of the expanding mechanism. But even here the plant, which one is almost tempted to endow with human intelligence, knows a way out: the opening remaining in the covering membranes is reduced to two minute openings by either a swelling of the placenta (the so-called placental tubercle) or, as in the case of the most highly evolved types such as Ruschia, by two ridges on the covering membranes. What influence do the covering membranes have on the ejection of the seeds from the capsule? Garside, Lockyer and Volk have investigated this problem in detail, and have come to a surprising conclusion. They found that in the case of the complicated capsule the seeds are ejected twice as far, which means that seeds will be distributed over an area four times as large. This effect seems to be due to a jet action, which is caused as follows:

The large, relatively slow-falling raindrops possess a high kinetic energy. When they fall in the centre of a capsule, they are split up into a number of smaller drops. The high kinetic energy of the large drops is passed on to the smaller drops, imparting to them a high initial velocity, which enables them to carry seeds along with them. As one can well imagine, the whole mechanism can only work satisfactorily if the covering membranes can bend inwards. This is only possible if they do not form a continuous membrane but consist of two halves, each of which can fold inwards. It thus becomes clear how important it is for the covering membranes to be neatly separated in the centre which, as has been mentioned, is brought about in a rather complicated fashion with the aid of the valve-wings.

The second effect of the closing mechanism is even more interesting: the time required for the complete ejection of all seeds is much longer in the case of complicated types of capsules than in the simpler types. Volk found that, in the more complicated types, a considerable number of seeds was still left in the capsule after 24 hours, while the simpler types were nearly emptied within a few minutes. We are thus confronted by the paradoxical condition that one and the same adaptation, namely the presence of covering membranes and other closing mechanisms, encourage dispersal in space while, on the other hand, it inhibits dispersal in time.

We have seen that many Mesembryanthemum capsules are so constructed that a portion of these seeds is only released after a long delay. The adaptations described so far do not, however, seem to suffice. There are a number of Mesembryanthemum species in which the capsules contain special pockets in which individual seeds are detained. These seeds are released only after the capsules have disintegrated entirely. This can some-times happen only after several years or, particularly after continuous rains.

The seed pockets of the Mesembryanthemums were discovered and described in some detail by Schwantes in 1929. Since then many new cases have become known and investigated. It appeared that the different types of more or less perfectly formed seed pockets are by no means morphologically homologous. All the pockets have only one thing in common: almost invariably one or both of the false septa somehow play a part in their formation.

In this short paper it would be impossible to describe all the types of seed pockets found in the Mesembryanthema. So far three types among the species with central placentation, have been mentioned, and three types among the species with parietal placentation, which are of fundamentally different construction (Straka 1955, Ihlenfeldt 1959). Moreover, the conditions are so complicated that it would be very difficult for anyone who is not well acquainted with the capsule structure of the Mesembryanthemums to understand their construction. I have chosen two types of pockets that occur in the group with parietal placentation. These examples have the advantage that one can even today trace different stages of evolution from a simple primordial type to the fully developed seed pocket which releases its seed only after it has disintegrated.

It is necessary to refer again to the structure of the ovary. We saw that the cavity in the tubular carpel is divided lengthwise into two halves by two false septa (fig. 1). In the case of the examples, which I would like to enlarge on, only one false septum is involved: the one that raises the placenta. As far as we know, it is only in the group with central placentation that the second type of false septum participates in the formation of seed pockets. As we shall see, the whole complex of the seed pockets is greatly complicated by the fact that the formation of seed pockets is almost invariably accompanied by a total alternation of the structure of the fruit.

First, let us consider the most primitive type of pocket formation. Let us look at a longitudinal section of a fruit of Micropterum which has a five-locular capsule (fig. 3). On the

Fig. 3: Longitudinal section of the fruit of Micropterum. Abbreviations as in fig. 1-2. s = seed.

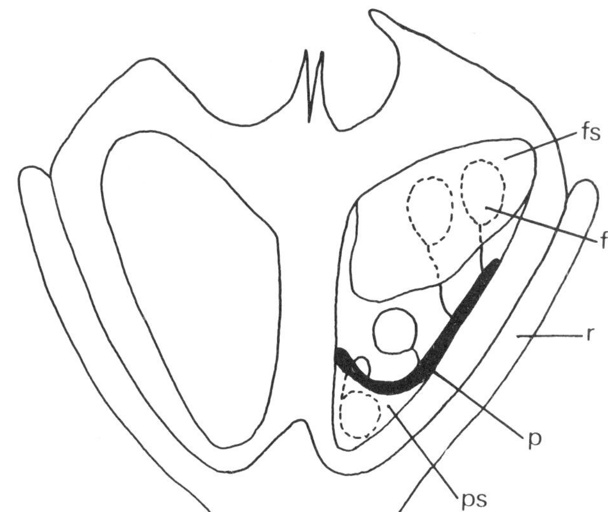

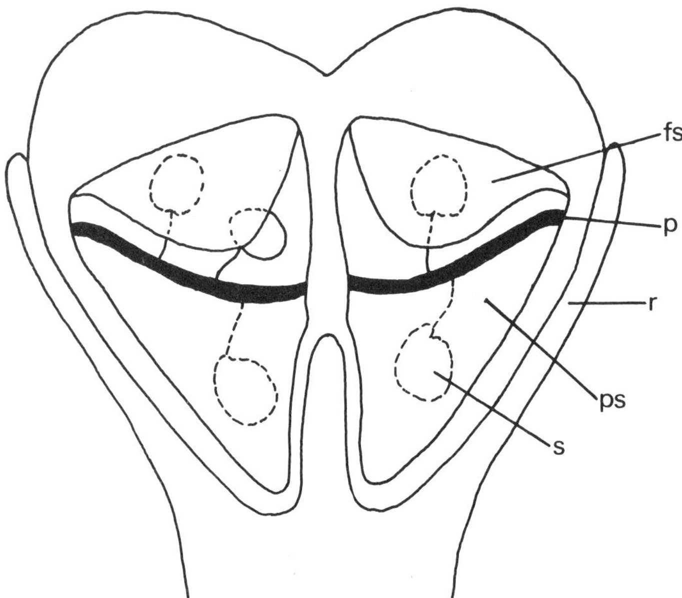

Fig. 4: Longitudinal section of the fruit of Malephora. Abbreviations as in fig. 1-3.

left we have cut through a true septum, while on the right the section is through the centre of the locule. On closer examination, we find that the lowest part of the placenta has been slightly raised by a false septum which, in this case, is only poorly developed. In this way two pockets are formed to the left and right of the lowest part of the placenta in each locule. In these pockets one generally finds one or sometimes two seeds. As the placenta is somewhat swollen, the opening of the pockets is contracted at the top. The opening is, however, still wide enough for the ripe seed to escape. These seed pockets are therefore still imperfect. Nevertheless, the seeds formed in this part of the locule are strongly inhibited in their dispersal. If one inspects an apparently empty capsule of Micropterum, one almost invariably finds seeds in these pockets at the base of the locule.

In the case of Malephora, however, the construction of these pockets is much more perfect (fig. 4). The genus has a multi-locular capsule. The longitudinal section shown in our diagram therefore passes through the middle of locules both on the left and on the right. In the genus Malephora, or at least in some of its species, the pockets occupy almost half the locule. For two reasons the seeds in the pockets of Malephora cannot leave the capsule in the normal way: the locules are very narrow due to the great number of locules present in each fruit, and the ripe seeds are very large. We can therefore in the case of Malephora speak of true seed pockets. It is interesting to note that the halves of the carpels break apart relatively easily and can be readily detached from the receptacle. The capsule of Malephora is thus no true capsule, but can almost be regarded as a schizocarp. The single fruitlets, which are formed when the fruit breaks up, consist of half carpels and must therefore be regarded as clausae.

The fruit of Malephora was found to be an intermediate type; we shall now consider a series of genera illustrating different stages of development from a hygrochastic capsule to a true schizocarp. This morphological series, which also appears to be a phylogenetic series, is formed by the four genera Carpanthea, Apatesia, Conicosia and Herrea (Straka, 1955). Their capsules are all multilocular. Carpanthea has a typical hygrochastic capsule. When the fruit of Carpanthea ripens, the central portions of the carpels become detached from the columella of the receptacle (fig. 5). This phenomenon, together with the splitting of the true septa, are the first indications of a schizocarpic fruit. Carpanthea has distinct seed pockets. As in the case of the other genera in our series, the outer (abaxial) portions of the placenta are raised, in contrast to Micropterum in which the seed pockets are formed by a placenta of which the inner (adaxial) portions are raised. In the fruits of the genera of our series, two seed pockets containing a few seeds are formed in each locule between the raised placenta and the walls of the locule. The openings of the pockets in the fruits of Carpanthea are large enough to allow the seeds they contain to escape when the fruits are ripe. Carpanthea therefore has no true seed pockets.

In the case of Apatesia, Conicosia and Herrea, the placental septum is much better developed; moreover, lateral ridges are formed below the placenta, which narrow down the entrances to the pockets to the extent that the ripe seeds can no longer leave the pockets. These genera thus have true seed pockets.

Fig. 5: Longitudinal sections of the fruits of Carpanthea (fig. 5a), Apatesia (fig. 5b), Conicosia (fig. 5c) and Herrea (fig. 5d). Abbreviations as in fig. 1-3. co = columella.

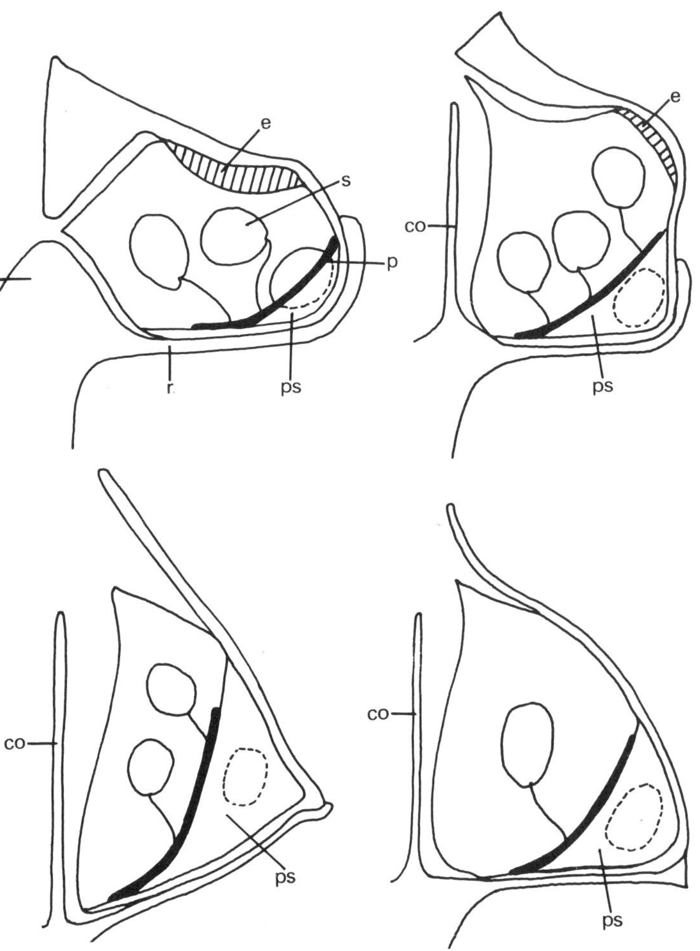

Within the series Carpanthea-Herrea the number of free seeds, seeds not contained in the pockets, decreases. At the same time, the fruits show more and more pronounced tendencies to become schizocarpic; while the fruits of Carpanthea show the first signs of breaking up, the capsules of Apatesia and Conicosia break up easily into mericarps. These mericarps consist of the two halves of neighbouring carpels and two seed pockets, each of which contains one seed. For these mericarps Straka (1955) suggested the name double clausae. Finally, in the case of Herrea, we have a true schizocarp breaking up into double clausae which in turn split to form two clausae.

Parallel with the tendency of the fruits to become schizocarpic, we find a reduction of the expanding mechanism: Carpenthea has fully functional expanding keels; Apatesia too has functional keels but they are considerably shorter. Conicosia and Herrea have none at all; not even rudiments of keels can be traced with certainty.

Let us now consider the dissemination biology of the Carpanthea-Herrea series in some detail. We saw that in the most evolved type of capsule, a large portion of the seeds (about 50 per cent) is enclosed in perfectly developed seed pockets from which they can escape only after the surrounding tissues have disintegrated. This must be regarded as a case of maximal inhibition of dispersal. At the same time, however, something unexpected happens: The hygrochastic capsule gradually changes into a perfect schizocarp. The dispersal unit is now no longer a single seed which is ejected from the capsule that has opened in the rain. It is a light mericarp containing the seeds and, being easily transported by the wind, it is much more effectively dispersed than single seeds. We now have the paradoxical case that a development tending to produce a highly efficient adaptation for the inhibition of dispersal, namely seed pockets, simultaneously produces a mechanism which encourages dispersal.

So far I have only mentioned the relation between capsule structure and the biology of seed dispersal of the Mesembryanthemums. But there is another, completely different factor which has a bearing on the seed dispersal of the Mesembryanthemums: the nature of germination.

The classical biology of seed dispersal never included germination behaviour in its field of study. This was not justified, because without knowledge of the germination, the picture remained incomplete. This becomes clear when we consider that even the fact that a seed germinates very rapidly and anchors itself in the ground by its radicle, must be regarded as an example of inhibition of dispersal. So far only very few and unsatisfactory observations have been published on the nature of germination of Mesembryanthemums. It is, in fact, a completely unexplored field.

The brief statements which follow must therefore be accepted with reservation, particularly as the seed material on which the findings were made was limited, and as it ripened in Europe where climatic conditions are radically different from those in South Africa.

If we make a diagram of the course of germination of a plant species, we generally get the following picture, which for the present we shall call normal (fig. 6). On the ordinate we mark the number of seeds that have germinated up to the present

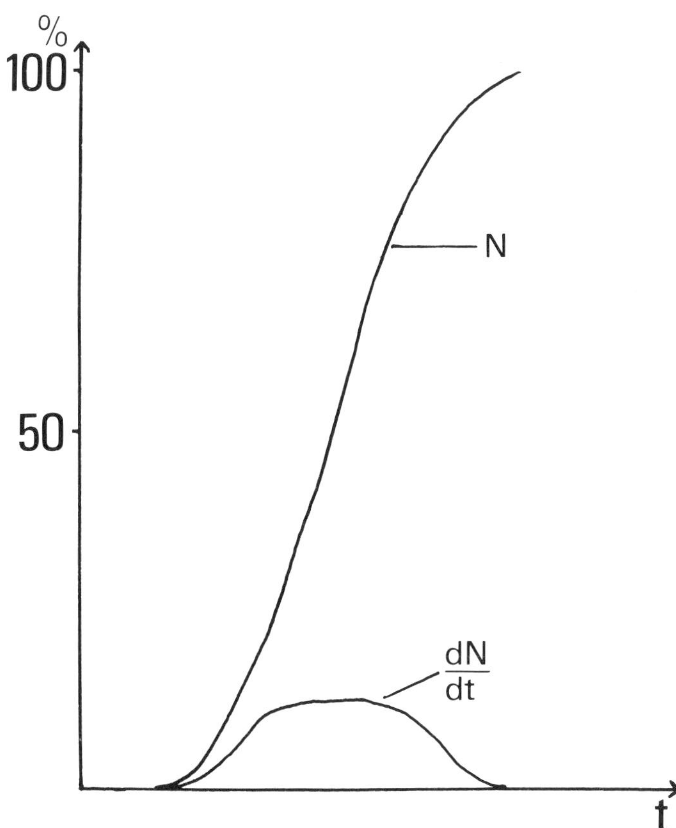

Fig. 6: Diagram of "normal germination" of a plant species. Ordinate: total number of seeds germinated as a percentage; abscissa: time. $\dfrac{dN}{dt}$ = curve of differential-quotient according to time, showing "normal distribution" with only one peak.

moment, and on the abscissa we mark the time that has elapsed since the seeds started to swell. The following are the findings on interpreting the curve: the bulk of the seeds germinate within a fixed period, which is usually very short; only a small number of seeds germinate before or after this period.

If we plot the number of newly germinated seeds per unit of time against time, we get the above curve (fig. 6). This well-known curve the mathematicians call the "curve of normal distribution".

In the case of Mesembryanthemums which I have investigated so far I have, however, found completely different relationships (fig. 7). On interpreting this curve we find that a considerable portion of the seeds germinates within a short time; this is followed by a period in which only a small number of seeds start to germinate. Finally, we have another period of time in which numerous seeds germinate. If we calculate the differential-quotient according to the time, as we did before, we obtain a completely different curve, one with two distinct peaks (fig. 7).

The space of time in which germination takes place can be very prolonged. In the case of the species of Trichodiadema, which I have studied most intensively up to now, the first seeds begin germinating after only twenty-four hours, but it takes twenty-eight days for ninety-eight per cent to germinate; the potential rate of germination in this particular species is ninety-

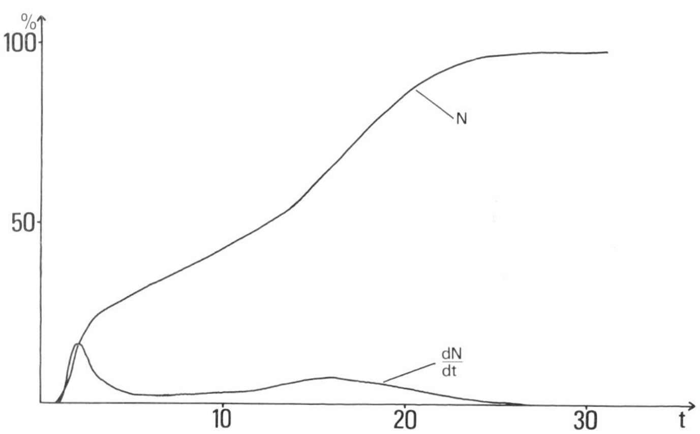

Fig. 7: Diagram of the germination of a Trichodiadema species at a constant temperature of 25°C; abscissa: time in days. The curve of the differential-quotient shows two distinct peaks.

nine per cent. This fact, incidentally, is not at all new; gardeners have known for a long time that many species of Mesembryanthemums germinate badly and "irregularly". This means that after long periods, during which nothing happens, odd seeds still come up.

If we consider this finding in connection with the natural habitat conditions of the Mesembryanthemums, we see once more how marvellously they are adapted to their surroundings in all their characteristics. We know that the rainy season begins with isolated showers, which may be followed by prolonged dry periods. As we have seen, a certain percentage of the number of seeds germinates rapidly, while the rest lie dormant for a long time. In this way the plant caters for all eventualities: the rapidly germinating seeds can make full use of the growth period, should the next shower follow soon afterwards. The seeds germinating after a long period of rest safeguard the species against the catastrophe which would occur if the first showers were followed by a long dry spell during which all newly germinated seeds perished.

As we have seen, the Mesembryanthemums are provided with a double protection against a dangerous dry period which may follow the first rains, or against showers which may occur at an "inopportune" season. The first protection is given by the structure of the capsules. We saw that in the majority of Mesembryanthemums provision is made to prevent all seeds from leaving the fruit during the first rains. In this respect the formation of seed pockets must be regarded as the highest degree of perfection. The second protection was found in the nature of seed germination: seeds which germinate after a short time, and seeds which germinate after a long interval.

During investigations into germination, phenomena were observed which are both difficult to explain and extremely intricate. I do not intend dealing with them at length, as they have as yet not been sufficiently verified. One matter I would like to mention, is so simple that no-one appears to have given it much thought; it is nevertheless most difficult to explain. We saw that many seeds leave the capsule with difficulty and thus after a long delay; on the other hand, we saw that at least a portion of the seeds germinates very rapidly after becoming wet. One should therefore often find seeds which have germinated in the capsules. So far, however, seeds of Mesembryanthemums which have already germinated in the capsule, have very rarely been observed, and only in very old, nearly decomposed capsules.

LITERATURE CITED

BERGER, A., 1908: *Mesembrianthemen und Portulacaceen*. Stuttgart.

GARSIDE, S. and LOCKYER, S., 1930: Seed Dispersal from the Hygroscopic Fruits of Mesembryanthemum. Ann. Bot. 44 (1930): 639–655.

IHLENFELDT, H.-D., 1959: Über die Samentaschen in den Früchten der Mesembryanthemen. Ber. Dtsch. Bot. Ges. 72 (1959): 333–342.

SCHWANTES, G., 1929: Biologisches und Systematisches über die Mesembryanthemeen. Mitt. Inst. Allg. Bot. Hamburg 8 (1931): 161–167.

STEINBRINCK, C., 1883: Über einige Fruchtgehäuse, die ihre Samen infolge von Benetzung freilegen. Ber. Dtsch. Bot. Ges. 1 (1883): 339–347, 360.

STOPP, K., 1958: Die verbreitungshemmenden Einrichtungen in der südafrikanischen Flora. Bot. Studien 8. Jena.

STRAKA, H., 1955: Anatomische und entwicklungsgeschichtliche Untersuchungen an Früchten paraspermer Mesembryanthemen. Nova Acta Leopoldina N.F. 17, 118. Leipzig.

VOLK, O. H., 1961: Flowers and Fruits of the Mesembryanthemums. In H. Jacobsen, *A Handbook of Succulent Plants*, 3: 945–950. London.

ZOHARY, M., 1937: Die verbreitungsökologischen Verhältnisse der Pflanzen Palästinas, I. Die antitelechoren Erscheinungen. Beih. bot. Cbl. (A) 56 (1937): 1–154.

On the Interest of the Seedlings and the Epidermis of the Mesembryanthemaceae *by Dr S. Dupont*

The beauty and the strangeness of the shapes of the flowers of the Mesembryanthemaceae have long aroused the interest of the amateur as well as that of the botanist, who wishes he could penetrate the mystery of their taxonomy and of their evolution.

Quite a number of problems are being raised by the delimitation of the numerous genera and species constituting this large family, and by their rational classification.

The morphological characters of the vegetative organs cannot be considered as the only guide to such classification, owing to both the phenomena of convergency, which are very important in this group, and the great variations which occur according to the ecological conditions of the environment. Their use has, moreover, led to an inordinate increase of the genera and species.

The characters of the reproductive organs, though they give more accurate information, do not, however, allow the researcher to make a detailed classification.

In order to make a contribution for the systematics and the evolution of this family, I have studied for the last fifteen years the development of numerous species, from the germination of the seeds to their fructification.

In the course of my observations, I have noticed that in a very definite genus all the species have the same type of seedling at cotyledon stage and the same stomatic types from the life primordia. These characters can therefore help to delimitate the genera within a group of species having the same capsule; they also enable the researcher to spot species which have been wrongly classified within a genus.

Thus the *Lampranthus verruculatus* has seedling which is unusual in the Lampranthus genus, and I had the pleasure to read that L. Bolus classified it as a Scopelogena; likewise, the seedling of the *Lampranthus comptonii* is not the same as that of the Lampranthus, neither is the seedling of the *Ruschia maxima* the same as that of the Ruschia. The systematic classification of these two species has to be revised.

I want to insist on the case of the *Delosperma algoense, lehmannii, taylori*. In these three species, the seedlings and the epidermis are quite different from those of the Delosperma, and they must be withdrawn from this genus. I have suggested that they be classified in the Corpuscularia genus, which would thus be recreated and would, for the moment, include only three species.

I think it necessary to divide the Delosperma genus, since all its species do not have the same type of epidermis, whereas Aloinopsis and Nananthus, which have the same type of capsule, of seedling and of epidermis, might be grouped together.

The seedling of the Mesembryanthemaceae which have been observed can be divided into four groups, and the epidermis into three. These characters can be associated with those of the capsules to distribute the genera into subtribes, tribes or more important groups. This led me to modify to a certain extent the classification given by Schwantes in 1960; more particularly, I have dissociated the tribe of the Malephorinae, brought the Jacobsenia nearer the Monilaria, withdrawn the Carruanthus from the Ruschiinae and the Vanheerdea from the Leipoldtiinae.

The Hymenogynoidea and all the Mesembryanthemoideae observed have the same type of seedling, a type which is found in a few Ruschioideae: Aethephyllum, Carpanthea, Conicosia, Dorotheanthus, Micropterum.

All the Mesembryanthemoideae have the same type of epidermis, which is to be observed also in the genera previously mentioned, as well as in some Delosperma, in the Drosanthemum, Jacobsenia, Mitrophyllum, Monilaria, Trichodiadema.

Moreover, the first observations I made on the pollen grains enabled me to notice that the pollen grains of the Mesembryanthemoideae have the polar axis longer than the equatorial diameter, as do the pollen grains of the Carpanthea and Conicosia whereas, for the greater part, the pollen grains of the Ruschioideae have the polar axis shorter than the equatorial diameter. The morphology of the exine, which varies from one genus to another (as is shown in the photographs of *Aptenia*

Aptenia lancifolia L. Bol., X 4400. Stereoscan Cambridge, Paris.

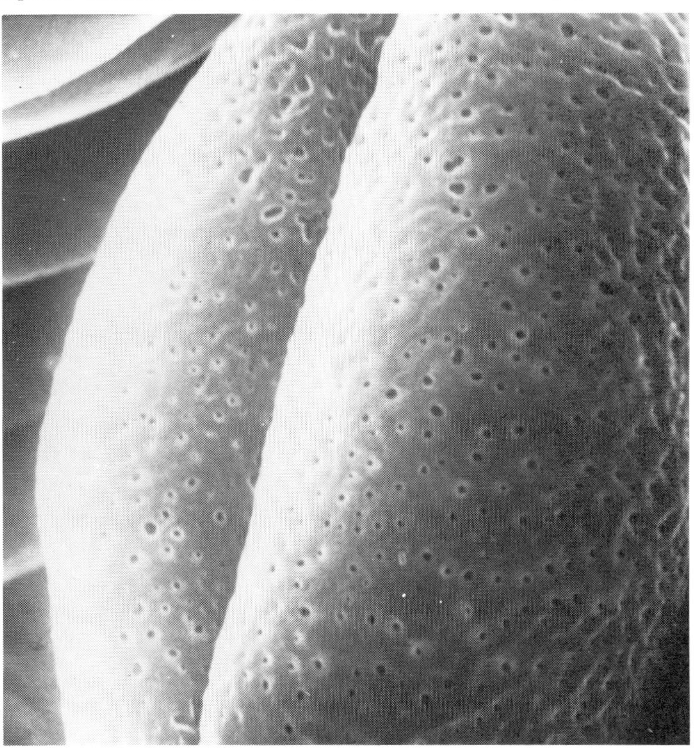

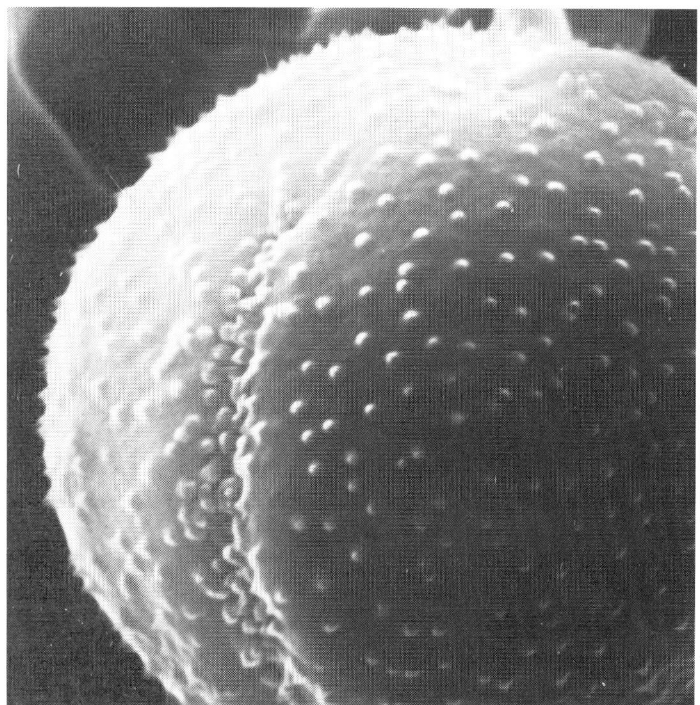

Lithops aucampiae L. Bol., X 5200. Stereoscan Cambridge, Paris.

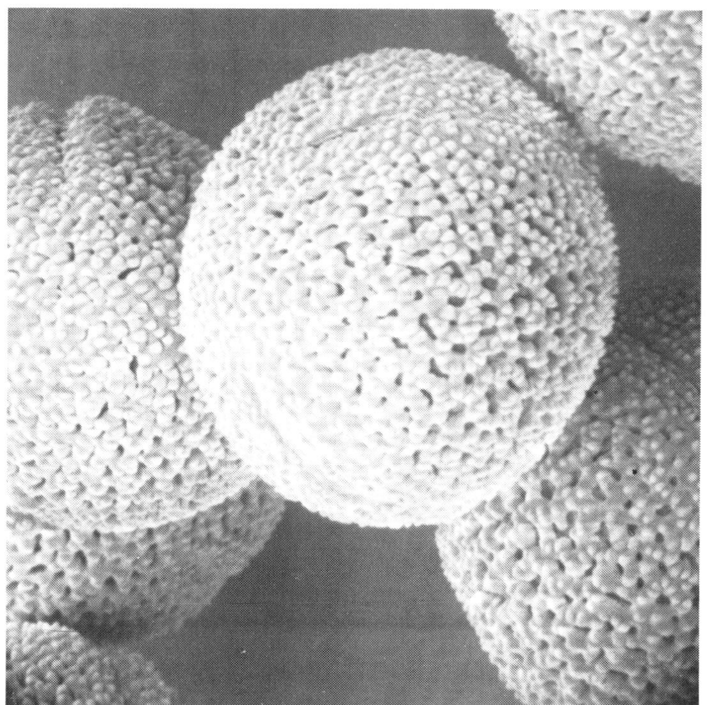

Ruschia perfoliata (Mill.) Schwant., X 3000. SEOL, Paris.

lancifolia, Lithops aucampiae, Ruschia perfoliata, taken by a scanning electron microscope), may also be used in the systematic classification of the family.

The large groups established by the study of the seedling, of the epidermis, of the placentation and of the capsules cannot be superimposed; the transition from one to another occurs at different instances according to the character taken into consideration. In many genera, evolutionary advance has not affected all parts of the plant to the same extent and, as a result, advanced features are often combined with primitive ones.

We are thus led, naturally enough, to speak of evolution. Evolution and adaptation are narrowly linked and must be explained together. The general trend of evolution is toward greater suitability to the conditions of existence: life in a hot and dry environment. By natural selection, favourable variations having survival value are preserved and accumulated through successive generations, thus bringing about greater adaptation to the environment. Succulence is one of the characters which best reveal this adaptation.

Some genera, such as the Conophytum, the Lithops, the Oophytum, show great changes in their vegetative organs, linked to a particular physiology. These changes are fixed by heredity. The capacity of adaptation of these plants seems to be very weak. Their range of tolerance is rather limited. When I saw those stone plants, this thought of Vandel came to me: "Any adaptation, even when it looks perfect, is a cause of the senescence of the issue."

The morphological variations observed in the course of the development of the species differs widely in the various genera.

There are practically no variations for most succulent plants, but in genera which have been less transformed by their adaptation, the variations are important. For instance, the *Mesembryanthemum crystallinum* shows in the course of its development the passing from slightly connate leaves to not connate leaves, from opposite, petiolate leaves on the sterile adult to isolated, sessile leaves in the vicinity of the reproductive organs; in the Conicosia, the flower-bearing stems are long, whereas the sterile adult has very small internodes.

If we start on these principles of Gaussen which have been so often verified: "Sexual influence is an ancestral influence . . . the flowers and their neighbouring parts are the organs which will give us the ancestral characters. A character capable of evolution seems more advanced in the juvenile stage than in the adult. The young plant points to the direction of the future evolution. When in the species the character has reached its maximal evolution, the juvenile stage has nothing to reveal," we can see that the general evolution of these species, for those to central placentation as well as for those to parietal placentation, tends to a general decrease of the size, bringing about an extreme reduction of the internodes, the formation of more and more connate leaves, an increase in succulence.

The genera with very weak ontogeny variations, which are also the most succulent, seem to have reached their maximal evolution for most of the characters. On the other hand, the genera with large ontogeny variations are more primitive but also more capable of evolution.

The family of the Mesembryanthemaceae, in spite of the extremely elaborate adaptation of many of its species, is far from having exhausted all its evolutive potentialities.

Poisonous Mesembryanthemaceae *by H. Herre*

As with the cacti, there are poisonous Mesembryanthema. Amongst the cacti there are the well-known poisonous plants in the genus Lophophora of Mexico, which were sought after long before the White man came to that country. For centuries they were collected by the natives and dried and eaten at certain religious festivals. At first they cause headaches, then sleep, in which one sees the finest, most delicate colours and colour patterns. This is caused by mescaline, an alkaloid. As far as I know, these plants were never cultivated for this purpose, but always collected in their natural habitat, Mexico, and one is surprised that they are still plentiful. Besides mescaline ($C_{11}H_{17}O_3N$), it contains eight other alkaloids.

Our poisonous plants occur in Namaqualand, the Karroo and the Little Karroo. The genus Sceletium, of which the leaves whilst decaying show the typical skeletonising before they are entirely destroyed, are the most important plants in this respect and to a lesser degree the *Mesembryanthemum* (L.) emend. L. Bolus, like *M. crystallinum* and its relatives. Unfortunately there is, besides the poison, also a strong solution of different salts in its cells, which hinders the investigation considerably, especially as there is much less poison present. Thus the main source of the poison is to be found in Sceletium, of which until now 21 species and two varieties have been discovered. All look very much alike, also as far as their flowers are concerned, which are mostly pale yellow. Since these plants belong to the group of Mesembryanthema with axile placentation, the capsules are also very similar. All the species form prostrate plants which grow along and under shrubs and bushes, which are the main grazing of the sheep, goats and other livestock. These creeper-like plants catch all the seeds that the wind blows along from the various shrubs, and are therefore of great value to the farmers in the regeneration of these valuable shrubs. The farmers therefore do not want these plants to be collected in large quantities when one searches for the poison.

Long before the White man came to South Africa, the Hottentots used to collect these plants; they wadded them into a vessel so that fermentation was caused. At the right moment, this process was interrupted and the dark and wet material was dried and chewed. Later it was chewed as a substitute for roll tobacco. The plants therefore became known in Afrikaans as "kougoed", and is still used. Its smell and appearance are not attractive to Europeans. It is nevertheless used as a remedy for infants' stomach troubles in the form of tea, and it is said to be effective. The chewing helps the chewer to bear thirst and hunger and, according to the Hottentots, makes him tough. (In this respect it is similar to the chewing of plants containing cocaine [$C_{17}H_{21}NO_4$], e.g. the leaves of *Erythroxylon coca*, by the natives of South America.) To this day there are store-keepers in Namaqualand who buy this plant from the natives and re-sell it to others.

The scientific investigation of the poison was started in 1914 by a dissertation of E. Zwicky at the Technical School in Zürich, Switzerland, on material sent to him by the late Prof. R. Marloth. He found, as main ingredient, an alkaloid which he named "mesembrine". About ten years ago the well-known firm of C. F. Boehringer & Söhne of Mannheim, Germany, tried hard to find its chemical formula, which is $C_{17}H_{23}O_3N$, and according to them it is related to the orinan group of the *Amaryllidaceae alkaloides*. This result was later on confirmed by the firm of S. B. Penick in New York, D.C. But these investigations are still continuing. In South Africa, as in the U.S.A., some Chemistry Departments of universities are trying to find out more about these poisons, especially as some species of Sceletium contain more alkaloids and in quantities which make it possible to discover their formulae too. One or other species may yet prove to have medicinal value.

In Germany, and probably also in other European countries, no mesembrine is formed at all, but in the U.S.A. it is formed.

Up to now the following Mesembryanthema have been investigated for their content of mesembrine and have been found to contain some: *Aptenia cordifolia, Delosperma cooperi, D. ecklonis, D. lehmanii, D. subincanum, Drosanthemum floribundum, D. hispidum, Glottiphyllum lingueforme, Lampranthus glomeratus, L. scaber, Mestoklema tuberosa, Aridaria splendens, A. umbelliflorus, Oscularia caulescens, Prenia relaxata, Ruschia congesta, R. multiflora, R. rubricaulis, R. tumidula, Trichodiadema intonsum, T. stellatum.*

The History of the Discovery and Botanical Introduction of the Mesembryanthemaceae, with appropriate Biographical Notes

Botanical research on the South African flora commenced with the arrival of Jan van Riebeeck on 6 April 1652. Although the Commander never attempted to climb Table Mountain, he made a close study of the indigenous trees and shrubs and other plants. This was necessary in connection with his work: the founding of a half-way refreshment station at the Cape.

Carpobrotus edulis (L.) N. E. Br., drawn by Paul Hermann and published in 1687 in *Horti Academici Lugduno-Batavi Catalogus* on p. 245.

PROFESSOR PAUL HERMANN

As far as this early botanist is concerned, I cannot do better than to follow the life sketch given in that fine book, *The development of the Gardens of Leiden University, 1587–1937* by the Hortulanus H. Veendorp (Haarlem, 1937).

Hermann was born on 30 June 1646 at Halle, Saxony. It seems that even as a child he had a great love of plants. He attended several universities, among which was Leiden, and obtained his M.D. at Padua in 1670. After a tour of Italy he went to Holland, where he became known as a naturalist and botanist. In this capacity he was sent on a trip to India, Africa and Ceylon; the trip was financed by the Government. He was the first botanist who visited the Cape, and arrived about 25 years after Jan van Riebeeck. After spending eight years in Ceylon, where he was a physician of the Dutch East India Company, he was recalled to the chair of Botany at his old university of Leiden, Holland, on 21 November 1678. His reply, written on 23 October 1679 in Colombo, stated that he would be glad to take the first ship to Holland. On his return he was appointed to the chair on 24 August 1680. In the autumn of the same year, the Secretary of Curators spoke to Hermann about the many plants in his conservatory belonging to private persons, friends and acquaintances of either the director or the gardener. There were so many of these plants that there was not enough room for the academic collection. A large conservatory, in which plants could be kept in winter, was a great temptation to amateurs, and is even to this day, and Hermann, who had proved too lenient in this matter, was urged to use the conservatory only for the plants growing in the Gardens. The cultivation of plants was impossible without glass houses, and from the dedication of Hermann's catalogue of 1687, it appears that the Gardens actually boasted of this most necessary equipment, built between 1680 and 1687. The new succulents introduced from the Cape of Good Hope in particular found good shelter there.

Of great interest is the large illustrated catalogue of plants cultivated in the Leiden Gardens between 1681 and 1686. It was published in 1687. Schuyl's catalogue of 1668 showed 1 827 names and this number seems to have increased within eighteen years to 3 029. After a long period of slow accumulation, a great number of important acquisitions, especially of East Indian, Virginian and South African plants, were cultivated. The following particulars about the South African ones are given: Since Schuyl's time the number of Mesembryanthema increased from one to twelve. Hermann calls these species *Ficoides seu Ficus Aizoides Africana*, because of the resemblance of their fruit to that of the genus Opuntia. The English common name is still Fig Marigold. It appears that the collection contained the "Hottentot Fig": *M. edule* L.; *M. acinaciforme* L.; *M. barbatum* L. (the latter grown from African seed in the garden of Mr Heemskerck).

Further, *M. glaucum* L.; *M. aequilaterale* Haw. (which flowers all summer); *M. nodiflorum* L. and *M. pemeridianum* L. – ten species of Pelargonium were mentioned, and also *Stapelia variegata* and various Crassula, and quite a number of South African shrubs. Two other South African plants in these Gardens deserve to be mentioned: *Zantedeschia aethiopica* Spreng., the common Calla Lily, and the African Lily, *Agapanthus umbellatus* L'Herit., the latter listed under the strange name of *Hyacintho affinis africana*.

Jacob Breyne, a Danzig merchant, whose son was a friend of Boerhaave, also sent some plants. From Breyne's hand we find the description of remarkable plants grown at Leiden in his *Prodromus fasciculi rariorum plantarum*, published 1680, and also in the second volume of 1689.

Hermann's directorate had been fruitful in all respects. A more systematic organization of the Gardens, according to the principles of Robert Morison of Oxford, with a concomitant increase of area, was recorded in 1685. The year 1682 was one of important travels for Hermann: he visited England, especially the Gardens of Oxford and Chelsea, on behalf of the hortus. He gathered and brought back almost 200 living plants on this trip. The following year he became Secretary of the Senate. In April 1685 the Curators taking into account the vast increase in the number of species grown in the Gardens, thanks to the energy of the Professor of Botany, ordered the catalogue to be printed and distributed by their own board. In 1687 the Curators offered a grant of 315 guilders to Hermann for his catalogue. Later that year he received a generous donation of 700 guilders for his intended trip to Italy, which actually became a trip to France.

Hermann died on 25 January 1695, only 49 years old, and was survived by his wife and four young children.

He was undoubtedly one of the greatest botanists of his century. His herbarium is now in the possession of the British Museum, but no portrait of him has survived.

H. B. OLDENLAND

Heinrich Bernhard Oldenland was the head gardener at the Cape from 1692–93 until his death in 1697. This was in the time of Simon van der Stel. Oldenland was born in 1663 at Lübeck, Germany, and was the first *baastuinier* at the Cape who had any scientific education; he had studied medicine for three years under Prof. Paul Hermann at Leiden. Oldenland and his successor, Jan Hartog, came to the Cape on the same boat. They sent much herbarium material and many seeds to Holland, especially to Professor Hermann. Jan Hartog, a brother of the hortulanus at Leiden at that time, succeeded Oldenland and stayed on until 1715, when he returned to Holland. He worked under W. A. van der Stel (Governor from 1699–1707), who very often ordered Hartog to send plants and seeds to Holland. Among these were Mesembryanthemum seeds. Hartog undertook many journeys inland, usually joining expeditions sent to locate and barter for cattle and sheep belonging to the natives.

Jacob Breyne

As a result of the collections by Hermann, Oldenland and Hartog, the unique flora of the Cape became known in Europe.

Nearly a hundred years later Linnaeus wrote the following lines about the few days which Hermann had spent at the Cape: "The Cape of Good Hope, situated on the farthest point of Africa where no botanists have ever collected. . . . O Lord, how many, what beautiful plants were there seen in these few days which Hermann was at the Cape. In those days Hermann collected more new South African plants than all the botanists on earth before him!"

According to the botanist A. Sparmann, the great Linnaeus often regretted having refused the offer he had had to visit the Cape.

JACOB BREYNE (also Breyn and Breynius)

He was born in 1637 at Antwerpen (now Belgium). His interest in plants often took him to Holland in search of new ones. Later he lived at Danzig (now Poland), and became a rich businessman. In his work *Exoticarum plantarum centuria prima* (1678), 48 Cape plants are illustrated. Among these is the oldest illustration of a Mesembryanthemum, today known as *Herrea fusiformis* (Haw.) L. Bol., which was originally collected at Robben Island. These 48 plants were sent to him between 1663 and 1670. Thus he came to know our "vygies" and noted that the flowers opened about noon; he therefore named these plants "Mesembrianthema", midday flowers, in the second edition of his *Prodromus fasciculi rariorum plantarum* (1689). The first edition of it appeared in 1680. In his *Species plantarum* (1752) Linnaeus took over this name out of Dillenius' *Hortus Elthamensis* (1732), where it was incorrectly spelt with a "y" instead of an "i". This name is actually a more accurate one for the whole family than Ficoideae. Of course, we know now that there are also Mesembryanthema which flower during the evening or at night. Breyne died at Danzig on 25 January 1697. His son Philippus (1680–1764) became a botanist and in 1739 published a reprint of the two parts of his father's *Prodromus*.

RICHARD BROADLEY *by Gordon D. Rowley*

He was born before 1688. He is believed to have been well-born and initially a man of means, although he died in poverty. No portrait has survived, nor do we know where he lived. He was elected a Fellow of the Royal Society in 1712 and became the first Professor of Botany at Cambridge University in 1724, apparently on the strength of a promise to provide a botanic garden, which he never did. A course of twelve lectures prepared soon after his appointment survived in manuscript, and he published *A Course of Lectures upon the Materia Medica* in 1730, but the carrying out of most of his professional duties seem to have been left to John Martyn, so that Broadley could proceed with the writing of books, which was his main source of income. He became the most prolific writer on gardening and agriculture of his day, and his works must have made a lasting impact by sheer number, if nothing else. It is to these books therefore, that we must turn in order to learn more of this curious and controversial figure.

He is credited with having published the first horticultural journal in 1721–22 and the first botanical dictionary (1728). Other works on agriculture appeared under Broadley's name as editor, such as the translations of Agricola (1721, 1723, 1726) and Xenophon (1727). He was a keen observer of nature and often in advance of his times. He visited quite a number of English Gardens and undertook trips to France and Holland in 1714.

His *History of Succulent Plants* appeared belatedly in 1716. It was the first book in any language devoted exclusively to these plants. It seems that the idea of a monograph on succulents originated at least as early as 1706. The pamphlet advertising it in 1710 informs us that he then planned a book with fifty plates and separate chapters on cultivation, published as a limited edition to subscribers only. There can be little doubt that the venture received a new stimulus in 1714 when Broadley visited Holland and saw the wonderful collections of newly introduced Cape plants growing at Amsterdam and Leiden. Many of these he brought back to England for the first time, and they figured prominently in his book: no less than thirty Mesembryanthemaceae as compared with only seven American cacti and an Agave.

It is said that 330 copies of the *History of Succulent Plants* were published, of which fourteen were hand-coloured, but since purchasers often coloured their own, it is not easy to say which are the originals. Broadley himself is said to have done at least some of the colouring. There was a spate of misprints and transpositions, which have caused lasting confusion to later botanists. The book was cited by many celebrated botanists: by Dillenius, by Linnaeus (thus providing the types for some of his species) and by many subsequent workers from De Candolle to Britton and Rose. Now, no copies of the *History* I have examined are bound exactly alike. A new edition of 1739 (Kew gives the date as 1734) consisted of only 150 copies. All Broadley's succulents can be recognized from the drawings and all are in cultivation today.

LIST OF LITERATURE
Advertising pamplet of 1710.
Historia Plantarum Succulentarum, 1716–27.
New Improvements of Planting and Gardening, 1718.
A Philisophical Account of the Works of Nature, 1721 (Chapter 3 entitled "Of Plants and Super Plants".
Re-published in 1964 by the Gregg Press Limited, 37 Catherine Place, London, S.W.1 (*Collected Writings on Succulent Plants*, with an introduction by Gordon D. Rowley).

BARON NICHOLAS JOSEPH DE JACQUIN

He was born on 16 February 1727 in Leiden, Holland, where his father Claudius Nicholas owned a cloth and velvet mill, founded by his grandfather towards the end of the 17th Century when he emigrated from Paris, France. His mother, (*née* Von Heyningen) belonged to an old, rich and aristocratic family. His father had had an excellent education and was an enthusiastic

Baron N. J. de Jacquin

lover of the great classical writers. Hence it was his desire that his son Nicholas, whom he had wanted to become a businessman, should be introduced to the beauty of classical literature. He therefore enrolled him at the same famous gymnasium which he and many other Roman Catholic Dutchmen had attended. But before gaining entry, the young man attended the gymnasium of the Jesuits at Antwerp together with his friend Gottfried van Swieten. He left in 1744 with excellent testimonials. Unfortunately, his father had suffered heavy losses, and had died after his return from Portugal, where one of his customers had fraudently claimed insolvency. Jacquin's studies, originally regarded as an extension of study for its own sake, now became a science from which the young Nicholas had to make a living. First, he matriculated in 1745 at the University of Leuven, and later he returned for further study in Leiden. At first he was deeply interested in the classics, but in course of time he preferred botany. He was especially fond of the showy flowers of exotic plants. One of these, *Costus speciosus* (*C. arabicus*, named by Jacquin) took the first place in his famous work, *Icones plantarum rariorum* (1781–1786). Later he studied medicine in Rouen and Paris. The first years of his studies were very difficult, as his family wanted him to study theology. He did, but when he informed his mother in 1748 that he did not want to become a clergyman, his mother told him that he was old enough to earn his living, and that she did not want him to return home. As a result, lack of money was a pressing problem. It was impossible for him to pay the fees for his medical course. He confided his problems to his old friend, Gottfried van Swieten, who was quite willing to help him in every way, if he would come to Vienna. On 20 July 1752, Jacquin arrived in Vienna, but needed some time to settle and to learn the language. In spite of all his work, he never forgot the botanical lessons he had had from Anton and Bernhard Jussieu in Paris, and he often visited the Gardens at Schoenbrunn, of which two fellow-countrymen were in charge.

The Emperor Franz Stephan was very interested in plants and animals, and wanted to send a botanist to America. Through van Swieten he asked if Jacquin would undertake such a journey. Jacquin was quite willing and immediately began to study natural science and art so as to be able to make good botanical drawings. On 9 December 1754 he left Vienna for the West Indies. His base was Martinique, but he also visited Curaçao, San Domingo and Jamaica and, on the continent, Venezuela and Colombia. He had many adventures but his health suffered and he contracted yellow fever. He left Cuba on 4 January 1759 with many plants and animals and reached Vienna on 17 July 1759. He introduced the first platinum to Austria.

Through van Swieten he then became in 1765 Professor of Chemistry at the College for the Science of Mines at Schemnitz, and remained until 1768. In 1768 he became Professor of Chemistry at the University of Vienna and later also Professor of Botany and Director of the Botanic Garden. In 1771 he published his fourth volume of his *Observationes botanicae*; in 1770 the first, in 1772 the second, and in 1776 the third volume of his splendid edition *Hortus botanicus vindobonensis*. In 1773 he started another excellent book, *Florae Austriacae*, of which the fifth and last volume was published in 1778. On the subjects of Chemistry and Pharmacology he published other works. All in all, he published about 30 volumes. Especially worthy of mention are his splendid *Icones plantarum rariorum* (1781–1793) with 648 illustrations reproduced from copper plates; among these illustrations are many of Mesembryanthema.

Another fine book is *Plantae rariores horti Caesarei Schoenbrunnensis* (1797–1804). He also published a Guide for the identification of plants according to the method of Linnaeus, which became the standard work for the study of Botany at the University of Vienna for the next 50 years, and this book earned him the title of "the Linnaeus of Austria".

There is no doubt that Jacquin was first and foremost a botanist, and that his most important works are on botany. Nevertheless, he was also an excellent chemist and one of the most distinguished scientists of the University of Vienna towards the end of the 18th Century. He had an excellent classical education and could express his thoughts in correctly construed Latin hexameters, as in the case of the description of the strange qualities of the Stapelieae. He also did quite a number of excellent and artistic botanical drawings.

He often played the flute while wandering in the woods, and he became a friend of the von Schreibers family, and he later married Katharina von Schreibers. He had a convivial nature, and musicales were common in his home. Mozart was a familiar guest at his home. Some of Mozart's compositions are based on the merry events which occurred in Jacquin's home. Mozart was a friend of Jacquin's youngest son, Gottfried; they died almost simultaneously. Mozart was also the music teacher of Jacquin's very musical daughter, Franziska. In 1774, after his book *Pharmakopoea Austriacoprovincialis* was published, he was granted a title, and in 1806 he was made a Baron.

In 1791–92 his son, Joseph Franz, took over the professorship in Chemistry and at the end of 1796 Nicholas Joseph retired. At the end of 1808, at 81 years of age, he was elected Rector of the University of Vienna. These were troubled times, as Vienna was occupied by the French for some months. Jacquin was an elected Fellow of numerous scientific societies in Europe. He died on 26 October 1817 and his last question was: "Is there not yet a Stapelia in flower?"

The botanical drawings in his first books were all done by himself and were also partly coloured by him. Later only the best artists did the drawings, according to his sketches and under his supervision. There are about 3 000 coloured copper plate engravings and among them many of Mesembryanthema. (The original plants were raised from seeds sent from South Africa by his collectors, Booth and Scholl. The latter collected in South Africa for about 10 years, and found the famous *Fockea crispa* named by Jacquin. It is still alive and has flowered every year in Schoenbrunn since about 1798.) The coloured copper plates are jewels and immensely expensive today. Merely to look at them is a pleasure one will never forget.

I acknowledge with many thanks my indebtedness to the fine contribution of Dr Wilfried Oberhummer "Die Chemie an der Universitat Wien in der Zeit von 1749 bis 1848 und die Inhaber des Lehrstuhles für Chemie und Botanik" in *Studien zur Geschichte der Universitat Wien, Band III*, 1965, p. 126–202, Verlag Hermann Böhlaus Nachfolger, Graz-Köln. With this as my source, I have given a condensed version of this most interesting account of the life of Baron Nicholas Joseph de Jacquin.

Adrian Hardy Haworth *by William T. Stearn*

A. H. Haworth was born on 19 April 1768 at Hull, Yorkshire, England. Both his parents belonged to prosperous land-owning and merchant families associated with the great port of Hull. His father was Benjamin Haworth (1728–1790) of Hullbank (or Haworth) Hall near Hull. His mother, Anne, who died in 1784, was the daughter of John Booth of Hull and the heiress of her uncle John Booth of Killingholme. Through her Adrian was descended from Henry Booth, Admiral of the North in the reign of Henry VI. Haworth took his second name from the mother (*née* Francis Hardy). His uncle, Dr Richard Hardy, who died in November 1800 at the age of 92, studied at Leiden in 1737–39 and was the last surviving pupil of Boerhaave. Adrian's grandson, Benjamin Blaydes Haworth, 1823–1901, assumed by Royal Licence in 1869 the additional surname and arms of Booth. The family thus became Haworth-Booth; Haworth Hall remained in its possession until 1936. The family is represented in present day horticulture by Michael Haworth-Booth, the grandson of B. B. Haworth-Booth and author of horticultural books.

Adrian Haworth attended the Hull Grammar School and was then articled to a Hull solicitor; his father's intention was

A. H. Haworth

that he should enter the legal profession. His legal qualifications acquired, Haworth quickly abandoned law for the study of natural history, which was much more to his taste. His father died in 1790 and his elder brother, Benjamin Blaydes Haworth (1763–1836), inherited the family estates. Benjamin was a keen horse-breeder. He seems to have been a merry and sporting squire, very unlike the quiet and studious Adrian, to whom however, he granted an income adequate for his modest needs. Benjamin never married, but Adrian did, three times. On Benjamin's death in 1836 the family estates passed to Adrian's second son, also named Benjamin (1796–1868). Freed from the necessity of earning a living as a lawyer or merchant, Adrian was able to devote his life to botany, entomology and gardening. His mother, judging from her portrait, was an intelligent and beautiful woman, and she probably stimulated and encouraged her son's interest in living things, for he became "a gardener practically at seven years old".

The Haworth family dower house was at Cottingham near Hull, and here he lived for several years, collecting specimens and compiling "A botanical arrangement and scientific descriptions of all the plants discovered wild in the ancient Parish of Cottingham and some account of the antiquities, zoology and natural history of that place", probably inspired by the works of Gilbert White. He never published this work, but it served as sound training. In March 1792 he married Elise Sidney Cumbrey, by whom he had four children. She died in 1803, aged thirty-four. Some time after his marriage, Haworth left Cottingham and settled at Little Chelsea. "In 1792 (although I had long before collected Aloes) I first beheld the rich Gardens at London and Kew," he wrote in 1827. He lived at Chelsea until 1812. A Chelsea friend was the botanist and entomologist William Curtis (1764–99), founder of the *Botanical Magazine*. In 1794 Haworth published the first part of his *Observations on the Genus Mesembryanthemum*, followed by other works about Lepidoptera, and about other plants. He entomologized all over England. Haworth's first publication on succulent plants, the *Observations*, remains one of his most important works. It was published in two parts, at his own expense. The first part is mainly concerned with cultivation and general characteristics. The second part describes the individual species and tabulates their synonymy. For that time, it was a remarkably thorough and comprehensive monograph. Haworth here distinguished 132 species, as well as 30 of doubtful standing. When he published his first work on Mesembryanthema, Haworth was twenty-six years old. Even then he had in mind the preparation of "A general account of succulent plants". Mesembryanthema remained his favourite group however, and he revised it in 1803, 1812 and 1821, and published new species up to 1831. But Aloe – taken in a broad sense, i.e. including Haworthia, Astroloba (*Apicra* Haw.) and Gasteria, Narcissus, Saxifraga and Crocus – shared his interest and attention.

In 1812 Haworth was able to describe no fewer than 206 species of Mesembryanthema. His earlier descriptions were relatively detailed; many of his later ones were essentially diagnoses in concise Linnaean style. N. E. Brown observed in 1926, "He was the first to make a systematic classification of its species, and as he described from living plants, chiefly cultivated by himself or at Kew, and had a thorough knowledge of them;

he made very few mistakes as to species." (As early as 1828 this work was out of print.) A very large proportion of the known species are represented in the Kew Herbarium by a series of excellent coloured drawings, made from the type plants; the majority can be correctly identified.

In 1812 Haworth returned to Cottingham, taking with him the greater part of his natural history collection, and he lived there for about five years. His second wife died about twelve years after the first, leaving him with another child. His third wife outlived him. While at Cottingham, he helped to found and arrange the Hull Botanic Garden and entered into correspondence with the Dutch scientist Martinus van Marum (1750–1837) and the German prince Joseph zu Salm-Reifferscheid-Dyck (1773–1861), a fine artist and keen student of succulent plants, who was already collecting material for his magnificently illustrated monograph, *Generum Aloes et Mesembryanthemi* (1836–63) The first part of this did not appear until several years after Haworth's death. He left Cottingham between January 1817 and March 1818 and took up residence at Chelsea. Here he lived until his death.

In the autumn of 1814, while at Cottingham, Haworth gave up the cultivation of succulent plants. He sold his collection to an amateur of Liverpool, who was no botanist. A good number of the plants perished on the journey or shortly afterwards. The end of the Napoleonic Wars in 1814–15 had again enriched the Gardens with new plants. Thus in 1821 Haworth had built a small greenhouse and had begun to assemble a new collection of succulent plants. In this he was helped by gifts from van Marum and Salm-Reifferscheid-Dyck, but probably most of the plants came from the Chelsea Physic Garden, thanks to the generosity of the curator, William Anderson (1766–1846), whom Haworth described in 1818 as being "now our chief cultivator of these plants". He also received plants from William Townsend Aiton (1766–1849) of Kew. He was so successful that a German professor, Joseph August Schultes (1773–1831) of Landshut, who visited him at Chelsea in 1824, found no fewer than 200 Aloes, 362 Mesembryanthema and 90 Crassulaceae in "dem Gärtchen des genialischen Haworth". Schultes attested to his amiable character: "Herr Haworth ist höchst mitteilend, und ein sehr gutmütiger, lieber Mann!" His garden was very small, "some 30 square yards only". On his death in August 1833, the collection comprised "above 160 species of Aloe, 330 Mesembryanthemum, 25 Cotyledon, 20 Cacalia, 12 Rulingia or Anacampseros, 21 Haworthia, and numerous species of Crassula, Mammillaria, Sedum, Sempervivum, Echeveria, etc., amounting in the whole (including duplicates) to nearly one thousand pots." Like his successor, N. E. Brown, he had a keen eye for minute differences. Haworth was one of the last of Britain's 22 000 victims of Asia's cholera. "He was enjoying his usual health and watering his favourite plants at seven o'clock in the evening of August the 23rd (1833), when he was seized with malignant cholera, and died between three and four o'clock in the afternoon of the 24th."

His collection of living plants was sold, then sold again but, according to N. E. Brown, plants reputed to be from Haworth's collection were still cultivated by William Wilson Saunders (1809–79) at Reigate in about 1870. Saunders employed Brown as curator of his natural history museum until 1873, and these plants introduced that keen-eyed man to the study of the group in which he became as pre-eminent as Haworth had been a century earlier.

Haworth died at a time when horticultural taste was changing. The cultivation of Cape and Australian Proteaceae, Ericaceae, Iridaceae and succulent plants, which had reached its zenith between 1800 and 1830, had begun to decline, and tropical and subtropical plants from Central and South America and the East Indies began to take their place in the new water-heated greenhouses that replaced the old "stove" ones. Orchids gradually became popular. Had Haworth then begun instead of ended his botanical studies, Mesembryanthema would hardly have been the group of his choice, and a group of succulent South African Liliaceae (Haworthia) would not have borne his name.

(From Adrian Hardy Haworth: *Complete Works on Succulent Plants*, Volume 1. Biographical and bibliographical introduction by William T. Stearn. *Observations on the Genus Mesembryanthemum*. 1794–95. Gregg Press. 1965.)

Adrian Hardy Haworth (1768–1833) was the leading English authority on succulent plants during the first part of the nineteenth century, and the author of many books and papers about them. He was from his youth a gardener-botanist.

These plants cannot be studied and described satisfactorily with the aid of dried material only. To understand them, one requires an intimate acquaintance with the living plants, usually to be gained only by cultivating many of them side by side and observing them constantly. This was Haworth's procedure.

Such plants are the despair of professional botanists concerned with them simply as part of routine duties. On the other hand, they offer an attractive field of study to the patient amateur enthusiast who is both botanist and gardener. England has produced a succession of such gardener-botanists, the most notable being Haworth himself. Probably N. E. Brown (1849–1934) and H. W. Pugsley (1868–1947), among his successors in the critical study of his favourite groups, have been closest to him in attitude and method.

Unfortunately he was unable to study the fruit characters which have been used since 1921 to justify division of Mesembryanthema into minor genera, although in 1821 he provisionally proposed the separation of Conophytum, Gibbaeum and Glottiphyllum, all formally adopted by N. E. Brown a century later.

Readily available information relating to the man himself is meagre.

Joseph Maria Franz Anton Hubert Ignaz zu Salm-Reifferscheid-Dyck Reichs-und Altgraf, later Fürst von Dyck by *W. T. Stearn*

Prince Salm-Dyck was an enthusiastic cultivator and student of succulent plants. He was born on 4 September 1773 at Schloss Dyck near Düsseldorf in the Rhineland of Northern Germany, and died at Nice on 21 March 1861. He lost his father, Franz Wilhelm, when he was three. His mother saw to his education, first under private tutors, then at the Jesuit College in Cologne, and at the age of eighteen he married a young countess, Marie Therese von Hatzfeld, whom he divorced

Prince Joseph Salm-Dyck

in 1801. When the French revolutionary armies overran the Rhineland in the seventeen-nineties, Salm-Dyck offered no resistance. He lost his sovereign powers but, by extending hospitality to Generals Jean Baptiste Kléber and Jean Baptiste Jules Bernadotte, later King Charles XIV of Sweden, he managed to retain his possessions.

This friendship with the French determined his future. He made frequent visits to Paris, where he met the French author, Constance Maria de Théis (1767–1845). She had divorced her husband in 1799, and Salm-Dyck married her in 1803, and met such distinguished botanists as Desfontaines, Antoine Laurent de Jussieu, Thouin and Thuillier, and that great botanical artist, Pierre Joseph Redouté (1761–1840). Some years earlier, Redouté had made paintings of succulent plants difficult to preserve in herbaria; he prevailed upon a young Swiss student, Augustin P. de Candolle, to write descriptions to accompany them. Under the title *Plantarum succulentarum, ou Histoire naturelle de Plantes grasses*, they were published in parts between 1798 and 1829. These plates probably did more than anything else to interest Salm-Dyck in succulent plants. From Thuillier he had acquired a good knowledge of botany. He had a natural talent for drawing and painting, and lessons from Redouté made him an expert in the drawing of succulents. At Schloss Dyck he built glass houses and cultivated plants so successfully that his collection of cacti, aloes and Mesembryanthema soon became the finest in Europe. By 1835, according to James Forbes, the Duke of Bedford's gardener, who in that year made a tour through Germany, Belgium and part of France, the Prince's glass houses needed rebuilding, but they contained many rare species. Forbes' journal testifies that even then a zeal

for the cultivation of cacti and other succulents was widespread in Germany.

The Prince was no mere collector. He studied his plants carefully and produced many publications. His most important work is the *Monographia Generum Aloes* and *Mesembryanthemi* (Düsseldorf, 1836–1842; Bonn 1849–1863), a collection of lithographed and partly hand-coloured plates accompanied by Latin descriptions. This beautiful work was issued in seven parts (fascicles). Salm-Dyck published only six parts, to which he contributed both text and plates. After his death in 1861, seventeen unpublished plates of Aloe were found. Salm-Dyck divided both Aloe and Mesembryanthema into numerous sections, distinguished by numbers. The accuracy of Salm-Dyck's naming has been much questioned in recent years, but not the excellence of his draughtmanship. His beautiful illustrations can never become out of date, but many of the names used by him have now been superseded by others and no doubt this process will continue, according to W. T. Stearn in *The Cactus Journal*, Vol. 7, Nos 2 and 3 of 1939.

Unfortunately, the prince did not see the ripe capsules of most of his described Mesembryanthema, as these are usually not formed in Europe. He was therefore not able to study these; the same is true for Haworth. Today the capsules denote the main difference between the various genera.

Quite a few of the species are named in honour of the prince, also the three genera, Salmia, Reifferscheidtia and Dyckia.

From his *Hortus Dyckensis, Catalogue des Plantes Cultivées dans les Jardins de Dyck*, Düsseldorf, 1834, we give the following extract: "The Castle and Gardens of Dyck lie on both banks of the Rhine on the Düsseldorf-Aachen route. The country is flat. The soil is good and well-suited for the lay-out of gardens. The gardens were started in 1800, a difficult time, as the armies of France occupied the left bank of the Rhine and devastated the beautiful gardens. Especially the conservatories were completely destroyed, so there was nowhere to keep a good collection of Aloes and Cacti. I tell this because I wanted to devote myself to the cultivation of succulents which I was able to get in France, Belgium and Holland. I also exchanged plants with many public gardens in places such as Paris, Brussels, Ghent and Haarlem. In 1808 I visited Madrid and brought back with me a fine lot of plants. In 1809 my collection consisted of 335 species, but they were badly named. In 1814 and 1815 I visited the gardens of Vienna and Berlin. The baronets Jacquin, father and son, Messrs. Boos and Antoine, the Chefs of the Imperial Gardens at Vienna and Prof. Link and Mr Otto of the Botanic Gardens at Berlin let me see all the plants they had in their care. I was able to enrich my collection considerably and I was able to see with my own eyes the fine species of plants described by the Jacquins and Prof. Willdenow. In 1817 my collection of succulent plants contained 550 species and varieties, including 114 species and varieties of Aloes. The study of these plants enabled me to put right some faults and to publish some new species in the Catalogue. The species and varieties of Aloes I published during that time.

"Afterwards I became associated with the English Gardens, especially with Mr Haworth who already had a collection of succulents at Chelsea and who let me have fine plants of his, also others such as Mr Aiton, the Director of Kew Gardens.

In 1822 I already had a collection of about 900 species and varieties of succulents. This collection was not a large one, but had the advantage of including mostly plants of the botanists who had described them. In 1820, 1821 and 1822 I was able to publish in *Observations botaniques* some species and varieties which were overlooked by them. In 1829 my collection consisted of 1 150 species and varieties and I am grateful to the Botanic Gardens of Berlin, Munich, Geneva, Karlsruhe, Bonn, Hamburg and Leiden who have helped me in getting these plants.

"The collection included plants of the following families: Portulacaceae, Crassulaceae, Ficoidées, Cactées, and also the genera Aloe, Begonia, Euphorbia and Stapelia. I also collected Aroideae, Scitamineae, Dracaena, Yucca, Iris, Saxifraga, Paeonia, and some trees and shrubs.

"Of the monographical works about the genera Aloe and Mesembryanthemum the first parts will be published soon."

(Taken from the *Journal of the Succulent Society of Great Britain and Ireland*.)

According to a letter of the Princess of Salm-Reifferscheid, there were still plants of the collection of Prince Joseph in cultivation up to the beginning of the first World War. Owing to the lack of fuel, all died later on.

My grandfather, Adolar Herre (1826–1910), went to see the Prince's collection when it was at the height of its glory and, as he himself was a collector and connoisseur of succulents, he was full of praise for this fine collection.

ALWIN BERGER

German horticulture and botany sustained a great loss when, on the evening of 20 April 1931, Alwin Berger died at Cannstadt, near Stuttgart, only a few months before the end of his sixtieth year. Born in 1871 at Möschlitz, a small village in Thuringia

Alwin Berger

where his parents farmed, he commenced his career as a gardener at the castle of Ebersdorf. Afterwards he worked successfully at the Botanical Garden at Dresden, in Fratelli Rovelli's nurseries at Pallanza, Italy, and in the Botanical Garden at Greifswald. My long and intimate friendship with him dated from this period (1895–96). From Greifswald, Berger went to the Palmengarten at Frankfurt, which at this time was one of the best horticultural institutions of Central Europe. In 1899, if I remember rightly, Berger became curator of the famous Hortus Mortolensis, Sir Thomas Hanbury's world-renowned garden at La Mortola, Italy. Here he spent nineteen successful years until, during the Great War, he was forced to leave the place he dearly loved. He became director of the gardens of the King of Württemberg, and lived in Stuttgart. In 1919, Berger had to resign his office and, following an invitation from the Experimental Station at Geneva, New York, he went to the U.S.A. to undertake some scientific work. After three years' absence from Germany, he came back to Stuttgart, and took a position as the head of the Botanical Department of the Museum of Natural History. Berger's chief contribution to horticulture and botany lies in his studies of succulents and cacti. Besides a great number of contributions to horticultural and botanical magazines, he published some very important books. In 1907, he started a series of small handbooks on succulents, beginning with his *Sukkulente Euphorbien*; in 1908 followed his second book on *Mesembrianthemen und Portulacaceen*. In 1910, Volume III of the series on *Stapelien und Kleinien*. His monograph *Die Agaven* (1915), and his best book on cacti, *Die Entwicklungslinien der Kakteen* (1926), prove him a most careful and far-sighted investigator of difficult phylogenetic problems. In 1929, Berger published a more popular book on cacti, and his last contribution to botany is the monograph on Crassulaceae in Engler-Prantl's *Natürlichen Pflanzenfamilien*. For Parey's *Blumengartnerei*, a comprehensive horticultural encyclopedia, Berger also described the succulents. Besides these books, his *Hortus mortolensis* (1912) ought to be mentioned. "His sudden death after a painful illness was a great shock to his numerous friends all over the world. We shall remember him as a kind and modest man, and a keen, indefatigable worker on all problems he was interested in. He left a widow and two children. His only son is in every respect like his father, and works at present as a foreman with Mr Hertrich in the Huntingdon Gardens at San Marino, California," Camillo Schneider wrote in *The Gardeners' Chronicle* of 9 May 1931, page 363.

In his book *Mesembrianthemen und Portulacaceen* (1908) Berger wrote: "It is in vain that Haworth searched for characters to divide this genus into more genera with fewer species." He divided this large genus into eight sections and 72 sub-sections, where in he placed all the 315 species treated in his book. His sub-sections do not correspond at all with the actual genera of today.

He knew more about the capsules, and experimented with them as far as the distribution of the seeds was concerned, but otherwise he did not realize the differences of the capsules and their worth for the division of this big group into natural genera. It was the late Dr N. E. Brown who discovered this and made use of it.

Dr N. E. Brown (1849–1934) became chief assistant in the Herbarium of the Royal Botanic Gardens at Kew in England, where he had worked since 1873. In 1921 he began the more detailed sub-dividing of the Mesembryanthema, basing his work on the characteristics of the fruits. Most of these studies were published in the *Gardeners' Chronicle*, London. He was a painstaking worker whose genera and species are well considered. "He was never too busy to help others too," the late Alain White wrote in his monograph of the Stapelieae. I had the same experience, but it was not always like that. Dr L. Bolus and Prof. Schwantes had quite another impression of him. Dr Brown who was older than either of them, and looked upon the Mesembryanthema which he first began dividing as his own plants. When Prof. Schwantes and later Dr L. Bolus started to do the same, it was too much for him and towards Prof. Schwantes in particular, he was not always fair. In German gardening journals they had many quarrels; sometimes the reason was only a misunderstanding, as with the two genera, Herrea and Conicosia. Dr Brown mistook the first one for the latter and thus the controversy arose. Prof. Schwantes had great respect for Dr Brown and his work, and was touched by this unhappy incident. In London he one day passed the house of Dr Brown and would very much have liked to go in and greet him, but he thought he would not be welcome and did not enter. If he had done so, I am sure each would have changed his mind about the other. Dr Brown was accurate and science owes him much for the fine descriptive names he gave to his new genera.

Alain White in his book, *The Succulent Euphorbias of Southern Africa*, gives a good description of him: "He was a quaint-looking, rather hunched little man with a narrow face and a short goatee beard, somewhat deaf (at least in old age), but endowed with a keen sense of humour and a kindly disposition that made him ever ready to share his vast knowledge of South African plants. He was endowed with remarkable eyesight, so that he could, for instance, distinguish the individual strands in cotton with the naked eye. Doubtless this sharp eyesight helps to explain the very critical nature of his botanical work. After he had retired in 1914, he continued to work at the Kew Herbarium almost to the time of his death."

Hence his great work on the Mesembryanthema was all done after his 65th year. Mr A. Beer, Ex-Curator of the Botanic Garden at Innsbruck, Austria, who lived for some time in London and translated Prof. Schwantes' German articles on the Mesembryanthema for Dr Brown, as well as Dr Brown's replies, confirmed wholeheartedly the kindly nature of Dr Brown described by the late Alain White.

Dr H. M. Louisa Bolus

Dr L. Bolus, Ex-Curator of the Bolus Herbarium of the University of Cape Town, did valuable research work on the Mesembryanthemaceae. From 1908 and especially since 1927 she diagnosed in her *Notes on Mesembryanthemum and Allied Genera* an enormous number of new species and arranged them in genera, in some cases in new ones of her own creation. In 1958 she published in her *Notes* a key to the genera of the Mesembryanthema and also a number of keys for species, groups and sub-groups, and she continued this work in the *Journal of South African Botany*.

She was born in 1877 at Burgersdorp in the Cape Province. Her father, William Kensit, was born in London. Her mother, Jane Stuart, was Scottish. She was married twice and had 14 children, of whom Louisa was the thirteenth. Her mother died when she was two years old and her father did not marry again, but himself looked after six of his children. Later Louisa had the privilege of looking after her father for about 26 years till he died in his 89th year.

At first she attended the Erica School in Port Elizabeth, and later the Girls' Collegiate School, where she matriculated in 1898. In the following year she began her studies at the South African College in Cape Town, now the University of Cape Town, where in December 1902, she received her B.A. degree in Greek, Latin, English, French and Philosophy. The studies were made possible by the generous help of an uncle, Harry Bolus, who was a merchant who later became interested in botany. For his outstanding work in connection with botany he received, on the same day on which she graduated, an honorary doctor's degree.

During her years of study she helped her uncle whenever she had time with the preparation of plants for the Herbarium. It was a pity that she could not accept his offer to study botany at the University of Cambridge; she had to look after her father. Her uncle then appointed her as Herbarium Assistant, in which capacity she began work in January 1903. When Harry Bolus died in May 1911, she became the Curator of the Herbarium which he had left to the South African College.

In 1912 she married Frank, the younger son of Harry Bolus. At first her husband helped with fine drawings of orchids and other plants, but in later years he became an ornithologist. After the death of Harry Bolus, she first published the third

Dr N. E. Brown

Dr H. M. L. Bolus·

continued in three books. Later the Latin descriptions were illustrated. Superb colour plates of the Conophytum were published in Volume 3 of her Notes, while in Volume 1 colour plates of excellent drawings had been published. She continued with the publication of her Notes in the *Journal of South African Botany* up to her 91st year. She published in other journals too, such as the *South African Museum Annual, Journal of Botany* and *Kew Bulletin*. All in all she published 1 445 species of Mesembryanthema, but she did not regard all these as true new species. Her ambition was to bring together as much material as possible, so that a later author of a monograph would have ample material at his disposal. On most of the herbarium sheets there are extra pockets with flower and fruit material. There the flowers are laid out on gummed paper, so that it is possible to take out specimens for examination when necessary. Without the collection of the Bolus Herbarium, the Mesembryanthema cannot be researched comprehensively. It is truly a treasure trove.

She had the privilege of visiting Kew and its institutes no fewer than seven times, and also the Herbaria of Linnaeus and Thunberg in Sweden, and of Jacquin in Vienna. She also visited Germany, Zürich in Switzerland, and many more places. Thus she was able to erect the new and modern Bolus Herbarium in the complex of the University of Cape Town on the slopes of Table Mountain. In spite of her active life with its great responsibilities, she retained her health and fitness and was able to continue her beloved work for many years, and for this we are truly thankful.

PROFESSOR MORITZ KURT DINTER

He was born 10 June 1868 at Bautzen, Germany. After his botanical studies at Strasbourg and Dresden, he was appointed Curator of the famous Botanic Gardens of Sir Arthur Hanbury at La Mortola, Ventimiglia, Italy. Later, he spent some time at Kew Gardens near London, where he became co-editor of the famous journal, *The Gardeners' Chronicle*. In 1897 he went to South West Africa, where he was employed by the farmer Gessert at Inachab, not far from Lüderitz, to help him with the planting of shrubs to stop the encroachment of the huge sand dunes on his farm. They tried several trees and shrubs, but Dinter did not stay long enough to complete the task. During his stay he collected as much herbarium material as possible and sent specimens to Prof. Dr H. Schinz in Zürich, and also to Prof. Dr A. Engler in Berlin-Dahlem. At about that time, a great deal of livestock died from eating poisonous plants, and someone who knew enough about these plants was needed to help the farmers with advice and investigation. Dinter was appointed Government Botanist in 1900. He studied the grasses and poisonous plants and started at Okahandja an arboretum of trees and shrubs introduced from abroad. During the Herero revolt of 1903 he lost all his herbarium material. In 1905 he took leave and became engaged in Germany. He returned, and his fiancée followed him some time later. They were married by Dr H. Vedder, who became a lifelong friend.

With his wife, Dinter started botanical collecting trips. The fine genus Juttadinteria was named in honour of his wife, and also a good number of other plants. Accounts of his first

volume of his *Icones Orchidearum*. Later she worked with the Ericaceae, the largest family of flowering plants in South Africa, with more than 500 species. With this work, she published the results of other studies about the Orchidaceae, Gramineae, Iridaceae and of the Mesembryanthemaceae. After the death of Prof. H. H. W. Pearson, she had to undertake the editor's work on the *Annals of the Bolus Herbarium* from April 1918 to April 1925, i.e. parts of volume 2 and volumes 3 and 4. In this she published her descriptions of new and unknown plants, illustrated with the best botanical drawings which were published at that time. These were done by the late Miss Mary Page. More than 40 fine drawings by her of the Mesembryanthema were published in *Flowering Plants of Africa*.

When she started her work at the Herbarium in 1903, her uncle had told her: "Look here, you will realize that the plants which really need to be worked on in South Africa are the Mesembryanthema and their relatives, and if you start with this work now and work until you are 80, you will realize how little you know about them!" At 83 she wrote to me: "And it came true!" She complained that she never had the time to give all her attention to this large family, as there was much other work to be done in the management of so an important an institution as the Bolus Herbarium. Her first description of new species was published in the *Transactions of the Royal Society of South Africa* (1908).

In the *Journal of South African Gardening and Country Life* she then started her "Notes on Mesembryanthemum and Allied Genera" (1927, *et seq.*); this work was subsequently

47

Prof. Kurt Dinter

journeys were published in the *Tägliche Rundschau*. In 1908, his first book was published, *Forst & landwirtschaftliche Fragmente*, followed in 1912 by *Veldkost* and in 1914 by *Neue und wenig bekannte Pflanzen Südwestafrikas*, which contained with fine illustrations. In this he treated succulents for the first time. The succulent flora of South West Africa amounts to about 12 to 15 per cent of all the flora. In his garden at Okahandja, he planted and cultivated succulents. The First World War found him on leave in Germany, but a longing for his second homeland brought him and his wife back in 1922. Now they collected privately and travelled in a wagon pulled by ten oxen; later they were given a lorry by the authorities. In Fedde's *Repertorium* he published in "Sukkulenten-forschung I," the result of his studies. On the farm Lichtenstein (near Windhoek, in the Auas Mountains), which belonged to his friend E. Rusch, he laid out a fine succulent garden which was still there in 1968. Another journey (1923–5) took him to the southern part of the country, and after it he published his *Sukkulentenkunde 2*. From 1897–1925 Dinter travelled about 12 000 km, partly by ox-wagon, for the most part collecting specimens. About 8 000 different species were collected. He undertook an especially interesting journey in 1929 when he was given a permit to collect in the diamond area. There he collected the very rare *Lithops optica* var. *rubra*, which is more precious than diamonds. From 1933 to 1935, Dinter was in his beloved country for the last time. This journey was the best and most successful one. As a result of the heavy rains in 1934, the variety of plants was superlative and there were many he had never seen before. Twelve thousand

species of herbarium material alone were collected, and his findings of some were published in his *Diagnosen neuer Südwestafrikanischer Pflanzen*. Since my late friend Ernst Rusch, Jun. was his partner on this journey, I was able to take with him the same route nearly 30 years later. In 1935 Dinter returned to Germany, 67 years old, and worked in Neukirch in Saxony, mainly on the identification of his plants. It is a pity that he could not finish his Index. He died in Neukirch on 16 December 1945.

A great idealist and an agreeable, tireless and hard-working man was lost to the world on his death. He was especially fond of the Mesembryanthema, and all those he found and suspected of being new species were sent to his friend, Prof. Dr G. Schwantes. Schwantes was then in Hamburg and later in Kiel, and many species were named in collaboration with him. The fine genus Dinteranthus was named in honour of Dinter; many other species commemorate him. His wife, Jutta Dinter, outlived him for a few years.

One has to read his most interesting books to learn more than the indefatigable collector of so many Mesembryanthema and to discover his character and to follow the pattern of his life. Hardships could never daunt him. His name will always be remembered in connection with the fine flora of South West Africa.

PROFESSOR DR GUSTAV SCHWANTES

He was born near Hanover on 18 September 1881. Even in his early years he was interested in nature. Later he became interested in burial grounds containing prehistoric urns, as well as in succulents. He selected a particularly interesting and attractive group of plants for his research, the Mesembryanthemaceae. In 1923 he became Doctor of Philosophy, working in the faculty of Prehistory at the University of Hamburg, and in time he obtained a degree in Botany. His first work was on the paraspermous Mesembryanthema; it appeared in 1929 in the "Proceedings of the Institute for General Botany at Hamburg". In 1929 he was appointed Professor of Prehistory at Kiel. His large collection of Mesembryanthema was brought to Kiel and the greater part of it was accommodated in the conservatories of the Botanic Garden of the University of Kiel. The botanist there, Prof. G. Tischler, gave him all the necessary assistance, and in H. Jacobsen, who had been appointed Curator at Kiel, he found a skilful and adaptable gardener, who quickly acquainted himself with the details and dodges of the cultivation of these desert plants of Southern Africa.

The founding of new genera on the morphology of the fruit (capsule) of the Mesembryanthema had to wait till the nineteen-twenties, and is the work of Dr N. E. Brown of Kew and Dr L. Bolus of Cape Town. Dr Rappa of Italy had as early as 1913 begun the study of the fruit for the purpose of classification. And so it happened that Schwantes was the author of many of the specific and generic names and that many other plants were renamed by him. Many of his friends who are of importance in the scientific world are commemorated, especially Prof. K. Dinter, with whom he worked on some of his new genera and species. Illness in his family, many obligations and the difficulties of the war and post-war periods interrupted his work on succulents, but after he became Emeritus Professor in 1945 he began to publish again. Then, for the first time, he published a complete presentation of the classification of the Mesembryanthema, in English; through it these plants became widely known.

Prof. H. Straka, who knew him well, said of Schwantes in an obituary: "Although he held firmly to the Origin of Species, the Darwinian explanation of the further development of living creatures did not satisfy him. Could all the wonderful forms of leaf succulence and all the many varied shapes of the Mesembryanthemum capsules have arisen only by selection through the struggle for existence? He could not believe this and sought for an 'idea' in Goethe's sense of an inherent plasticity within the plants themselves. Not all scientists, naturally, could follow him along this road, but he calls Chapter VIII of his book: *Flowering Stones and Mid-day Flowers:* To the Limits of Knowledge and Beyond."

Schwantes had no students in the usual sense, but he stimulated the interest of a large number of people. H. Jacobsen's articles and books, well-known amongst people interested in succulent plants, are based on Schwantes' work which also inspired the cytological work carried out at Kiel Institute of which Prof. G. Tischler was the Director (Wulff, 1940, 1944; Heldt, 1951). Anatomical, evolutionary researches directed to systematics have appeared (Straka, 1955; Ihlenfeldt, 1958,

1959, 1960). The last year of Schwantes' life was overcast by tragedy. His wife and daughter died within a short time of each other. He himself had been ill for some time and, though he bore his illness with patience and displayed a great will to live, he followed them soon after.

From 1916 to 1952 he published some 107 contributions on Mesembryanthema, as given in a list published in his booklet entitled *Cultivation of the Mesembryanthemaceae,* a supplement to the *Cactus and Succulent Journal of Great Britain* (1952/53).

Dr Schwantes was a singularly gifted man. Normally the post of Professor of Prehistory would occupy most people so fully that they would have little time for a hobby, yet he was able to accomplish much as a botanist too.

In his special field he published in 1908 a comprehensive work, *Deutschlands Urgeschichte,* of which the seventh edition was published in 1952. Later, at Kiel, he was very busy with the excavation of the rich Haithabu, the old Schleswig and the largest of the settlements of the Vikings, which occupied him for many years.

He was a likeable and much admired man. Though we corresponded regularly, I never had the good fortune to meet him, and this is a constant source of regret. (H. HERRE.)

PROFESSOR DR R. MARLOTH

He was born 28 December 1855 at Lübben in the province of Brandenburg. After his practical work as apothecary's assistant in Lübben and elsewhere in Germany and in Switzerland, he enrolled at the University of Berlin in April 1880. In 1883 he received his doctorate at the University of Rostock. Through a schoolfriend in Cape Town, Richard Müller, he received in December 1883 an appointment as chemist with the firm of Wentzel & Schleswig. Soon afterwards he started a herbarium. When he died, he left 20 000 sheets. In August 1884, he received the Government's permission to practise as a chemist and druggist. From November 1885 to February 1886, he ran an apothecary's shop at Kimberley and studied the flora of the Free State. In February 1886 he collected for the first time in Bechuanaland, and from April to June that year he visited South West Africa, with much success. He was an excellent botanical explorer and mountain climber, and founded the Mountain Club, of which he served for a long time as President. From 1889 to 1892 he was Professor of Chemistry at the old Victoria College at Stellenbosch. It was here that he met his future wife, Miss Mariana van Wyk of Clanwilliam, who was a teacher at Bloemhof Girls' High School. They were married in July 1891. They had three sons, one of whom also became a chemist, with a doctorate from Germany. In 1889 Dr Marloth started a laboratory for analytical chemistry and pharmaceutical research in Cape Town. Until 1904 he was a lecturer at the Elsenburg Agricultural College, Stellenbosch. He was an expert photographer and a very acute observer of plants in the wild and of all biological phenomena connected with these plants. On the many journeys he had to undertake for his excellent book *Das Kapland* (1908), he visited the Karroo and Namaqualand. The so-called mimicry among the Mesembryanthemum was of tremendous interest to him. Besides his comprehensive and valuable standard work, *Flora of South Africa,* with its many excellent

Prof. Dr G. Schwantes

Prof. Dr R. Marloth

colour pictures, he also wrote about this mimicry. Later he published a pamphlet in which mimic plants such as Didymaotus, Pleiospilos and *Titanopsis calcarea* were treated and shown. As he once told me, he found the Titanopsis with his feet and not with his eyes. When I arrived at Stellenbosch in 1925, we used to work together in his garden at Cape Town every Saturday afternoon, and I learned much from his wide experience. He also helped me to obtain suitable conservatories for the cultivation of succulent plants. Apart from his work on mimicry, he conducted experiments to observe the water-absorbing qualities of the green parts of certain succulents (e.g. Crassulaceae, Mesembryanthemaceae, Portulacaceae, Liliaceae). In November 1911 he received the Prussian title of Professor. On his seventieth birthday he received an honorary doctorate from the University of Stellenbosch (1922) also from Heidelberg and Cape Town in 1929. In spite of his many activities, he was always willing to help others, and he had many friends. When I met him for the first time, he already suffered severely from arteriosclerosis. A second stroke caused his death at Caledon on 15 May 1931. He was buried at St. Peter's cemetery in Mowbray, Cape Town. General Smuts, one of his students, who he helped with his work *Holism and Evolution*, was one of the speakers at the funeral. On his tombstone one reads: "He worked as long as there was day-time." His wife outlived him for more than 30 years; she died at 93.

PROFESSOR DR GERT CORNELIUS NEL

In "Nel", in *The Gibbaeum Handbook*, Prof. P. G. Jordaan gives the following details of Dr Nel's life:

"He was born on a farm at Greytown, Natal, April 6th, 1885. In this Garden Province of South Africa he was first imbued with the love for the South African veld and flora. He matriculated at Franschhoek and took his B.A. degree at Stellenbosch. At the latter two places he came to know the botanically rich and varied flora of the winter rainfall area – a flora very much different from the subtropical and savanna flora he became acquainted with in Natal. He could not have realized at this stage that he would later spend several years in the Orange Free State in the grass steppes ('grasveld') of South Africa and that his future research would lead him to an intimate experience, love and knowledge of the vast, arid regions of Southern Africa, such as the Little Karroo, the Karroo, Namaqualand and South West Africa.

"After studying in Halle and Berlin he obtained his Ph.D. degree in 1914 on a thesis on the Amaryllidaceae-Hypoxideae embodying the results of his research under the well-known A. Engler. After filling responsible educational posts in Lindley and Bloemfontein in the Orange Free State, he was appointed in 1921 to the newly instituted professorship of Botany at the University of Stellenbosch – a post he held up to his death in 1950 – where he could build on the sound foundation laid by Professor R. Marloth and Dr A. V. Duthie.

"One of the first tasks to receive the attention of Professor Nel was the founding of a botanical garden, and through his endeavours the Botanical Garden of the University of Stellenbosch came into being, only two years after he assumed duty as Professor of Botany. He had the good fortune of getting the services of Mr H. Herre as head gardener and curator who, to a great extent, was responsible for building up at Stellenbosch a collection of succulents of international repute and who collaborated closely with Professor Nel. In the Botanical Garden Prof. Nel could study under close observation living plants that were collected in nature and could so exclude the danger of studying plants that had been unintentionally hybridized by cross-pollination under cultivated conditions.

"Administration and teaching, especially during the first years of renewed expansion in the Botany Department, took up much of the time of Professor Nel and it was only about 1930 that he could find time for research. He started off by studying the succulent Euphorbias and Stapelias and described many new species and varieties. After the publication of the book of White and Sloane on the Stapelieae (1937) and the book of White, Dyer and Sloane on the *Succulent Euphorbias of Southern Africa* (1941), in both of which his authority and research on these plants are acknowledged, he switched to the succulent Aizoaceae – or 'vygies' as they are called in Afrikaans – and for the rest of his life he spent on them the time he had available for botanical research.

"During the last three decades very many new genera and species of the succulent Aizoaceae were described. Many of these descriptions were based on inadequate material, without taking the phytogeographical and ecological variation of these plants into consideration. Furthermore, no keys for the identification of the species were published. In this way a chaotic state of affairs came into existence and very few, perhaps nobody, could again identify all the plants from the descriptions – often lengthy descriptions in Latin – occurring in a great variety of books, periodicals and pamphlets published all over the world. To put the taxonomy of the Aizoaceae on a sounder basis, a better knowledge of the different species and varieties in their natural environments was essential, as well as keys to the different genera and to the species of each genus. Professor Nel was more attracted to a study of related species and stressed the importance of taking a genus, studying its species carefully and compiling a key. He started with the highly succulent genus *Lithops* (University Publishers and Booksellers, Stellenbosch). *Gibbaeum* was the second genus to receive his attention and, if it were not for his sudden death, revisions of several other genera could have been expected, as he had made extensive observations on several other of the highly succulent genera such as Dinteranthus and Titanopsis.

"Most of the succulent Aizoceae occur in South Africa, where they exist in nature under conditions quite different from those in Europe where much of the research on them had been and is being done. One of the urgent needs in furthering the scientific knowledge of succulent plants is experience of these plants in their natural environment. This need Professor Nel clearly realized and one of his great contributions to the study of South African succulents is the field knowledge of these plants which he put on record and tried to incorporate in his taxonomic studies. To study these succulents he made many expeditions and excursions into the arid parts of southern Africa and in this way obtained a field knowledge of these plants, especially of Lithops, Gibbaeum and other highly succulent genera, of which few [people], if any, possessed. Much of this knowledge was lost to botany on his sudden death. Reading his descriptions, as in his book on Gibbaeum, is not only enjoyable, because these descriptions transfer the reader to the natural habitats of these plants, but his reasoning, attempting to show that many of the described species merge into one another in nature, is an intellectual pleasure. His books, *Lithops* and *Gibbaeum*, the first and only two monographs on genera of succulent Aizoaceae, are hallowed monuments because [they emphasize] the importance of knowing these plants under natural conditions and may well be an inspiration and model for further study of and research on the numerous remaining genera. Not only his two books, but also the genus Nelia, the number of specific epithets bearing his name and the great number of plants described by him will make his name live on in botanical literature and in the history of the study of succulents.

"G. C. Nel was not only a lover and student of succulents and a botanist with a sound knowledge of the fundamentals of his subject, he also possessed a deep and intimate knowledge of human nature which grew from wide experience and an understanding of all sections and races which make up the population of South Africa. He had a sound philosophy of life, a very good sense of humour and a strong personality. He was kind-hearted, friendly, approachable and intimately interested in each member of his staff. In many respects he had his own ways of doing things, of which the unconventional make-up of his book on Lithops is proof."

Thus Professor Jordaan in his life sketch about Professor Nel for the *Gibbaeum Handbook*. Since we were both members of the staff of the late Professor Nel, I can endorse every word written by Professor Jordaan, and I think that it could not have been told better than he has done it. There is nothing to add.

Prof. Dr G. C. Nel

PROFESSOR J. A. HUBER

He was born at Landshut in Bavaria, Germany, on 8 September 1899. After the First World War, he studied Biology and Botany at the University of Munich. His thesis, *Morphology of the Mesembrianthemum*, was done in 1923 under the guidance of Prof. K. Goebel. From 1924 to 1932 he worked as scientific assistant, especially on genetics, at the Institute of Plant-breeding and Cultivation at the Technical School of Munich at Weihenstephan.

Later he was appointed Professor of Biology and Anthropology at the Phil-Theol. High School at Dillingen/Danube, a post which he held until he retired in 1966. From 1936 onwards he held an honorary post in Nature Preservation, which he did not relinquish until February 1968.

His last major work about the vegetation organs of the Mesembryanthema was done during the first years of the Second World War, with the kind assistance of the late Wilhelm Kesselring, at that time Curator of the University Botanic Gardens at Darmstadt, and Dr H. Jacobsen, Curator of the University Botanic Gardens of Kiel, who let him have valuable plants for this purpose. The results were published in the last

Prof. J. A. Huber

Dr A. Tischer

war-editions of 1943 of the German Cactus Society under the title *Morphological Studies on Mesembryanthema*.

In 1950 he was a foundation member of the International Organisation of Succulent Scientists (I.O.S.), and he is still a member of it.

Later works on succulents were mostly done on European Crassulaceae, especially Sempervivum, as he had not the time nor the opportunity for the cultivation of African succulents.

Dr A. Tischer

He was born in 1895 and studied Law and Commercial Science (Volkswirtschaft) and obtained his doctorate in 1921. Until 1928 he was occupied in the industrial field, and then he went into Government service, he later became "Oberregierungsrat".

From his early youth he was interested in natural sciences, especially Botany. While still a student, he became interested in cacti and other succulents. From 1920 he worked on the Mesembryanthema. From 1923 he worked in collaboration with the late Carl Schick and others on the most interesting highly succulent Mesembryanthema. From this time dates his special interest in Conophytum. Through the late Dr N. E. Brown and other workers in England, at the Botanical Gardens at Stellenbosch and at Kirstenbosch, as well as numerous South African collectors who knew him, he was able to build up a collection of about 1 500 Conophytum comprising about 200 species and varieties, probably the most comprehensive collection in the world. In later life, he presented his collection to the Botanic Garden of the University of Heidelberg, but he is still looking after it and uses it for further scientific research.

After working at it all his life, he finished a monograph about the genus Conophytum, but he has not yet been able to publish it, which is very regrettable.

Dr H. W. de Boer

Hendrik Wijbrand de Boer was born at Adorp in the Province of Groningen, the Netherlands, on 26 August 1885. He attended the schools in Groningen and later the University too, where he studied Pharmacology, and where he sat his last examination in 1912. When still a student he won a gold medal by answering an inter-academical prize question, written out by the University of Leiden. In January 1914 he received his doctorate at the Sorbonne in Paris for his thesis: *Etude micrografique de dix drogues végétales nouvelles de la pharmacogsee neerlandaise.* Later he qualified in food chemistry and became director of the food investigation service for some of the cities of the Netherlands. Finally he became director of the food investigation laboratory of the Province of Groningen. He also lectured on food chemistry at the university there. From 1938 he became interested in the cultivation of the Mesembryanthema, especially Lithops and Conophytum, and made it his hobby. Owing to the pressure of his work he was not able to make it more than a hobby. However, after his retirement in 1950, his hobby became a scientific study, although he continued to work at the university for several more years. In *Succulenta* he published some new Lithops; later he wrote also about Conophytum species and varieties, and of his views about the variability of the Conophytum, and the question of the Mitrophyllum-Conophyllum. Since 1961 he has published in

Dr H. W. de Boer

Dr Hermann Jacobsen

Succulenta his "Notes on Lithops", and in collaboration with Dr H. Boom his "Analytical key for the genus Lithops". He cultivated all its species and varieties and had a large collection of Conophytum as well. He also had an excellent collection of colour slides of the various capsule-structures, photographed by himself, which he used in connection with his studies of the different genera of the Mesembryanthemaceae, a very interesting part of botanical study. He died some years ago.

DR HERMANN JOHANNES HEINRICH JACOBSEN

Dr Jacobsen was born 26 January 1898. On leaving school, he served his apprenticeship in the private garden of the Freiherr von Donner at Breedeneek, near Preetz in Holstein, where Chief Gardener Hennig became his teacher. The large park with its fine old trees prompted his life-long interest in dendrology. There were conservatories and hot-houses, the plants of which were needed for arrangements in the castle. Since it was far away from any town, the young man could devote himself entirely to his studies. A few years later, he worked for nearly a year at the tree nursery of J. von Ehren at Nienstedten, near Hamburg. Later he worked for some time in the Court Garden of the Grand Duke of Mecklenburg-Schwerin. During the First World War he served for many months at the front in Flanders as well as in the north of France. After the war, in 1919, he joined the Botanic Gardens at Köln-Riehl. This was good training; in 1920 he started as assistant gardener in the Botanic Garden in Bonn under the Oberinspektor C. Wiesemann. Here he worked mainly with the fine collection of coniferous trees left by the famous L. Beissner,

a former head of the Garden. It is still an excellent collection, well worth inspecting. In the conservatories he worked mainly with the Bromeliaceae, and also for the first time in his life, with succulents. He also attended many lectures on Botany at the University. After he passed his examination as head gardener, he was appointed as the youngest first-class assistant gardener at the Botanic Gardens in Bonn. In Bonn he married, and after ten years he was appointed Garteninspektor at the Botanic Gardens of Kiel. In 1931 he attended the general meeting of the famous German Dendrological Society and won the confidence of the Count of Schwerin, its President, whose adjunct he later became.

In Kiel he was fortunate enough to cultivate the famous succulent collection of Derenberg-Dinter and Schwantes, which had been handed over by Prof. G. Schwantes to the Botanic Gardens of Kiel. Dr G. Schwantes had by this time been appointed as Professor of Prehistory. In 1933 Jacobsen published his first book, *Die Sukkulenten*, followed by another in 1938, *The Cultivation of Succulent Plants*. The first was translated into English before the Second World War. He assisted Prof. Schwantes with the further classification of the Mesembryanthemaceae. The genus Jacobsenia and also about eleven species are named after him. In 1954 he started with the German edition of his *Handbook of Succulent Plants*, of three volumes, of which the third one is entirely devoted to the Mesembryanthemaceae. In 1960 an enlarged English edition was published. In 1952 he published in Aachen a smaller book, *Kakteen und Andere Sukkulenten*. In 1950 he published in collaboration with H. Herre and Prof. G. H. Volk a small booklet about the Mesembryanthemum (Stuttgart,

Eugen Ulmer Verlag.). He published many articles on succulents in various journals all over the world, and also lectured on this subject. During the Second World War, the Botanic Gardens of Kiel were almost completely destroyed, but fortunately most of the succulents were no longer there; they had been distributed all over Holstein. With almost superhuman effort he built up his collections once more in new conservatories and constructed a new alpine garden. Nor did his work end there. He raised a new strain of Cyclamen with scented flowers which sold well all over Europe.

He was made honorary member of the German Cactus Society, Vice-President of the Cactus and Succulent Society of Great Britain and also of the African Succulent Plant Society in Great Britain. On 17 May 1963, the degree of Doctor rer.nat.h.c. of the Faculty of Philosophy of the University of Kiel was conferred upon him. He has served in many public capacities in Kiel, and is an able organizer and speaker. Like his forebears, he is fond of music and plays several instruments. In spite of all these activities, he is continuing with his work and, an enlarged *Handbook of Succulent Plants* is ready for printing. He is also engaged in the preparation of a lexicon of succulent plants, similar to the one of cacti by the late C. Backeberg, which appeared in December 1970 in Germany. An English edition will follow.

PROFESSOR DR HERBERT STRAKA

He was born at Brünn, Moravia, on 14 July 1920 and attended the grammar school there. Later he studied at the Universities of Vienna, Innsbruck, Bonn and Stockholm. He graduated at Bonn in 1951 with a dissertation on the palynology of the history of the late Quartaer vegetation of the volcanic Eiffel mountains. In 1951 he was appointed as Assistant at the Botanic Department of the University of Kiel, Germany. There he became interested in the Mesembryanthemaceae through Prof. G. Schwantes; his later work dealt with examination of the anatomical and historical development of the fruits of paraspermous Mesembryanthema (*Anatomische und entwicklungsgeschichtliche Untersuchungen an Früchten paraspermer Mesembryanthemen*). Still later, he published some palynological works on the flora of Madagascar. It was followed by two works about the Mesembryanthemaceae, the first one in collaboration with Dr H.-D. Ihlenfeldt, about the systematic position and organization of the Mesembryanthema (*Die systematische Stellung und Gliederung der Mesembryanthemen*). The second one was written in collaboration with Prof. Schwantes and Dr H.-D. Ihlenfeldt about the higher taxa of the Mesembryanthemaceae (*Die höheren Taxa der Mesembryanthemaceae*). He has been Professor of Botany at the University of Kiel since 1960.

PROFESSOR DR OTTO HEINRICH VOLK

He was born on 6 December 1903 in Richen near Heidelberg, Germany. His father, who was a pastor, died in 1913. Otto went to school at Heilbronn, where he matriculated at Easter 1923. Military service followed, and in 1925 he started with the study of Natural Science at the Universities of Munich, Vienna and Heidelberg. He obtained his degree in Botany in 1930. His examiners were Professors L. Joost and H. Walter. He also graduated in Zoology and Chemistry. During the years 1930

Prof. Dr H. Straka

Prof. Dr O. H. Volk

to 1938 he was an assistant of Prof. H. Burgeff at the University of Würzburg. In 1936 he became lecturer in General Botany and Plant Geography.

In 1938 he undertook a research trip to South West Africa. There he was interned, first at Windhoek, and in 1940 in Andalusia. He was repatriated in June 1944, and up to his release by the powers of occupation he was head of the Institute of Applied Botany at the University of Würzburg. Later he was involved in the pharmaceutical industry as well as with the cultivation and the procuring of pharmaceutical plant material.

In 1950 he married Miss Irene Rubov. In 1947 he had become a honorary lecturer, in 1949 he had become Professor. From 1950 to 1953, while on leave, he served as Professor of Botany at the University of Kabul, Afghanistan. He was invited in 1958 to join the Congress of South American Botanists in Peru. In 1960 he took part in a symposium on medical plants of the Middle East, organized by Unesco. He was the delegate for Afghanistan in Peshawar.

During 1956 he spent four months, and during 1963, six months in South West Africa, doing additional work in connection with his research which he had begun in 1938 and which had been interrupted by the war. From 1957 to 1963 he was head of the Academical Foreign Office by appointment through the Senate of the University of Würzburg. For some time he was also Chairman of the Board of the Faculty of Natural Science at the same university.

During his many journeys in connection with the examination of the behaviour of plants under arid conditions, he also studied the Materia Medica of the countries he visited. His main scientific interests are the Systematics and Chemotaxonomy of plants of medical and economic value, their infraspecific qualities, ecology and all their other vegetational problems. During the period that he was Assistant at the Botanical Institute of the University of Würzburg, he dealt with pharmaceutical industries and dispensaries interested in pharmacognostical aspects. His experience of lecturing for so many years was a great asset.

Since 1947 he has been Head of the Pharmacognostic Department of the Institute of Botany of the University of Würzburg. In 1964 he was appointed Professor of the newly created chair of Pharmacognosty of the Faculty of Natural Science at the University of Würzburg.

Professor Dr H. D. Wulff

He was born on 18 January 1910 at Hamburg-Altona in Germany, and studied at the Universities of Hamburg and Kiel. He obtained his Dr.Phil. at the University of Kiel in 1933 and in 1945 was Professor at the University of Kiel. From 1946 to 1951 he was Commissionary Director of the Institute for Pharmacy at Kiel, and he has been Professor-in-ordinary at the University of the Saarland, Saarbrücken, since 1957 and Director of the Institute of Botany and the Botanic Gardens, also from that date. From 1958 to 1960 he was Rector of the University of the Saarland. He produced the first work on the morphology of the Mesembryanthemaceae.

Dr Hans-Dieter Ihlenfeldt

He was born the 7 July 1932 at Kiel, the son of the Studienrat W. Ihlenfeldt and his wife, Bettine, (*née* Kickhefel). He matricu-

Prof. Dr H. D. Wulff

Prof. Dr H.-D. Ihlenfeldt

lated at Easter 1952 at Eckernförde in Schleswig-Holstein. From the summer semester of 1952 until the winter semester of 1958 he studied the Natural Sciences at Kiel. In March 1958 he graduated from the University of Kiel as Dr.rer,nat. (Botany, Zoology and Chemistry) with a thesis, *The History of Evolution, Morphology and Systematics of the Mesembryanthemaceae* (published in Fedde's Repertorium 63 (1960): 1–104). His promotor was H. Straka. On 28 July 1958, he sat the State Examination for Teachers in Grammar Schools (Biology, Chemistry and Physics). Since 1 February 1959, he was scientific assistant at the State Institute for Botany in Hamburg, working at the University of Hamburg. From October 1961 to May 1962 he was engaged in a research and collecting trip to South and South West Africa, and worked in Namaqualand with H. Herre and in the Transvaal with H. Schlieben. In March and April 1962 he undertook a similar trip into the Namib of South West Africa for the study of the flora there after good rain had fallen, and was accompanied by Dr de Winter and Mr D. S. Hardy. At present he is a university professor at the State Institute of General Botany at Hamburg. Latest work: *Systematics of the Mesembryanthemoideae, African Pedaliaceae; problems about the modern evolutionary morphology of the higher plants.*

Dr Suzanne Dupont

She was born on 20 September 1922 at Albi (Dept. du Tarn [81]) and went to school there. She studied at *L'Ecole Nationale d'Enseignement ménage agricole de Coetlogon* at Rennes (Ille-et-Vilaine [35]). She received the *Licence de Science naturelles* at Rennes. From 1946 she was Zoötechnical Assistant in the Faculty of Science at Toulouse (Haute Garonne [31]). In the course of her work there she began to study the Mesembryanthemaceae, as well as the Cactaceae; the studies of the latter took her three years. In 1962 she was appointed Chief Assistant. Her husband received his doctorate in 1960 and was appointed *Maitre de Conference*, and in 1963 became professor at Nantes. She changed her speciality and left Rennes and later also Toulouse, to be with her husband; she works in his laboratory. Their specialities are quite different; her husband is a phytogeograph and ecologist and is making a special study of the European Atlantic coast. They have two sons.

Dr H. Friedrich

He was born 29 December 1925 in Zittau, Saxony, where he also went to school. In 1943 he joined the Navy and served until the end of the war. In August 1945, after internment in Denmark, he was released and went to Bavaria as he had lost touch with his family still living in Zittau. But he continued to live in Bavaria even after he had re-established contact with his family. Pursuing his old desire of studying pharmacy, he set out to get a position as a practitioner, but was not successful. After waiting for more than two fruitless years, he decided to study the Natural Sciences, especially Botany, which was one of his hobbies. Before admission to the University of Munich, he was able to begin his studies with the intention of becoming a high-school teacher. But in 1950, while working for his doctorate under the guidance of Prof. Karl Suessenguth (at

Dr S. Dupont

Dr H. Friedrich

that time the Director of the *Botanische Staatssammlungen* in Munich) on a theme about the natural relationship of the Plumbaginales, Primulales and Centrospermae, he found the taxonomical work more interesting, especially as he was appointed Assistant at the *Botanische Staatssammlungen*. By this time he had already begun the preliminary work on what was later to be published as *Prodromus of the Flora of South West Africa*. He was entrusted with research on some families of the Centrospermae. In June 1954 he received his doctorate. He received a bursary from the German Forschungsgemeinschaft to continue with his work on the *Prodromus*. On 1 January 1955 he was appointed Scientific Assistant at the Botanic Garden in collaboration with the *Botanische Staatssammlungen*. Here, in the Botanic Garden, he had the opportunity to build up a fine collection of South West African succulents, especially of Crassulaceae and Mesembryanthemaceae. In 1956 he was promoted to Conservator. In the same year he married Dr Martha Holzhammer, also a student of Professor Suessenguth and collaborator on *Prodromus*. In 1965 he was promoted to Chief Conservator.

H. Herre (I.O.S.)

I, Adolar Gottlieb Julius (Hans) Herre, was born on 7 April 1895 at Dessau, Germany, where I attended the Fridericianum, and I served my apprenticeship at the famous Garden at Wörlitz near Dessau. Later I worked at Hamburg and Brugge (Belgium) and visited Paris and northern France, and also the north of Italy. During the First World War, I served in the north of France. Seriously wounded in the leg during the battle

H. Herre

of the Somme in July 1916, I returned to Germany for hospitalization. After my recovery I left the Army as an officer, with the Iron Cross, 1st Class. Early in 1919 I worked for some time in the famous University Botanic Garden at Munich-Nymphenburg until I began my studies of Horticulture at Dahlem near Berlin, where the well-known Prof. A. Engler and Prof. L. Diels were two of my teachers. On successful completion of the course, I accepted the position of Director of the Horticultural School for Women at Hermannstadt in Rumania; conditions in Germany were very bad. As the position there also worsened, I returned to Berlin and worked in the firm of L. Spaeth. Inflation aggravated matters and I had decided to emigrate to the United States of America, when I received the offer of laying out the Botanical Garden for the University of Stellenbosch in South Africa; I was approached by Prof. L. Diels to whom Prof. G. C. Nel had written. Some time after my application, I received the news that I had been appointed. Before leaving Germany in July 1925, I gained good marks in my last Diploma Examination. In August I arrived in Stellenbosch and immediately began laying out the Garden. It was of medium size and we resolved to collect succulents, since we could collect and grow a large collection in a small space. Because of the wet climate of the Western Cape, we had to built conservatories, and Prof. R. Marloth very kindly helped in this respect. In 1927 and 1930 the conservatories were built. As 1929 was a good year in Namaqualand, I collected there while the Rev. G. Meyer and his family accommodated me at Steinkop. In 1930, again a good year, I returned to Namaqualand and did more collecting. On these trips, we assembled ample material, and Dr L. Bolus, the Curator of the Bolus Herbarium of the University of Cape Town, came over with her artist and her staff and helped with the identification. As most of the collected plants were mesems, I had the chance to learn more about this large family. While interned during the first year of the Second World War, I whiled away the time by drawing many flower and capsule-sections and, as Prof. Volk of the University of Würzburg, Germany, showed much interest in this family, we decided to try to make a more or less phyllogenetic key for the 123 genera of this large family. We worked hard, and in June 1943 the work was finished, just before Prof. Volk was repatriated to Germany. I then started work on the distribution maps, which are published here. On my return to Stellenbosch, I continued with this work and went on with the investigation of the cross-pollination of the genera, a study which is still being continued. According to my findings, cross-pollination is by no means always necessary, but there are genera which will not form seeds without cross-pollination, such as Herrea, Conicosia, while others, such as Aethephyllum, are self-fertilizing.

Another task was to reduce the number of the genera. This is progressing, but it will take some years to complete it. There is other work to be done, as Dr S. Dupont has shown with her fine publication about the stomata and cotyledon of the Mesembryanthemaceae.

Mary M. Page

Mary Maud Page, botanical artist of the Bolus Herbarium, University of Cape Town, died in Cape Town on 8 February

Miss Mary Page

Miss Beatrice Carter

1925. She was the second daughter of Nathaniel Page, J.P. of Croydon, and was born on 21 September 1867 in London, where she lived till she came out to South Africa. She received her education at private schools in England, followed by a year in Paris at a finishing school. Later she spent six happy months at Grasse in the south of France, and learnt to converse easily in French.

Dogged by ill-health from early childhood, Mary Page's whole life was a brave struggle, and in spite of the many obstacles she had to overcome, she fought on till the end with indomitable courage. Anyone less imbued with a deep and lasting love of nature would have become saddened and would have given up. But she was comforted, restored, and inspired by her love, and this keen delight in the beauty of her surroundings, and the glorious sunshine, coupled with a never-failing sense of humour, bore her victoriously through the most painful periods of her life. Who can forget the merry twinkle of her blue eyes and the ready wit that refreshed us at every turn? She made friends and was beloved wherever she went.

The record of her activities is extraordinary. She worked at the School of Art (Caldrons) until her eyesight failed and she was obliged to give it up. Then she took a course of wood-carving, learnt to work with metals and enamels, and excelled in various kinds of needlework, embroidery and lace-making. She learnt Braille in order to help a blind friend, for whose use she translated many books. During the three years of her

father's mayoralty of Croydon – at the time of the Boer War, the death of Queen Victoria and the accession of King Edward – she did her full share of the duties which fell to so prominent a family.

In July 1911, soon after her father's death, Mary Page sailed for South Africa. Her health had become worse and after a serious operation it was hoped that a change to a warm climate would benefit her more than anything else. She first stayed at Dealesville in the Orange Free State. Early in 1912 she moved to Bloemfontein. In August 1912 she spent three months at Palapye, in the former Bechuanaland, and later visited Pretoria and Rhodesia, and also Basutoland, which she liked so well that she returned to it several times.

In January 1915, while she was spending a few weeks in the Cape Peninsula, Mrs L. Bolus met her for the first time and was at once struck with the beauty of the flower paintings she showed her. Mary Page had then not yet done any botanical drawing but with her usual enterprising spirit she was quite prepared to do her best. In October she spent three months with Mrs Bolus, and she learnt to discern the minute differences between the various species. Her enthusiasm was unbounded. This was the beginning of her connection with the Bolus Herbarium which lasted till her death. She made considerably more than 200 drawings of that most important South African genus, the Mesembryanthema; these constitute an invaluable contribution to our knowledge of that difficult group. Besides these, there are some 30 drawings of orchids and about 100 of other plants of the Cape Peninsula, and also many sketches of various other plants, especially in the family Iridaceae. It is indeed a splendid legacy and will immortalize her name in the annals of South African botany.

(*Done according to the obituary by Dr L. Bolus*)

BEATRICE ORCHARD CARTER

She was born in King William's Town in 1889 and was educated there and in East London. She received her first lessons in art in Queenstown, and later she attended the Art School at Cape Town. During and after her student days, she worked for Dr D. J. Wood, the oculist, making drawings for his ophthalmic work. In 1926 she was appointed as artist to the Bolus Herbarium, where she remained until her death in 1939.

Her death was a great loss to the Bolus Herbarium, as no successor was appointed and much of the work she began remained unfinished.

Prof. R. H. Compton, at that time the Director of the National Botanic Gardens at Kirstenbosch, paid her the following tribute in the *Journal of the Botanical Society of South Africa*, Part XXV, p. 4, 1939: "It is with great regret that we record the death of Miss Beatrice O. Carter in November 1939. Miss Carter had been for many years on the staff of the Bolus Herbarium, where she devoted herself especially to recording with her skilful pen and brush the details of form and structure of the Mesembryanthemaceae in connection with Dr L. Bolus' comprehensive studies of this vast group. We include in this Journal a coloured plate of drawings of various species of Conophytum, drawn by Miss Carter and giving some idea of her painstaking and accurate work."

I knew her too, during the days when she and other members of the staff of the Bolus Herbarium came to the Stellenbosch University Botanical Garden to investigate the flowering Mesembryanthema which I had collected on various trips to Namaqualand, and well remember her wit and kindness.

DESCRIPTIONS OF THE GENERA
OF THE MESEMBRYANTHEMACEAE

In the bibliography published with each genus, italic type has been used
to indicate: (*a*) the most important source of information on a particular genus; and
(*b*) a genus or genera that have been sunk and incorporated with another genus.

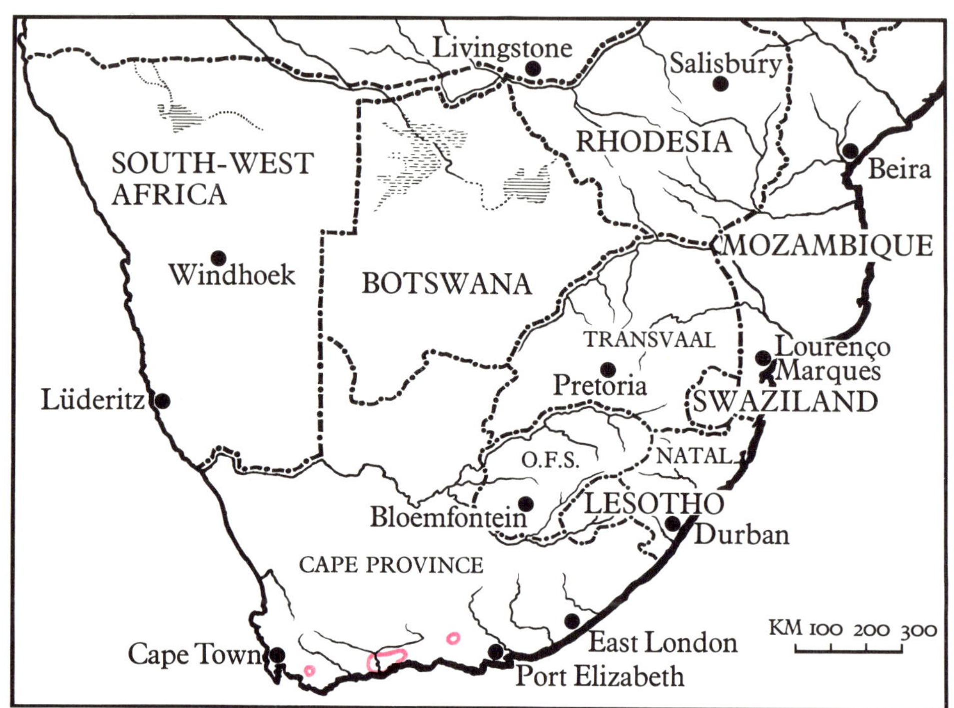

Acrodon

Acrodon N.E. Brown
From the Greek *akros* = point and *odus*
(genetivus of *odontes* = tooth).

N. E. Brown, Gard. Chron. 81: 12.1927; J. Bot. 66: 77 (in key). 1928. - Labarre (red.) Mesembr. 104; fig. 16, 17. 1931. – v. Pöllnitz, Aufteilung, 24. 1933. - Jacobsen, Succ. 89. 1933. - Pax, Natürl. Pfl. 211. 1934. - Jacobsen, Succ. Pl. 123. 1935, Verzeichnis, 11. 1938. - Goossens, Blomplante, 145 (in key). 1940. - Jacobsen, Herre, Volk, Mesembr. 72 (in key), 83. 1951. - Phillips, Genera, 319. 1951. - Jacobsen, Handbuch, 1156; fig. 1004. 1955. - Schwantes, Fl. Stones, 92. 338. 1957. - Jacobsen, Handbook, 952 (in system of Schwantes), 962 (in key of Bolus), 973 (in key of Herre & Volk), 977; fig. 1193. 1960. - Jacobsen, Lexikon, 356, t.143/1. 1970.

Plant branched from base into clumps about 10 cm high. Leaves fleshy, 3-5 cm long, opposite in rosettes, sheathing at base, crowded, spreading-reflexed, grey-green, smooth, 3-angled, acuminate, with scattered cartilaginous teeth on margin. Flowers solitary, about 4 cm in diam. opening midday. Sepals 5–6, nearly equal, ovate, sometimes with a membranous margin. Petals 2-seriate, free, spreading, white with reddish margin, narrowly lanceolate, emarginate. Stamens erect; filaments connivent, white, reddish above; anthers reddish, staminodes absent. Ovary conical; placentas parietal; glands dark-green, shortly denticulate; stigmas 5, short, plumose. Capsule 5-locular; valves ascending when wet; the keels awn-like pointed; loculi-roofs rather stiff; tubercle large, compressed, closing the loculi; seeds pear-shaped, rough, dark-brown.

A south-western Cape genus with 3 species. (Type: *A. bellidiflorus* (L.) N. E. Br.)

Acrodon bellidiflorus   63

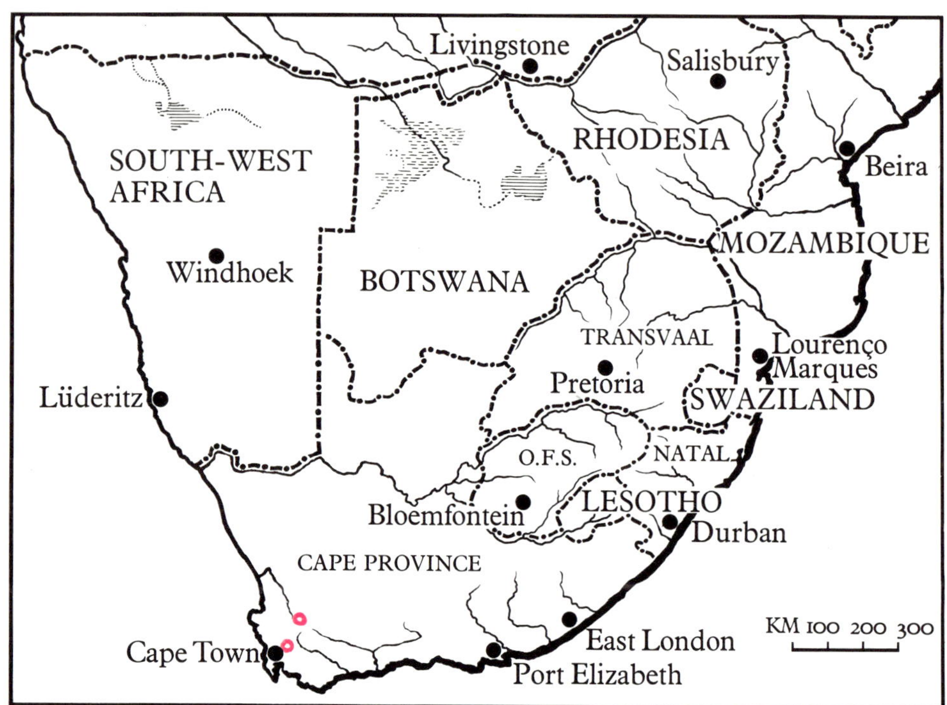

Aethephyllum

Aethephyllum N. E. Brown
From the Greek *aethes* = irregular, unusual and *phyllon* = leaf.

N. E. Brown, Möllers Gärtnerztg. 43: 400. 1928. - Bolus, Notes, 81: fig. 11 (capsule). 1928. - v. Pöllnitz, Aufteilung, 24. 1933. - Pax, Natürl. Pfl. 211. 1934. - Jacobsen, Succ. Pl. 123. 1935; Verzeichnis, 11. 1938. - Goossens, Blomplante, 149 (in key). 1940. - Jacobsen, Herre, Volk, Mesembr. 58 (in key), 83, 118; fig. 32 (capsule). 1950. - Phillips, Genera, 319, 1951. - Jacobsen, Handbuch, 1157, 1552. 1955. - Schwantes, Fl. Stones, 42, 340; fig. 6 (capsule). 1957. - *D. G. Winkler, Bydrae tot die kennis van Micropterum Schwant. subgenus Aethephyllum in J. S. Afr. Bot. 23: 19. 1957.* - Jacobsen, Handbook, 953 (in system of Schwantes), 959 (in key of Bolus), 967 (in key of Herre & Volk), 978; fig. 1194. 1960. - Jacobsen, Lexikon, 356, t.143/2. 1970.

Annual herb, prostrate, papillate. Leaves in outline oblong or ovate, lyrate-pinnatifid, irregular. Flowers solitary, dichotomous, pedicels without bracts, small, light-yellow. Sepals 5, unequal, largest lanceolate. Petals 2-3-seriate, linear-oblong, pointed. Stamens 1-seriate, only a few. Ovary vaulted; placentas parietal; stigmas 5, filiform, deciduous. Capsule 5-locular; the keels with membranous, broad, spreading wings; roofs reduced to a limb, without tubercles; seeds with very small tubercles. Cleistogamous plant!

A south-western Cape genus with 1 species. (Type: *A. pinnatifidum* (L.f.) N. E. Br.)

On page 22 the numbers 103–110 of the key to the genera of the Mesembryanthemaceae by H. Herre and Prof. Dr O.H. Volk were omitted.

103   Leaves of a pair more or less equal, more rarely one shorter than
the other but then not as with Glottiphyllum 104

–   Leaves of the same pair dissimilar, one of the leaves longer and
with an imprint of the shorter one, shiny, soft and pulpy, at the
apex either hook-formed and often obliquely keeled or rounded and
tongue-shaped, often distichous or obliquely decussate; capsules
pear-shaped, with 8–20 chambers, with roofs and big tubercles;
explanding-keels without wings, ending in membranous points or
awns . . . . **Glottiphyllum**

104   Leaves of a pair soon diverging 105

–   Young leaves of a pair for long time contiguous and somewhat
resembling a beak, during the resting period clothed with membranous white sheaths (remains of preceeding leaves), alternating
pairs often dissimilar; stigmas 8–20, plumose; chambers of capsules with roofs and large tubercles; expanding-keels diverging,
ending in awns, without wings . . . . **Cheiridopsis**

105   Leaves in transmitted light without hyaline spots 106

–   Leaves in transmitted light with hyaline dots; flowers with numerous staminodes; stigmas 6–10, slender . . . . **Khadia**

106   Leaves at least 3–4 times as long as broad or if shorter, then
flowering stalks with bracts 107

–   Leaves shorter than above or flowering stalks without bracts . 108

107   Leaves semiterete to terete. See no. 66 . . . . **Cephalo-hyllum**

–   Leaves keeled, sword-shaped, whitish-green, not dotted, often
mucronate; capsules with 5–15 chambers, with roofs and with small
tubercles . . . . **Machairophyllum**

108   Flowers deep yellow, large; 6–7–15 stigmas and chambers . . 109

–   Flowers white to violet of moderate size; stigmas in the flowers
and chambers in the ovary or in the capsules 5–25, usually 8–12;
leaves often very thick to ovoid, whitish-green, with more or less
edged keels, tuberculately toothed or entire; flowers with many
stamens with hairy filaments; nectaries united in a crenulated ring;
stigmas usually longer than the stamens . . . . **Juttadinteria** s.l.

A Flowering stalks without bracks . . . . B

–   Flowering stalks with opposite bracts; sepals 5, almost equal;
stigmas 8–12; chambers of capsules with roofs and tubercles;
expanding-keels diverging, dentated, without wings but with
awns; leaves irregularly 3-sided with a few teeth, rough but not
dotted . . . . **Dracophilus**

B Sepals 4; stigmas 5–11, usually 8; roofs of the chambers of the
capsules mostly reduced to a rim . . . . **Juttadinteria**

– Sepals 5, stigmas 9–25; roofs as above; leaves soft . . . . **Namibia**

109   Leaves short and very thick, keeled near the top, united for about
one or two thirds of their length; surface greyish-white, distinctly
dotted, minutely granulated . . . . **Dinteranthus**

–   Leaves stone-like, 3-sided with sharp edges, united at the base
only; surface smooth, whitish, not dotted; flowers solitary on a
compressed stalk; chambers of capsules with roofs, without tubercles; expanding-keels more or less diverging . . . . **Lapidaria**

Aethephyllum pinnatifidum   65

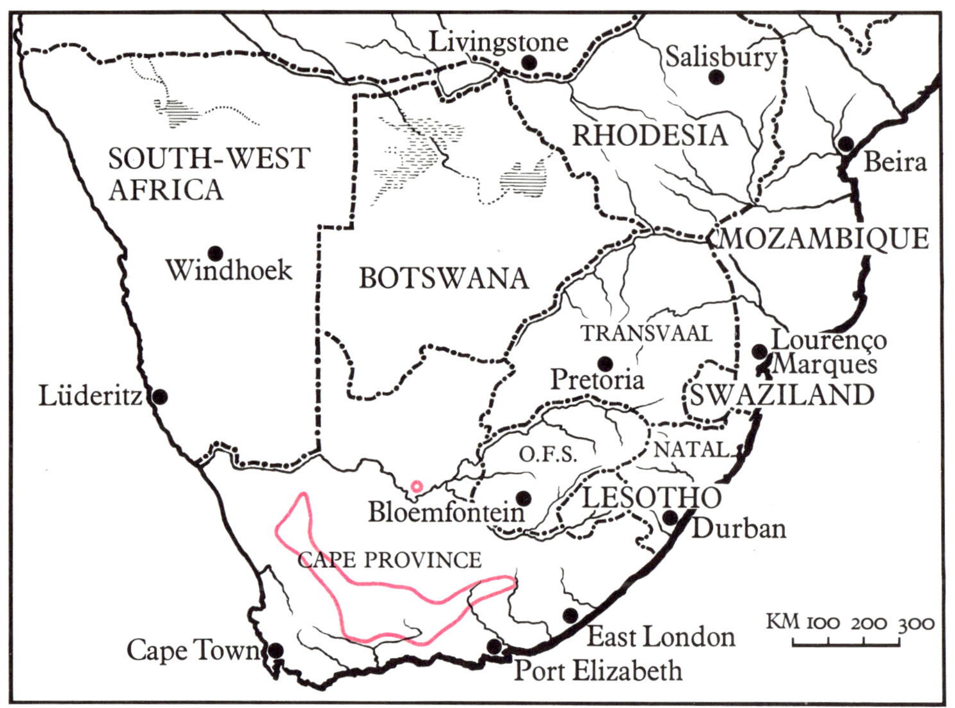

Aloinopsis

Aloinopsis Schwantes
From the name Aloe and the Greek *opsis* = appearance, on account of the similarity of the habit of this plant to an Aloe.

Schwantes, Z. Sukk. 2: 177. 1926. - Rendle, J. Bot. 66: 78, 1928. - v. Pöllnitz, Aufteilung, 24. 1933. - Jacobsen, Verzeichnis, 12. 1938. - Jacobsen, Herre, Volk, Mesembr. 83. 1950. - Phillips, Genera, 319. 1951. - Jacobsen, Handbuch, 1158. 1955. - Schwantes, Fl. Stones, 160. 1957. - *Bolus, Notes, 372 (key of species)*; *pl. 116-129. 1958.* - Jacobsen, Handbook, 953 (in system of Schwantes), 963 (in key of Bolus), 974 (in key of Herre & Volk), 979 (with key of species) fig. 1195-1202. 1960. - Jacobsen, Lexikon, 356, t.143/3, 144/1 & 2. 4, 6. 1970.

Dwarf tufted plants with tuberous rootstocks, perennial, herbaceous parts velvety pubescent or glabrous and then variously punctate. Leaves broadly spathulate, ovate, ovate-lanceolate, linear-lanceolate or sub-clavate; lower surface nearly flat, convex or obtusely keeled; the tubercles on the leaves similar, or some conspicuously larger than others, and sometimes differently shaped. Flowers solitary or rarely 2-3-nate; peduncles elongate or enclosed in the bracts, the lateral ones bracteolate, or as in *A. malherbei* the bracts are absent; 1,8-4 cm in diam., opening in the afternoon and closing at sunset, or opening towards evening and expanding after nightfall. Sepals 5-6, nearly equal, erect or recurved, obtuse, acute or acuminate, the membranous margins narrow. Petals 2-3-seriate, yellow, salmon, flesh-pink or rose, the yellow sometimes with a red central stripe. Stamens strongly inflexed or nearly erect, collected into a cone, or the outer sometimes erect and becoming somewhat diffuse; filaments papillate at the base, or from a little above it, up to about the middle; staminodes absent. Ovary flat, convex or subglobosely, or conically, elevated above; glands angular, crenulate; placentas parietal; stigmas 6-14, slender, shorter than, or equal to, or surpassing the height of the stamens. Capsule 6-14-locular, semiglobose, or convex, or obconic, or nearly flat below and above nearly flat, convex or semiglobosely, or conically elevated; keels often aristate; loculi completely roofed; tubercle absent, rudimentary or well developed; seeds broadly obovate, minutely tuberculate or nearly smooth, brown.

Species: 15 recorded; from Namaqualand, Bushmanland, the Karroo and Griqualand-West. (Type: *Aloinopsis spathulata* (Thunb.) L. Bolus.)

M.M.Page, del. ad vivam, IX, 1921.

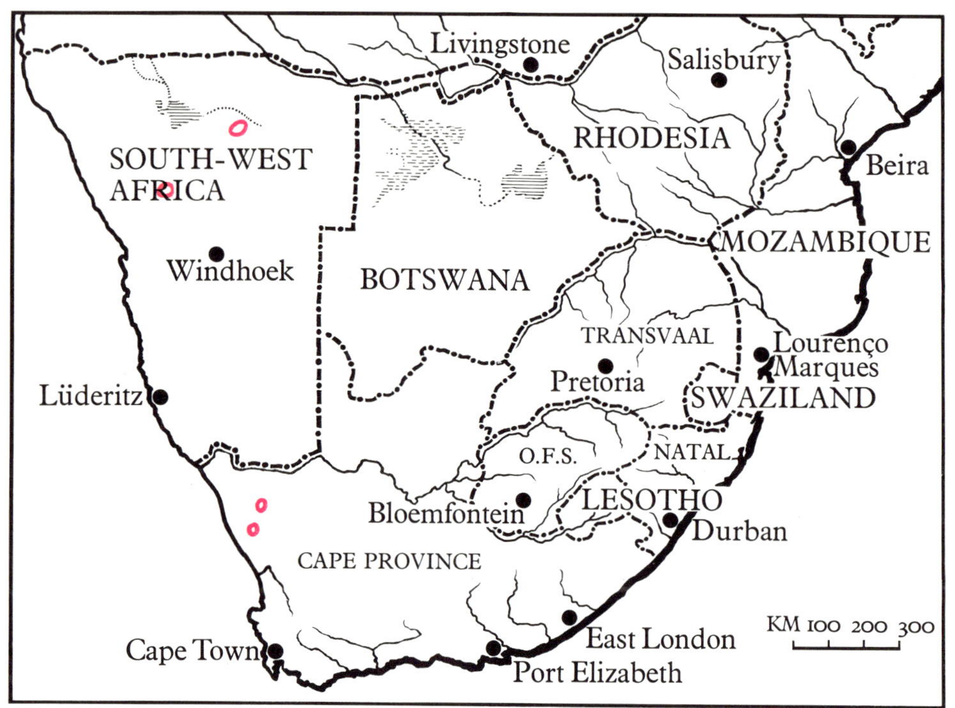

Amoebophyllum

Amoebophyllum N. E. Brown
From the Greek *amoebe* = change and *phyllon* = leaf, in reference to the alternate leaves.

N. E. Brown, Gard. Chron. 78: 433 (in key). 1926. - Phillips, Genera, 244. 1926. - N. E. Brown, Gard. Chron. 83: 419. 1928; J. Bot. 66: 79. 1928. - Bolus, Notes, 85; fig. 15A 1. 1928. - v. Pöllnitz, Aufteilung, 25. 1933. - Pax Natürl. Pfl. 211. 1934. - Jacobsen, Verzeichnis, 12. 1938. - Goossens, Blomplante, 142 (in key). 1940. - Jacobsen, Herre, Volk, Mesembr. 53 (in key), 84. 1950. - *Phillips, Genera, 295. 1951.* - Jacobsen, Handbuch, 1160. 1955. - Schwantes, Fl. Stones, 27, 36. 1957. - Jacobsen, Handbook, 954 (in system of Schwantes), 956 (in key of Bolus), 965 (in key of Herre & Volk), 984. 1960. - Jacobsen, Lexikon, 358. 1970. - Friedrich in Merxmüller, Prodromus, 85. (Mesembryanthemum). 1970.

Erect, stout, annual or biennial undershrubs, leaves and green parts with glittering warts. Leaves or at least those of the floral branches alternate, not connate, sessile, erect, sometimes spreading, later withering and spinescent or - with dried plants - only the hard leaf-base persisting, about 3 cm long and 2 cm broad. Flowers terminal, pedicelled in cymes with a few flowers, more or less light-violet, about 2 cm in diam. Calyx 5-lobed, short tubular-connate above the ovary, lobes pointed or with spine-like points, later spinescent. Petal short connate at the base. Stamens many, in different series, inserted in the tube. Ovary lengthened in the leaf-base. Placentas axile. Stigmas 4-5, short and strong, subulate. Capsule 4-5-locular, expanding-keels contiguous, with broad wings, pouchforming; loculi-roofs and tubercles absent.

Species 3 Northern Part of Namaqualand, Damaraland and N. South West Africa. (Type: *A. angustum* N. E. Br.)

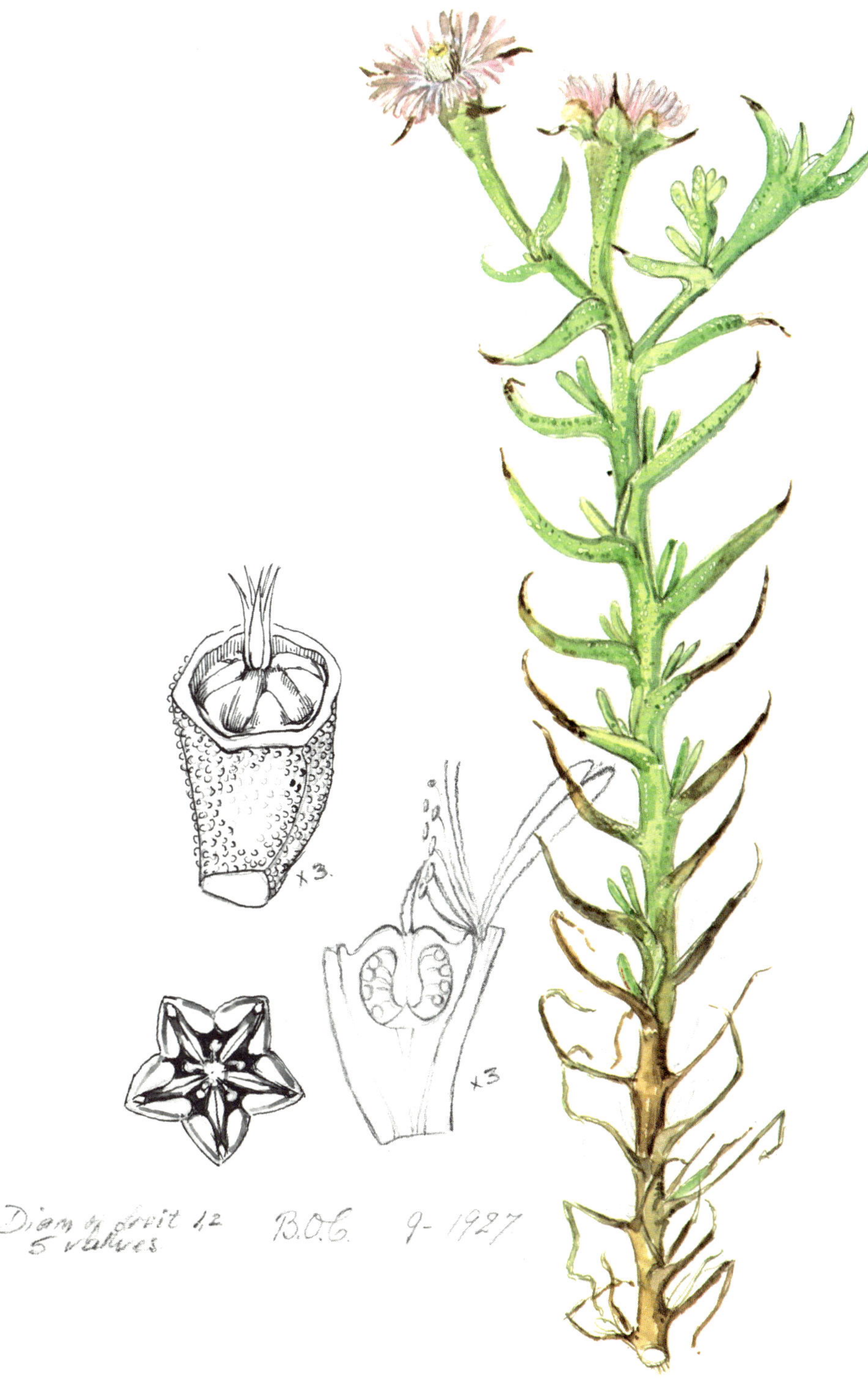
Diam of the 2.4 cm
8 - 9 - 1927.
Open
all
day
× 3.
× 3
Diam of fruit 12
5 valves
B.O.C.   9 - 1927

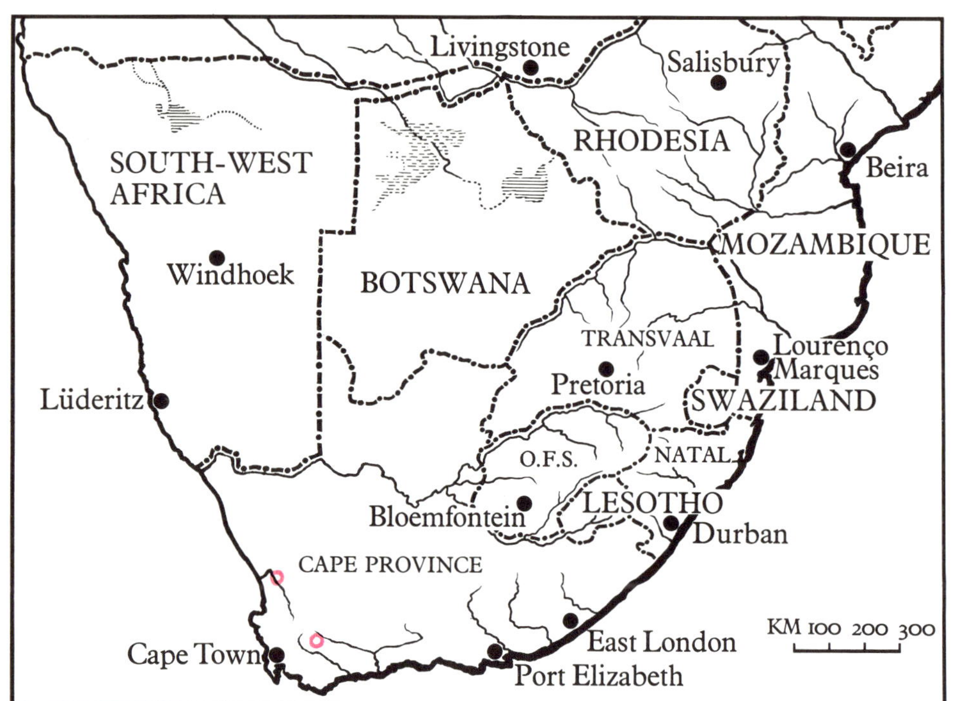

Amphibolia

Amphibolia L. Bolus
From the Greek *amphibolia* = uncertainty, doubt.

L. Bolus, J. S. Afr. Bot. 31: 169. 1965.
Stoeberia L. Bolus J. S. Afr. Bot. 28: 14.
1962. Jacobsen, Lexikon, 358. 1970.

Perennial shrubs like *Stoeberia* or *Ruschia*. Leaves ascendent, spreading, acute or subobtuse, bluish-green, the younger ones reddish. Flowers solitary or ternate, short pedicelled with bracts, pink. Sepals 5, nearly equal, sometimes with small membranous margins. Petals 2-seriate, obtuse or subacute. Stamens conical collected, staminodes mostly absent, flilaments often papillate. Ovarium slightly elevated; glands crenulate; placentas parietal, stigmas 5, subulate. Capsule 5-locular, obconical, valves with wings, keels erect, sometimes lacerated, loculi-roofs sometimes with 2 spur-like processes, tubercles small, sometimes deep into the loculus. Seeds: rotund-obovate, polished brown. As the capsule of *Stoeberia* opens when ripe and never closes again, this genus had to be created.

A Namaqualand genus with 3 species, also in the Ceres Karroo. (Type: *A. hallii* L. Bolus.)

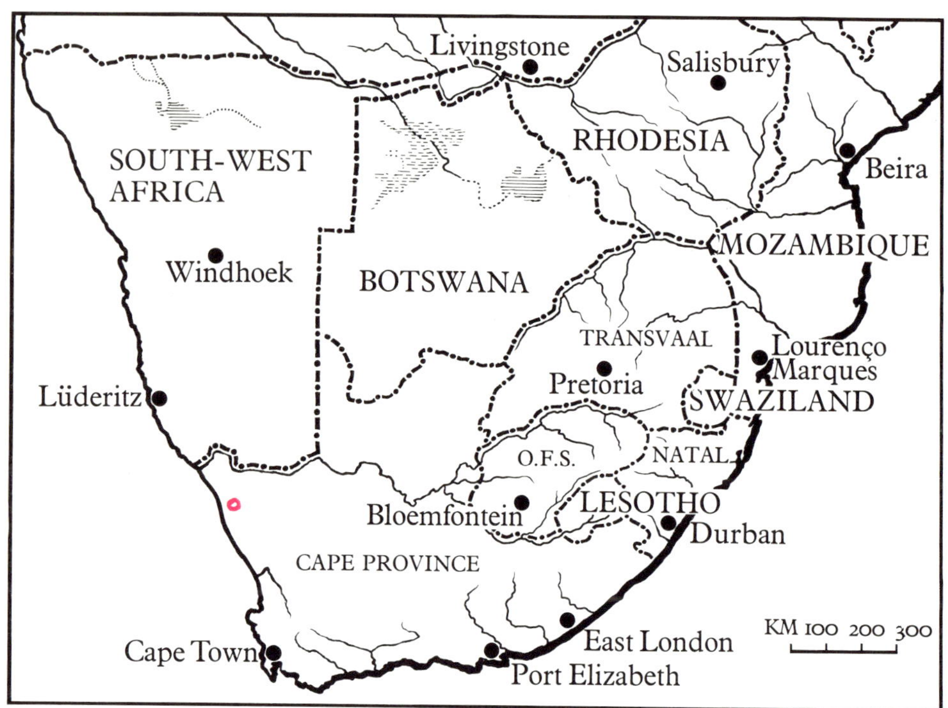

Anisocalyx

Anisocalyx L. Bolus
From the Greek *anisos* = unequal and *calyx* = calyx.

Bolus, Notes, 385. 1958. - Drosanthemum vaginatum L. Bolus Notes 2: 138. 1935. - Jacobsen, Lexikon, 359. 1970.

Small, compact, smooth shrublets, much branched; green parts minutely papillate. Leaves erect, diverging, keeled to the points, lower side roundish, upper side even, free part suboblong, obtuse or acute, seen from the side not acuminate, the points truncate or subtruncate, subconvex at the sides; taste brackish. Flowers solitary, open during day-time, pedicelled; the bract-like leaves at the base of the peduncle eventually produce axillary shoots and cannot therefore be regarded as true bracts. Sepals 6-8, unequal, 2 outer ones much compressed, acute, keeled, from the side nearly obtuse or roundish, the inner ones compressed with membranous margins. Petals perhaps 3-seriate (not well to distinguish), the inner ones only slightly smaller, obtuse or inwards only slightly acuminate, white. Stamens erect, connivent, pure white, the outer ones obscurely papillate at the base, anthers and pollen whitish. Ovary lobes slightly elevated, obtusely compressed; glands annular, around the lobes of the ovary undulate, finely crenulate; placentas parietal; stigmas 9–10, slender, upwards attenuate. Capsule 9-10-locular, lower side obconical, nearly angular, sutures compressed, fairly even above; valves with ample wings, wings changing to awns, intergrown with the keels, when old, free; keels erect, fairly thick, lacerate, the awns at the top attain nearly the valve, sometimes cruciate; loculi-roofs well developed; tubercle absent; seeds brown, about 2 mm long with lighter coloured navel.

A monotypic Namaqualand genus from Riethuis, district Kommaggas. (Type: *A. vaginatus* (L. Bol.) L. Bol.)

Anisocalyx vaginatus 73

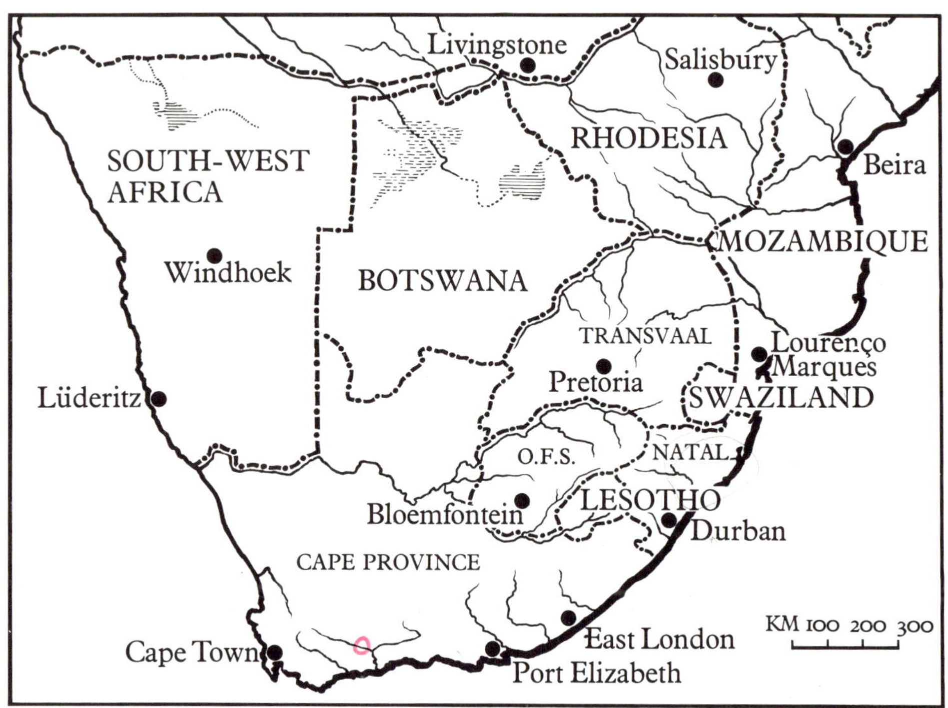

Antegibbaeum

Antegibbaeum Schwantes
From the Latin *ante* = before and
Gibbaeum, that is fore-*Gibbaeum*, a sort
of prefatory stage to *Gibbaeum*.

*Schwantes ex H. D. Wulff in Bot. Archiv.
45: 154. 1944.* – Jacobsen, Herre, Volk,
Mesembr. 80 (in key), 84. 1950. – Jacobsen, Handbuch, 1161: fig. 1006. 1955.
– Schwantes, Fl. Stones, 170, 340. 1957. –
Bolus, Notes, 235. 1958. – Jacobsen,
Handbook, 953 (in system of Schwantes),
962 (in key of Bolus), 976 (in key of
Herre & Volk), 985; fig. 1203. 1960. –
Gibbaeum fissoides (Haw.) Nel comb. nov.
in G. C. Nel, The Gibbaeum Handbook,
p. 81. 1953. – Jacobsen, Lexikon, 359,
t.144/5. 1970.

Plant branched from base into clumps
about 8 cm high. Leaves of one pair somewhat unequal and curved, upper surface flat or slightly convex, back surface
rounded, angled and compressed towards
the apex, surface grey-green or also reddish, smooth or somewhat rugose, blunt.
Flowers on short pedicels with 2 pairs
of bracts below the flower (distinction
from *Gibbaeum*) violet-red, 6 cm in diam.
Sepals 6, unequal, free to the ovary, 2
lateral lobes longer and often keeled as
also with membranous margin. Petals
1-3-seriate, free, often flaccid, linear. Stamens many, erect, bearded at the base,
with or without staminodes. Ovary
often raised above the body; glands annular, not separated as in *Gibbaeum*;
placentas parietal; stigmas 6–7 stout,
subulate, acute, plumose, style absent.
Capsule 6-7-locular, short, conical or obconical, often convex, with prominent
sutures; keels ending in large broad
wings; loculi-roofs present; tubercles
absent. Seeds fairly large, dark-brown,
ovate, with navel, short echinate.

A monotypic genus from Matjesfontein and the Western Part of the Little
Karroo. (Type: *A. fissoides* (Haw.)
Schwantes.)

Antegibbaeum fissoides   75

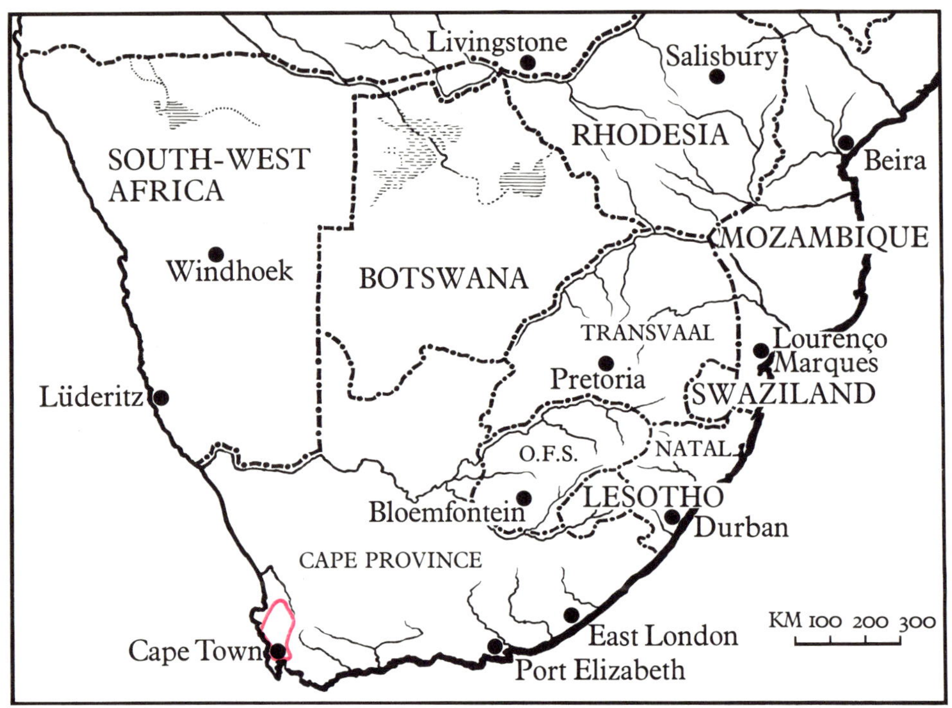

Apatesia

Apatesia N. E. Brown
From the Greek *apatesis* = deception,
from the likeness of this plant to *Hymenogyne* (before the flower is formed).

N. E. Brown, Gard. Chron. 81: 12 (in key). 1927. - J. Bot. 66: 80. 138. 1928. - Bolus, Notes, 16; pl. 4.G 1 & 2. 1928. - Schwantes, Mitt. Inst. Allg. Bot. Hamburg, 8: 166. 1929. - N. E. Brown, Gard. Chron. 91: 262. 1932. - v. Pöllnitz, Aufteilung, 25. 1933. - Jacobsen, Sukk. 90. 1933. - Pax, Natürl. Pfl. 211. 1934. – Jacobsen, Succ. Pl. 125. 1935; Verzeichnis, 13. 1938. - Goossens, Blomplante, 149 (in key). 1940. - Jacobsen, Herre, Volk, Mesembr. 58 (in key), 84. 1950. - Adamson, Salter, Fl. Cape Pen. 376. 1950. - *Phillips, Genera, 296. 1951.* - Jacobsen, Handbuch, 1162; fig. 1007. 1955. - Schwantes, Fl. Stones, 55, 341; fig. 43-47 (capsules); pl. 90, 91 (capsules). 1957. - Jacobsen, Handbook, 954 (in system of Schwantes), 958 (in key of Bolus), 967 (in key of Herre & Volk), 985, fig. 1204. 1960. - Jacobsen, Lexikon, 359, t.144/3. 1970.

Dwarf annual herbs with decumbent branches. Leaves petiolate, flat or concave like a spoon, not united in pairs at the base, dotless, petiole dilated below into an open sheath. Flowers solitary or axillary, on long pedicels. Calyx produced a little beyond its union with the ovary into a saucer-shaped limb, ovary part shallowly hemispherical. Sepals 5, unequal, narrowly spathulate from a broad base, the inner ones with broad membranous margins, the longest often longer than the petals. Petals numerous, arising from the saucer-shaped limb of the calyx, 4-5-seriate, free, widely spreading. Stamens numerous from the saucer-shaped limb of the calyx, all inflexed, bent down on top of the ovary and then upcurved around the style and stigmas, inner filaments bearded at the base; staminodes numerous, arising from the saucer-shaped limb of the calyx, hair-like or petal-like at the base and hair-like above, contorted, at first more or less bent over the stamens, afterwards becoming more erect. Ovary inferior, shallow, flat on the top and sunk below the rim of the calyx; placentas on the outer walls of the chambers; ovules, many in each chamber; stigmas 10-12 short, stout, 2 mm long. Capsule 10-12-locular, shallow, slightly convex beneath, circular in outline; valves horizontally spreading when expanded, narrowly deltoid; keels about half as long as the valve and subcontiguous into a central keel, adnate to the valve throughout their length, without free tips or wings; loculi open, without loculi-wings; tubercle none, seed globose, smooth without a nipple.

A south-western Cape genus with 4 species. (Type: *A. pillansii* N. E. Brown.)

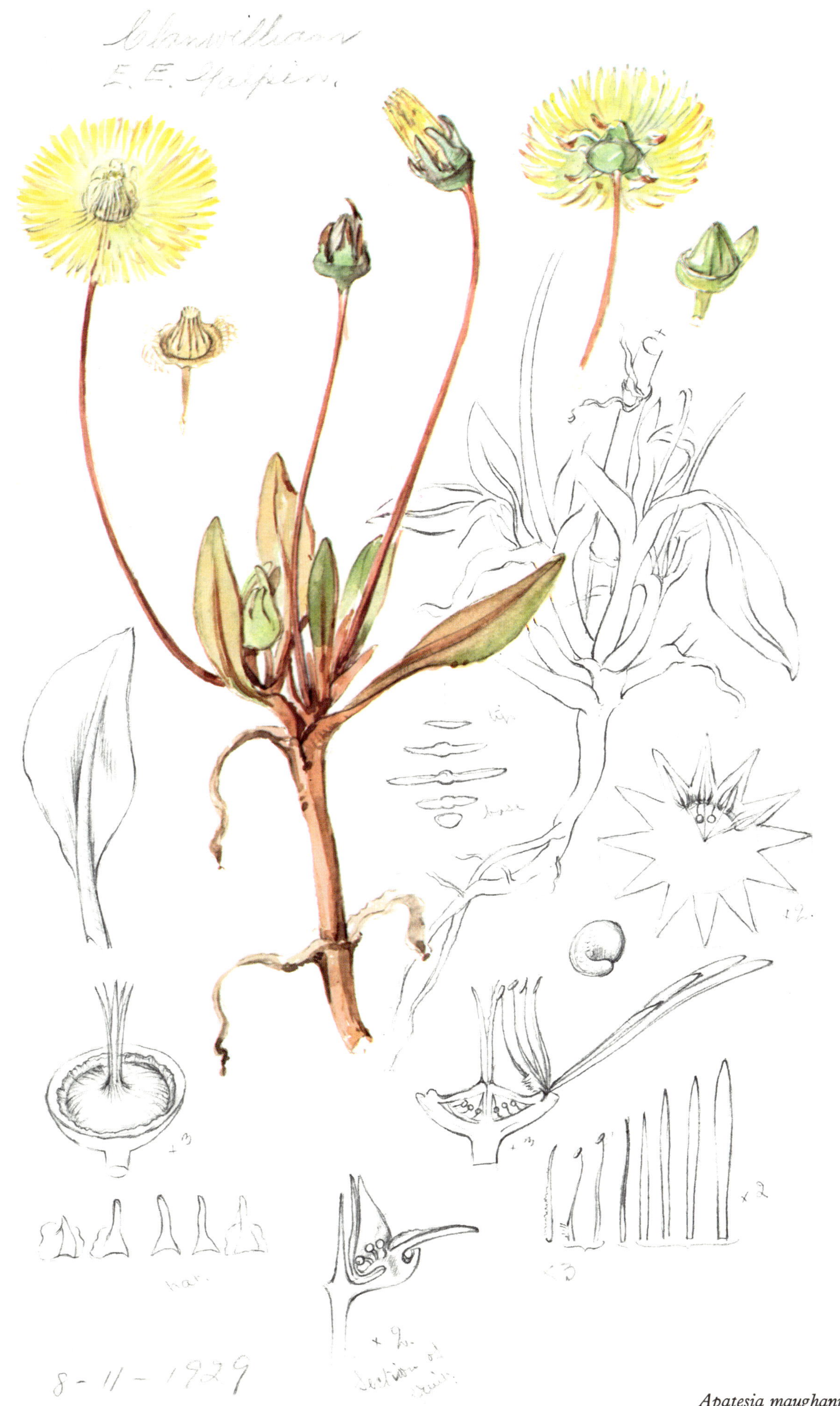

Clanwilliam
E. E. Galpin.
8 - 11 - 1929

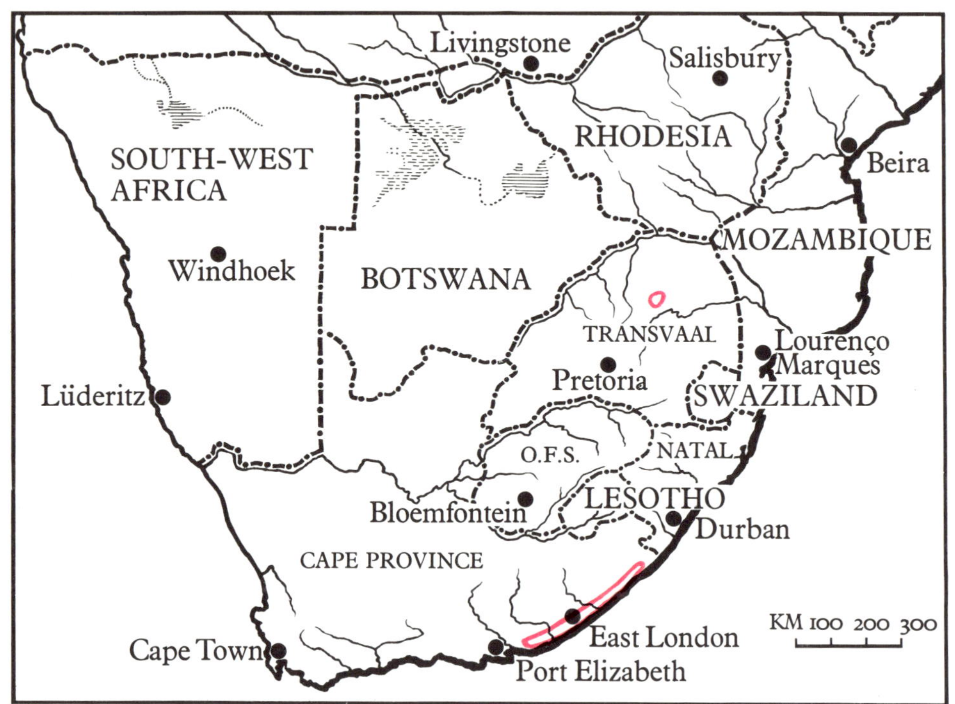

Aptenia

Aptenia N. E. Brown
From the Greek *apten* = wingless, since the valves of the capsule are without wings.

N. E. Brown, Gard. Chron. 78: 412 (in key). 1925. - Phillips, Genera, 244. 1926. - N. E. Brown, J. Bot. 66: 130. 1928; Gard. Chron. 84: 313. 1928. - v. Pöllnitz, Aufteilung, 25. 1933. - Jacobsen, Sukk. 90. 1933. - Pax, Natürl. Pfl. 211. 1934. - Jacobsen, Succ. Pl. 125. 1935; Verzeichnis, 13. 1938. - Goossens, Blomplante, 148 (in key), 1940. - Jacobsen, Herre, Volk, Mesembr. 55 (in key), 84. 1950. - *Phillips, Genera, 296. 1951.* - Jacobsen, Handbuch, 1162; fig. 1008. 1955. - Schwantes, Fl. Stones, 5, 34; fig. 1 (capsule); pl. 1a. 1957. - Jacobsen, Handbook, 951 (in system of Schwantes), 955 (in key of Bolus), 966 (in key of Herre & Volk), 987; fig. 1205. 1960. - Jacobsen, Lexikon, 359, t.145/4. 1970. - *Litocarpus cordifolius* (L.f.) L. Bolus Fl. Pl. S. Afr. pl. 261, fig. 2. 1927.

Perennial succulent herbs, but of short duration, prostrate. Leaves petiolate, flat, opposite, heart-shaped to oval, papillose. Flowers solitary in the forkings of the branches, pedicillate, without bracts. Sepals 4, free, 2 lobes often leafy and larger than the others. Petals numerous, united at the base into a short tube, shorter than the sepals. Stamens, many but not very numerous, arising from the corolla tube; filaments not bearded, surrounded by incurved, whitish staminodes. Ovary inferior; placentas axile, style none, stigmas 4, minute. Capsule 4-locular, valves broader than long, with the apical part so abruptly thickened that the basal termination of the thickening is quite vertical; keels forming 1 central keel with its end joined to the subvertical base of the thickened apical part of the valve, without marginal wings or flaps; loculi open; seeds compressed, circular in outline, tuberculate.

Species: 2, from the Eastern coastal districts of the Cape Province and from Bandolierskop near Pietersburg, Transvaal. (Type: *A. cordifolia* (L.) N. E. Br.)

When grown under bad conditions, it dies off soon (tendency towards annual plants).

M.M.Page. XII. 1922.

Aptenia cordifolia 79

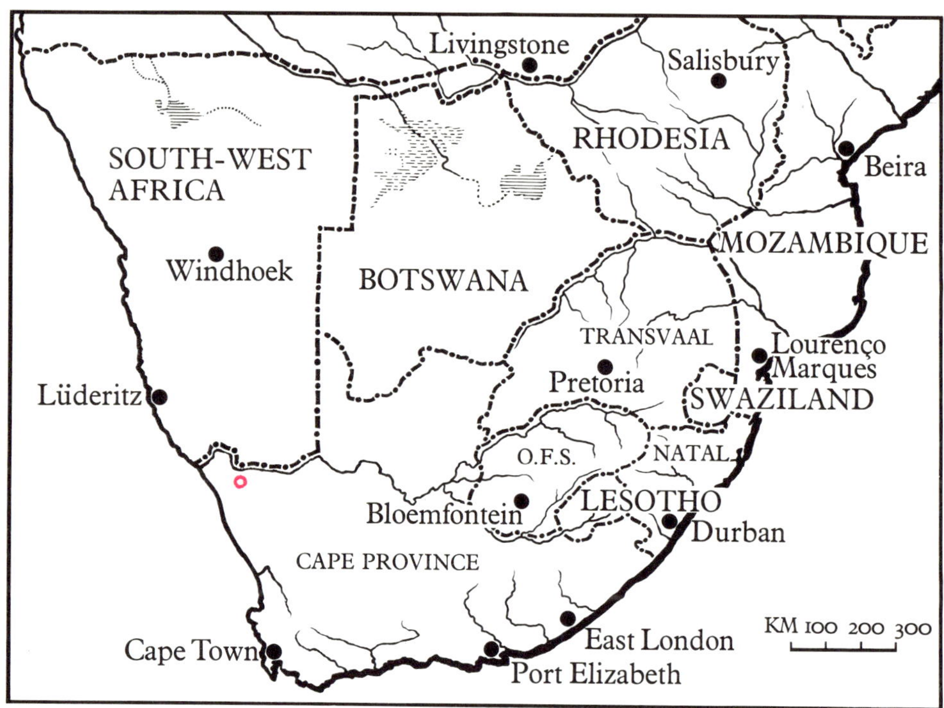

Arenifera

Arenifera Herre
From the Latin *arena* = sand and
ferre = to carry.

Herre, Jahrb. Schweiz. Kakt. Ges. Sukkulentenkunde 2: 35 together with 4 figures. 1948. - Jacobsen, Herre, Volk, Mesembr. 59 (in key), 84. 1950. - Jacobsen, Handbuch, 1163; fig. 1009. 1955. - Schwantes, Fl. Stones, 162, 338. 1957. - Jacobsen, Handbook, 952 (in system of Schwantes), 961 (in key of Bolus), 968 (in key of Herre & Volk), 987; fig. 1206, 1207. 1960. - Jacobsen, Lexikon, 359, t.145/1. 1970. - *Psammophora pillansii* L. Bolus *in Fl. Pl. S. Afr. 7: 280, 1927.*

Perennial, middle-sized shrub, erect, densely branched. Leaves opposite, somewhat connate at the base, the younger ones green and viscid, the older ones glaucous and rough, almost dry, covered with sand and dust, obtuse, somewhat convex, variable in shape: the longer ones up to 2,5 cm long and 5 mm wide and thick, somewhat curved and blunt above and along the sides, whilst the shorter ones are 0,6-1,5 cm long with a diameter of about 7 mm. The cover does not break up into pieces as in *Psammophora*. Flowers ternate, the later ones sometimes developing rather slowly; peduncle terete, central pedicel without bracts, open in the middle of the day, pale rose to light violet with an inconspicuous central band, about 2,2 cm in diam. Sepals 4, almost equal, the interiors with a broad membranous margin. Petals 2-seriate, recurved, linear, obtuse or subacute. Stamens at first conically collected, later diffuse, surrounded by a few whitish staminodes, which become rosy upwards, filaments sparingly bearded at the base, anthers and pollen whitish. Ovary slightly convex above; glands dark-green, annular, crenulate; placentas parietal; stigmas 8, long-subulate, acuminate, conspicuously fringed at the base. Capsule 8-locular, turbinate, about 0,8 cm in diam., woody, pedicel terete, sometimes with conspicuous traces of bracts, surface cupular, margins of the valves prominent, forming a small column in the centre, valves transversely ribbed, point of the valves with 2 conspicuous recurved teeth; valves with wings which are also connected with the keels, points of the wings free; loculi-roofs covering half the chamber, with spur-like projects at the mouth; tubercle conspicuous, middle-sized disc with a smaller top, pale, comparatively deep; seeds long quadrangulate, somewhat convex, navel along the shorter side, with produced micropyle, finely longitudinally ribbed, light yellowish brown.

Monotypic genus from the Richtersveld: Brakfontein near Lekkersing. (Type: *A. pillansii* (L. Bolus) Herre).

Arenifera pillansii 81

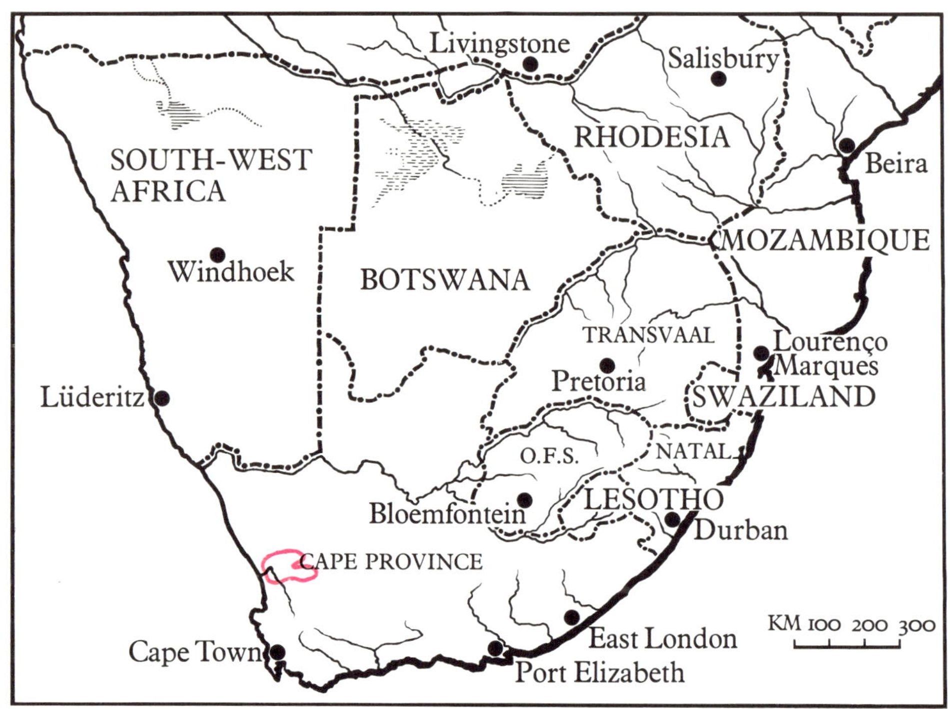

Argyroderma

Argyroderma N. E. Brown
From the Greek *argyros* = silver and *derma* = skin.

N. E. Brown, Gard. Chron. 71: 92, 105. 1922; 78: 413 (in key). 1925; 79: 268 with fig. 1926. - Phillips, Genera, 242. 1926. - N. E. Brown, J. Bot. 66: 265. 1928. - Tischer, Z. Sukk. 347. 1928. - Labarre (red.), Mesembr. 109; fig. 19-22, 35; t.2 fig. 6. 1931. - v. Pöllnitz, Aufteilung, 26. 1933. - Jacobsen, Sukk. 90; fig. 83, 84. 1933. - Pax, Natürl. Pfl. 211. 1934. - Jacobsen, Succ. Pl. 126; fig. 105, 106, 107. 1935; Verzeichnis, 14. 1938. - Goossens, Blomplante, 142, 144 (in key). 1940. - Jacobsen, Herre, Volk, Mesembr. 53 (in key), 84. 1950. - *Phillips, Genera, 296. 1951.* - Jacobsen, Handbuch, 1164; fig. 1010-1014. 1955. - Schwantes, Fl. Stones, 70, 339; fig. 19 (long section of flower); pl. 17. 1957. - Jacobsen, Handbook, 953 (in system of Schwantes), 956 (in key of Bolus), 965 (in key of Herre & Volk), 988; fig. 1208-1212. 1960. - Bolus, J. S. Afr. Bot. 31: 86 (key of subsections). 1965. - *Roodia N. E. Brown* in Gard. Chron. 78: 433 (in key). 1925; 79: 308 with fig. 1926. - Jacobsen, Lexikon, 359, t.145/2, 3. 1970.

Stemless succulent perennials, forming with age clumps with underground branches, each growth with 2 or, when a new pair is forming, 4–6 erect or spreading leaves which are united at the base; those with short and thick leaves, upper surface mostly flat or a little convex, back surface convex, often drawn forward over the upper side, silver-white to grey-green, immaculate, about 2,5 cm long, 2 cm broad and 1,5 cm in diam.; those with long leaves (formerly described as *Roodia*) several times as long as broad, more or less semi-terete, obtuse, immaculate, bluish-green often with red-violet tint, till about 8 cm long and 1,3 cm in diam. Flower solitary, terminal, those with short thick leaves more or less sessile, white, yellow, pink; light- to deep-violet red, about 1,5–4 cm in diam., those with long leaves with pedicels about 4 cm long, with bracts, mostly deep red-violet or reddish, about 3 cm in diam. and more. Sepals 6–8, produced above the union with the ovary into a short or distinct tube. Petals numerous, free, arising from the top of the calyx-tube. Stamens arranged in a dense ring, arising from the top of the calyx-tube. Ovary inferior, placentas on the outer walls or on the floor of the ovary-chambers; stigmas small, sessile on top of the ovary, entirely or faintly crenate. With this disc-like stigma ("circular pulvinus") this genus differs from all the other ones. Capsule 10-24-locular, broad and shallow, each valve with membranous wings, ending in membranous points, keels stout, ending in awns, without wings, only the long-leaved ones with wings; loculi roofed with rather rigid loculus-wings and the outer opening nearly closed by a tubercle; seeds small, globose to ovate, smooth with navel, brown.

Species: 48 recorded from the Vanrhynsdorp district, one from Calvinia (*A. orientale*). (Type: *A. testiculare* N. E. Brown.)

Argyroderma sp.   83

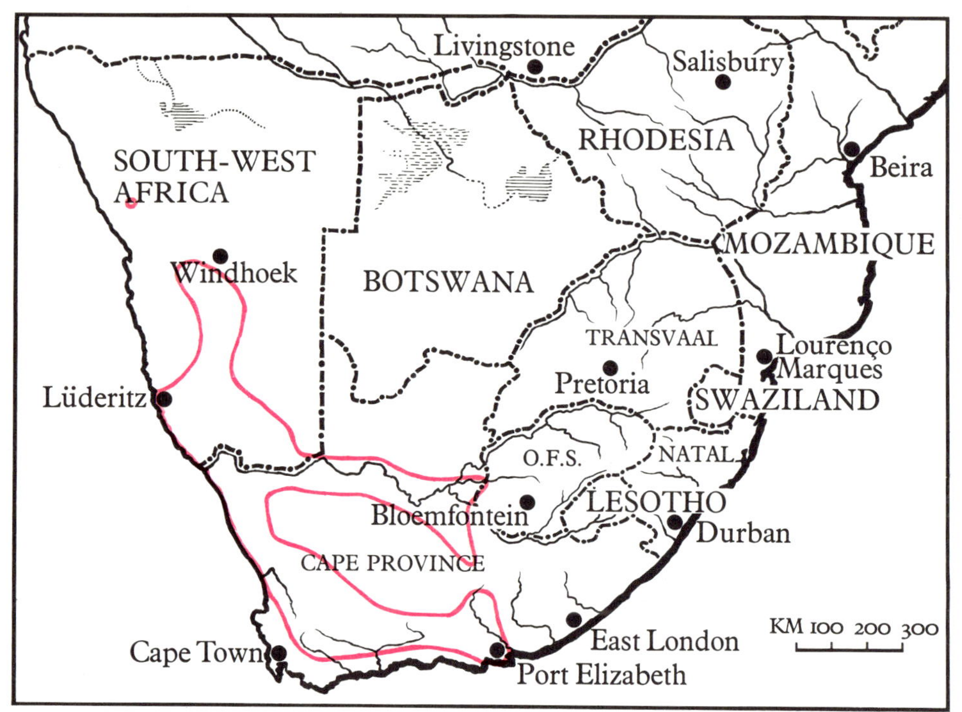

Aridaria

Aridaria N. E. Brown
From the Latin *aridus* = dry.

N. E. Brown, Gard. Chron. 78: 453 (in key). 1925; *J. Bot. 66: 140. 1928.* - Bolus, Notes, 98; fig. 23. 1928. - v. Pöllnitz, Aufteilung, 27. 1933. - Jacobsen, Sukk. 93. 1933. - Pax, Natürl. Pfl. 211. 1934. - Jacobsen, Succ. Pl. 129; fig. 108. 1935; Verzeichnis, 17. 1938. - Goosens, Blomplante, 142 (in key). 1940. - Jacobsen, Herre, Volk, Mesembr. 54 & 56 (in key) 84. 1950. - Adamson, Salter, Fl. Cape Pen. 374. 1950. - Phillips, Genera, 319. 1951. - Jacobsen, Handbuch, 1172, fig. 1015. 1955; Lexikon, 362, t.146/2. 1970. - Friedrich in Merxmüller, Prodromus, 16. 1970.

Nycteranthus Necker in Elem. 2: 82. 1790. - Schwantes, Fl. Stones, 31, 35. 1957. - Jacobsen, Handbook, 951 (in system of Schwantes), 957 (in key of Bolus), 966 (in key of Herre & Volk). 993, 1310; fig. 1528. 1529. 1960. - Bolus, J. S. Afr. Bot. 28: 228 (Nomenclature Congress Montreal 1960 decided that *Nycteranthus* was not a validly published genus). 1962.

Perennial, erect, decumbent or prostrate shrubs, small or high, leaves sessile, opposite, not connate or only slightly at the base, erect, ascending or spreading, terete or semi-terete or triquetrous, rarely flat and then fairly thick and fleshy, never flat and thin, often withering and deciduous, rarely papillose, up to 4 cm long and 0·8 cm in diam. Flowers solitary or of 3, white, yellow, pink or red, sometimes scented, 1,8-7 cm in diam. Sepals (4) 5-6, somewhat unequal, the 2 longer ones leafy, sometimes with a membranous margin. Petals in several series (3, 4, 5), narrow linear-lanceolate, obtuse or emarginate, shortly tubular, connate at the base. Stamens erect, sometimes connivent, proceeding to staminodes, glabrous, sometimes connate. Ovary conical; placentas axile; stigmas 4-5, subulate, glands deep divided, 6-angulate, crenate, dark-green (as far as reported). Capsule 4-5-locular, keels without processes, loculi-roofs and tubercles absent; seeds compressed, along the margins ridgelike tubercled.

Species 93. South-West Africa, Namaqualand, Bushmanland, Prieska, Calvinia, Ceres, Clanwilliam, Vanrhynsdorp, Cape Peninsula, Worcester, Calitzdorp, Oudtshoorn, Eastern Cape Province, Free State: Fauresmith. (Type: *A. noctiflora* (L.) Schw.).

Aridaria noctiflora   85

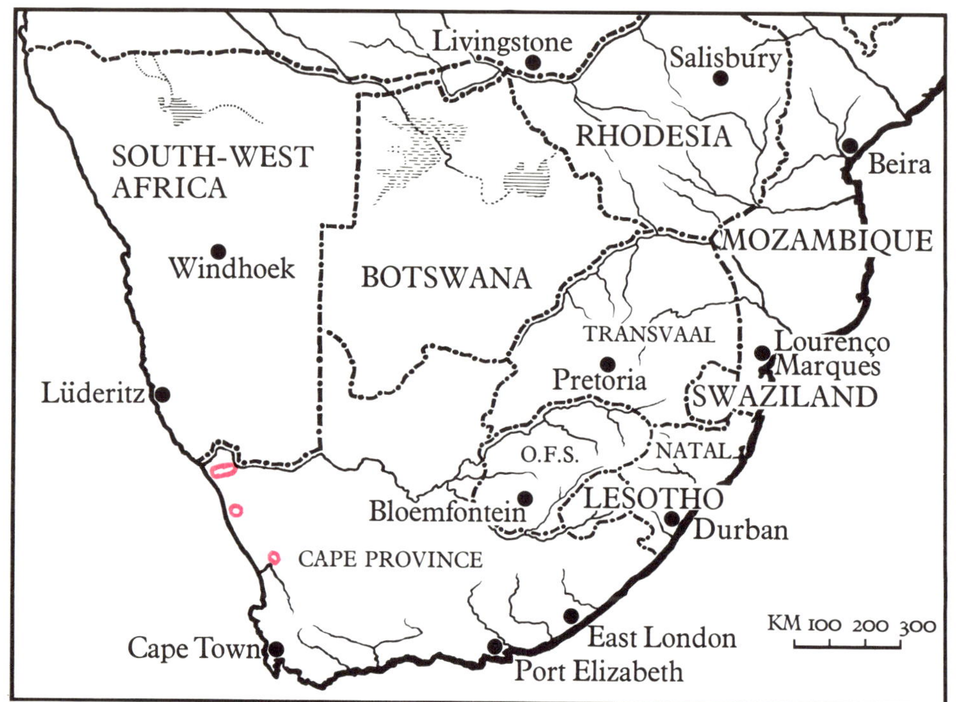

Aspazoma

Aspazoma N. E. Brown
From the Greek *aspazomai* = to clasp,
since the leaf sheath clasps the stem.

N. E. Brown, Gard. Chron. 78: 413 (in key). 1925. – Phillips, Genera, 244. 1926. – N. E. Brown, Gard. Chron. 83: 419. 1928. – v. Pöllnitz, Aufteilung, 27. 1933. – Pax, Natürl. Pfl. 211. 1934. – Jacobsen, Verzeichnis, 22. 1938. – Goossens, Blomplante, 142 (in key). 1940. – Jacobsen, Herre, Volk, Mesembr. 54 (in key), 85. 1950. – *Phillips, Genera, 297.* 1951. – Jacobsen, Handbuch, 1186. 1955. – Schwantes, Fl. Stones, 24, 35. 1957. – Jacobsen, Handbook, 951 (in system of Schwantes), 956 (in key of Bolus), 965 (in key of Herre & Volk), 996; fig. 1213, 1960; Lexikon, 365, t. 146/4. 1970.

A bushy branched succulent, perennial, about 15 cm high. Leaves more than 2 to a branch, some alternate and others opposite, semi-terete or subtrigonous always with a tubular sheath which embraces the branches of the sheath above and nearly or quite as long as the internodes, fine velvety pubescent. Flowers terminal, solitary, pale straw-yellow, about 4 cm in diam. Calyx produced above its union with the ovary into a distinct tube. Sepals 4–5 with 2 lobes leaf-like and the others shorter and more membranous. Petals numerous, linear, 2-3-seriate, united into a short tube at the base. Stamens and staminodes numerous, erect, at first conically collected. Ovary conical at the top; placentas axile; stigmas 4–5, subulate. Capsule 4-5-locular, loculi-roofs and tubercles absent, expanding-keels contiguous, with a broad inflexed or erect wing on each side. Seeds large, light-yellow, 2 mm long, compressed, rounded, rough. 21 into one large capsule.

A monotypic genus from the Richtersveld, where it is a characteristic plant of the Southern Part, also near Wallekraal in the Vanrhynsdorp district. (Type: *A. amplectens* (L. Bol.) N. E. Br.)

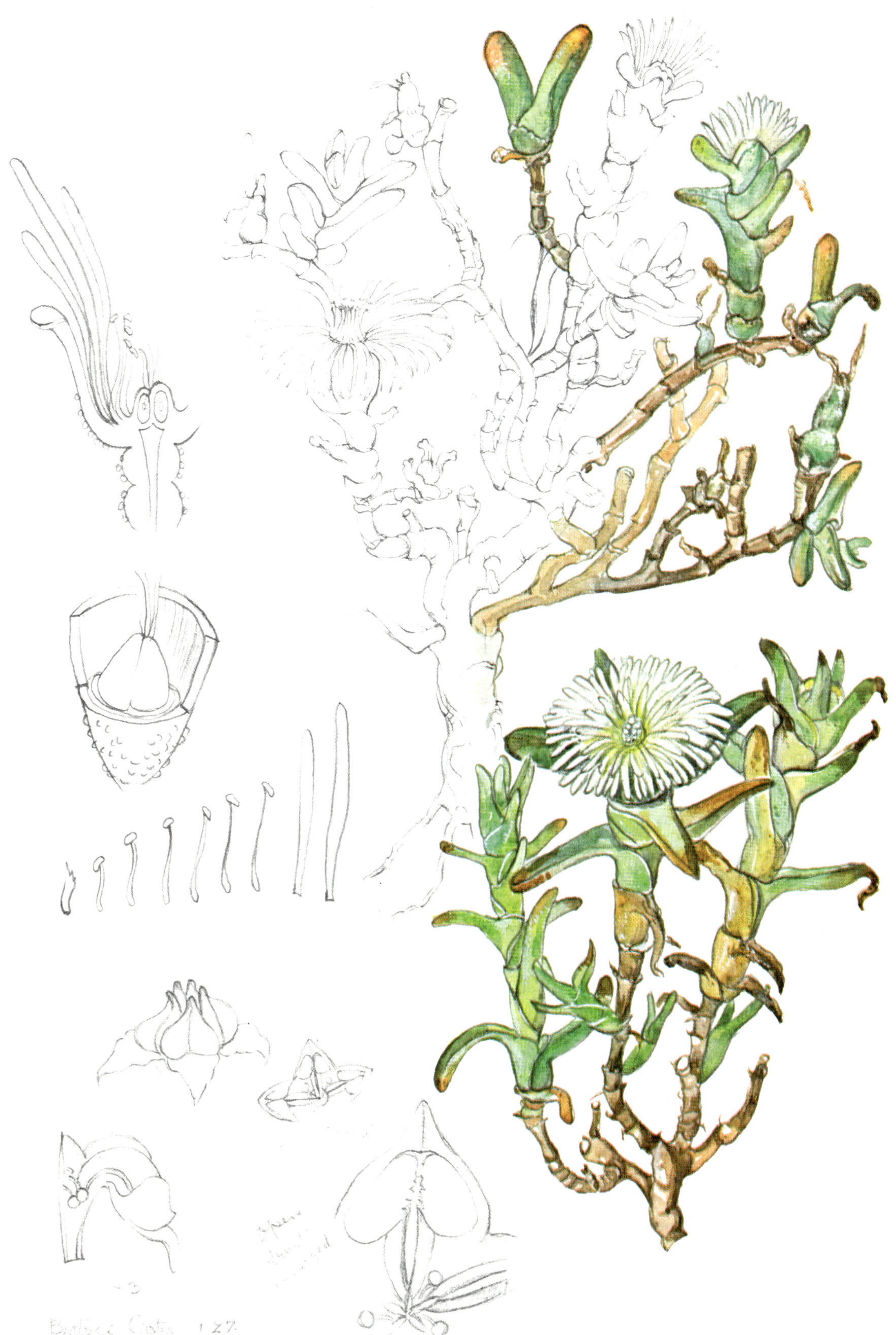

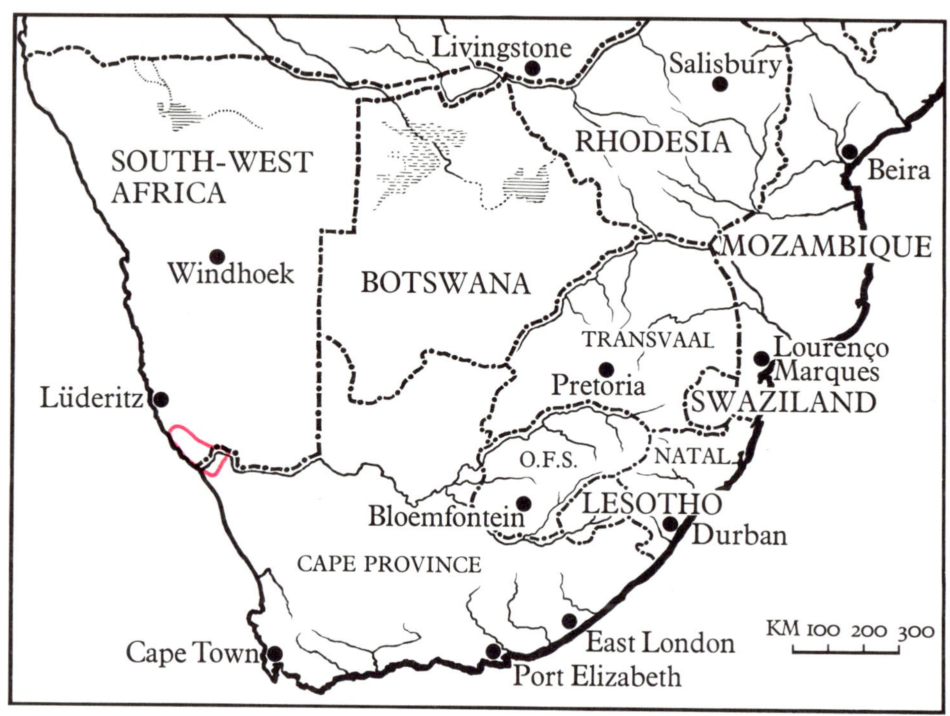

Astridia

Astridia Dinter et Schwantes
Named in honour of
Mrs. Astrid Schwantes.

Dinter, Gard. Chron. 80: 430, 447 with fig. 1926. - Dint. et Schwant. Z. Sukk. 16, 106. 1927. - N. E. Brown, J. Bot. 66: 266. 1928. - v. Pöllnitz, Aufteilung, 22 (in key of Schwantes), 27. 1933. - Jacobsen, Sukk. 93; fig. 86. 1933. - Pax, Natürl. Pfl. 211. 1934. - Jacobsen, Succ. Pl. 130; fig. 109. 1935; Verzeichnis, 22. 1938. - Jacobsen, Herre, Volk, Mesembr. 63 (in key), 85. 1950. - Phillips, Genera, 319. 1951. - Jacobsen, Handbuch, 1187; fig. 1016. 1955. - Schwantes, Fl. Stones, 68, 338; pl. 16. 1957. - Bolus, Notes, 298. 1958. - Jacobsen, Handbook, 952 (in system of Schwantes), 962 (in key of Bolus), 969 (in key of Herre & Volk), 996; fig. 1214. 1960. - *Bolus, J. S. Afr. Bot. 27: 171. 1961; 32: 233 (key of species). 1966.* - Jacobsen, Lexikon, 365, t.146/1 & 3. 1970. - Friedrich in Merxmüller, Prodromus, 17. 1970

Erect, robust shrublets 25-75 cm high, sometimes compact, herbaceous parts sometimes velvety pubescent or glabrous or very rarely, very minutely pubescent and glabrous in the same collection. Leaves nearly erect, ascending or widely spreading, closely attached to the branch and long-persistent, and what appears to be the extreme base still persisting after the rest of the leaf has eventually fallen, usually decurrent, sometimes sub-falcate, upper surface acute, or rounded at the apex which is sometimes recurved, the length being 2, 3-5 times that of the breadth in the middle of the leaf, obscurely obtusely, or rarely rather acutely, keeled, the sides slightly convex or rarely strongly compressed, usually 3-6 cm long, or rarely, in cultivation, up to 11,7 cm long and 1-3 cm in diam. (and thus among the largest in the perennial *Mesembrieae*), the sheath 2-6 mm long. Flowers solitary, or terminating a branch, 3-nate, the lateral ones developing much later than the primary one, diurnal, 4-7,5 cm in diam.; peduncle very short with the bracts near the apex and constricted there by their close attachment; bracts usually differing conspicuously in form from the leaves, broad and deeply concave below and embracing the receptacle, in side view obtuse, acute or acuminate and often more or less excavate near the middle of the margins, up to 2 cm long. Sepals 6, subequally, laterally compressed, obtuse or truncate in side view, inner ones subulate in the upper half marginated, receptacle shortly obconic or almost semiglobose. Petals: the outer ones in 1-2-series, the inner ones much narrower in 2-3-series, white, pink, violet-rose, red or rarely lemon-yellow. Stamens conically collected, staminodes usually present, ciliately papillate in the lower part, closely appressed to the stamens, equally or slightly exceeding them; filaments about 6-8-seriate, white, red or orange, the outer ciliately papillate in the lower part, as with the inner ones too but these are densely papillate near the middle, anthers and pollen golden. Ovary lobes distant, acutely or obtusely compressed, rarely reaching the nectary (disk); nectary (glands) inconspicuous, annular crenulate, 6-angled, the angles rounded; stigmas 6, very narrowly subulate and caudate, very rarely exceeding the height of the stamens. Capsule 6-locular, shortly obconic below, the sutures strongly compressed, expanded valves widely spreading, wingless; keels aristate; loculi-roofs well developed, the two processes on the underside of the membrane near the mouth obconical, there are capsules without these; tubercles well developed; seeds extending to the base of the loculus, covered with long brown papillate hairs (especially with *A. dinteri*), shortly echinate, tuberculate or muricate, brown or dark-brown, also pear-shaped to horseshoe-shaped.

Species 10, South-West Africa and the Richtersveld, along the Orange River. (Type: *A. dinteri* L. Bolus).

B.O.Carter. 29-2-1930
Astridia longifolia 89

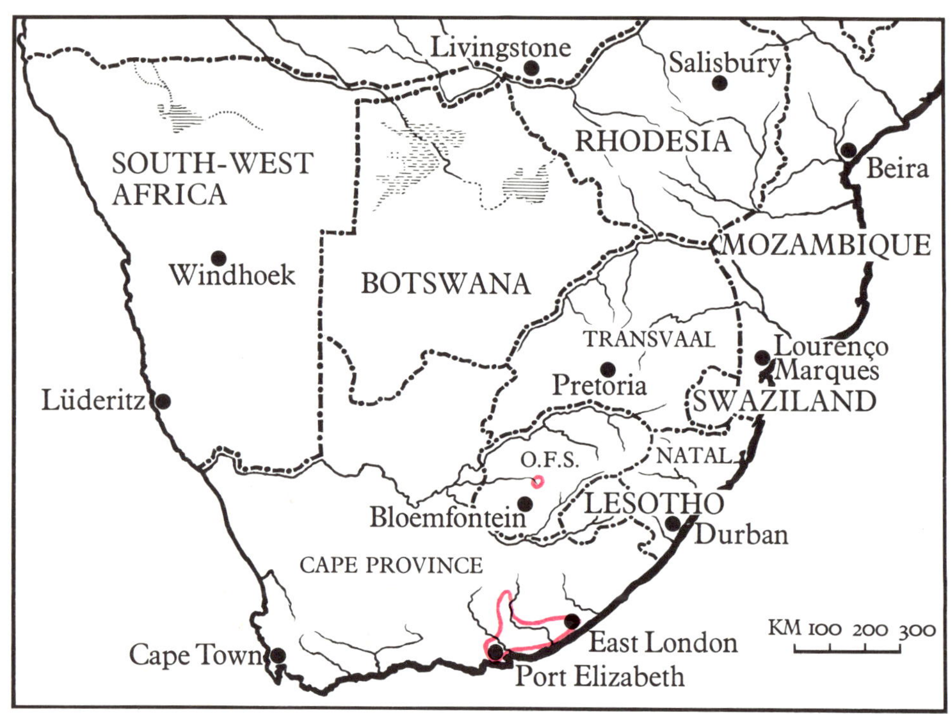

Bergeranthus

Bergeranthus Schwantes
Named in honour of the botanist
Alwin Berger, Stuttgart.

Schwantes, Z. Sukk. 3: 180. 1926; 3: 105. 1927. – N. E. Brown, J. Bot. 66: 266. 1928. – Bolus, Notes, 87, 88; fig. 16A. 1928. – Labarre (red.), Mesembr. 120; fig. 23, 24, 25. 1931. – v. Pöllnitz, Aufteilung, 27. 1933. – Jacobsen, Sukk. 94, 95; fig. 87. 1933. – Pax, Natürl. Pfl. 211. 1934. – Jacobsen, Succ. Pl. 131; fig. 110. 1935; Verzeichnis, 22. 1938. – Jacobsen, Herre, Volk. Mesembr. 53 (in key), 86. 1950. – Phillips, Genera, 319. 1951. – Jacobsen, Handbuch, 1187; fig. 1017, 1018. 1955. – Schwantes, Fl. Stones, 89, 338; fig. 22 (capsule); pl. 23b. 1957. – Jacobsen, Handbook, 952 (in system of Schwantes), 956 (in key of Bolus), 965 (in key of Herre & Volk), 997; fig. 1215, 1216. 1960; Lexikon, 366, t. 147/1. 1970.

Perennial clumped succulents with fleshy rootstock, nearly rape-shaped. Leaves opposite, shortly connate at the base, crowded, semi-terete, towards the apex more or less keeled, sometimes lower surface chin-like produced, upper surface more or less linear or broadened towards the midst, entire, oblong 3-angled to triquetrous, dark-green to whitish, more or less shining, darkly dotted, 13 cm long and about 2 cm broad. Flowers in dichotomous cymes, of which the axilliary flowers are sometimes suppressed, pedicels with sheathed bracts, about 10 cm long, yellow, outside reddish, about 5 cm in diam. There are species flowering during evening and night. Sepals 5, 3-angled or lanceolate, acuminate with a sharp spine, keeled, with membranous margins. Petals several series, linear-lanceolate, yellow, outside reddish. Stamens numerous, erect. Ovary flat-conical or conical with raised centre; placentas parietal; glands fairly large, sometimes yellowish-green, also small; stigmas 5, filiform or subulate, fairly long. Capsule 5-locular, keels dark-brown, very broad, with membranous points; loculi-roofs stiff; tubercle very large, more or less trigonous; seeds pear-shaped or round or trigonous with rounded edges.

Species: 11 from the Eastern Cape Province, and the Orange Free State. (Type: *B. scapiger* (Haw.) Schwant.)

M.M.Page. XII. 1922.

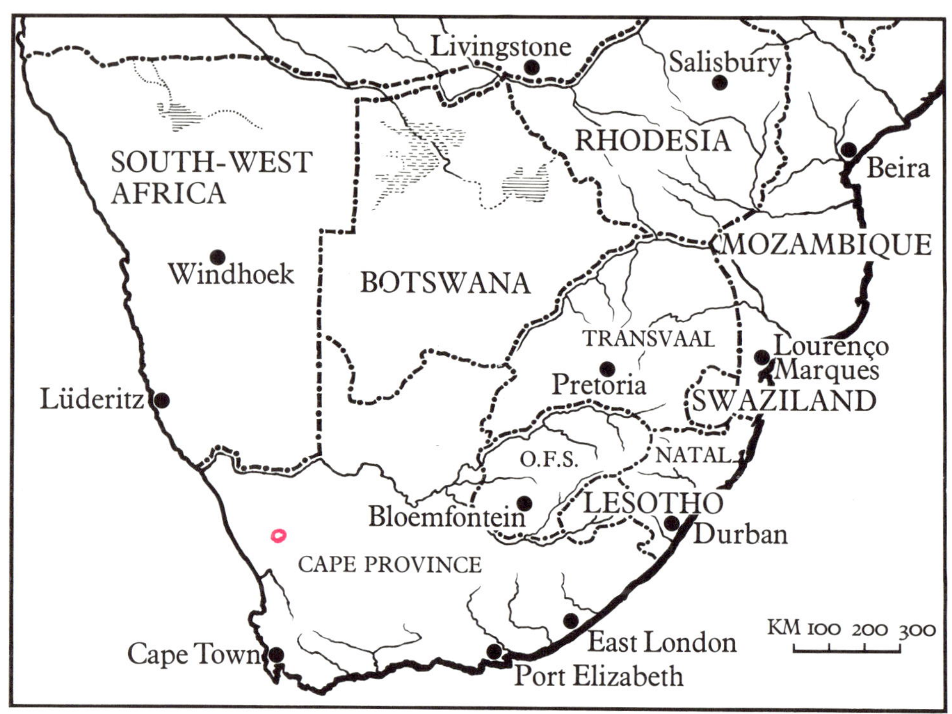

Berrisfordia

Berrisfordia L. Bolus
Named in honour of Mr. G. Berrisford
who discovered the plant.

Bolus, Notes 2: 313. 1931. - v. Pöllnitz, Aufteilung, 28. 1933. - Pax, Natürl. Pfl. 220. 1934. - Jacobsen, Verzeichnis, 23. 1938. - Jacobsen, Herre, Volk, Mesembr. 86, 90. 1950. - Phillips, Genera, 319. 1951. - Jacobsen, Handbuch, 1191, 1270. 1955. - Schwantes, Fl. Stones, 245, 340; pl. 67A. 1957. - Jacobsen, Handbook, 953 (in system of Schwantes), 956 (in key of Bolus), 999; fig. 1217. 1960; Lexikon, 367, t. 147/3. 1970.

Perennial very small plant, glabrous with very fine branchlets. Leaves in the midst just reaching the base of the calyx lobes, about 1,5 cm long, with a sheath of 1,1 cm length and about 0,8 cm in diam., at the base of the free part, up to 0,7 cm broad and at the apex about 0,9 cm in diam., the highest leaves are inside-to convex, almost completely hidden, 0,7-1,1 cm long with a sheath of 0,5-0,7 cm long, outside more or less warty with the warts crowded towards the apex, where they are nearly flat, so that the small leaves look as if they are dusted with flour, among them are some other larger ones, more round or conical in shape. Flowers with a small pedicel, about 0,4 cm long, with 1 or 2 pairs of bracts or these are completely absent, the flower is just exserted. Sepals 4-lobed, nearly equal, with membranous margin, 2 of them, keeled and towards the apex somewhat warty, about 0,3 cm long, connate to a 2-2,5 mm long tube 2,5 mm in diam., calyx 0,5 cm long. Petals 4-5-seriate, connate at the base for 3 mm, obtuse or emarginated or truncate, to 1 cm long and 0,2 cm broad; posteriors narrow, acuminate, mauve-pink. Stamens 4-5-seriate, inserted inside of the flower-tube, anthers about 0,2 cm long; staminodes absent. Ovary chambers less convex; glands distinct, thickly crenulated, yellow; stigmas 4, slender, towards the apex somewhat broadened and papillose, 0,8 cm long, longer than the stamens. Capsule 4-locular, small, like those of Conophytum without loculi-roofs and without tubercles; seeds pear-shaped, dark-brown, small, 1-1,5 mm long, with navel.

Monotypic genus from the Ezelskop near Garies in the Khamies Mountains, Little Namaqualand.
(Type: *B. khamiesbergensis* L. Bolus.)

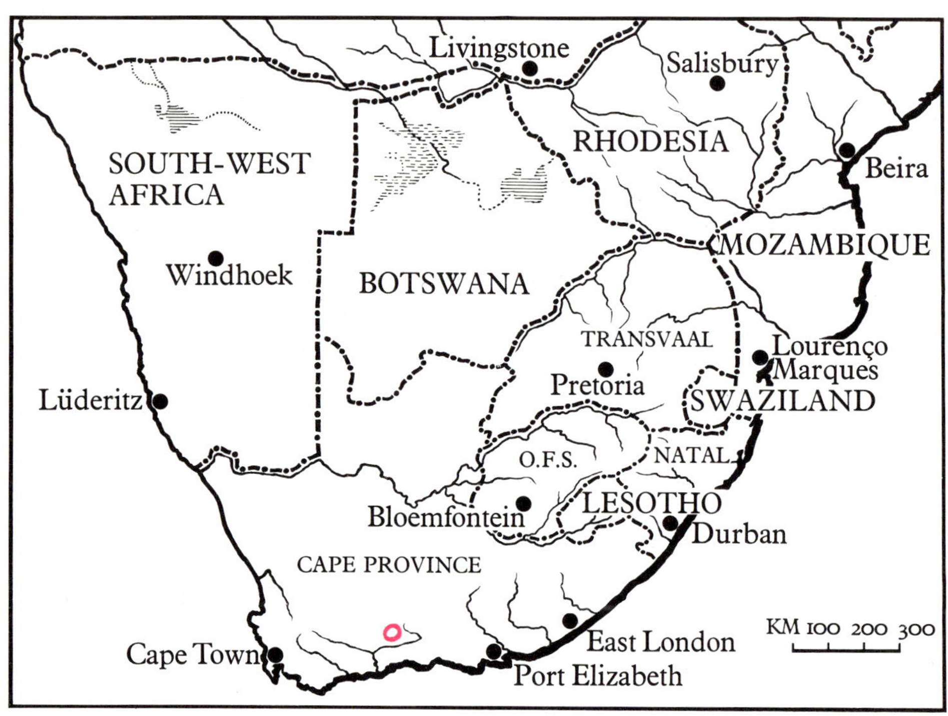

Bijlia

Bijlia N. E. Brown
Named in honour of Mrs D. van der Bijl.

N. E. Brown, J. Bot. 66: 267. 1928;
Gard. Chron. 85: 34. 1929. - v. Pöllnitz,
Aufteilung, 28. 1933. - Jacobsen, Sukk.
95; fig. 88. 1933. - Pax, Natürl. Pfl. 211.
1934. - Jacobsen, Succ. Pl. 132; fig.
111. 1935; Verzeichnis, 23. 1938. - Goossens, Blomplante, 145 (in key). 1940. -
Jacobsen, Herre, Volk, Mesembr. 73
(in key), 86. 1950. - *Phillips, Genera, 297.
1951.* - Jacobsen, Handbuch, 1191; fig.
1019. 1955. - Schwantes, Fl. Stones, 89,
338; pl. 23a. 1957. - Jacobsen, Handbook, 952 (in system of Schwantes), 962
(in key of Bolus), 973 (in key of Herre
& Volk), 1000; fig. 1218. 1960; Lexikon,
367, t.147/3. 1970.
Bolusanthemum tugwelliae Schwantes
in J. Bot. 268. 1928; Gartenwelt, 514.
1928.
Hereroa cana L. Bolus in S. Afr. Gard.
121. 1930; Gard. Chron. 34. 1929.
Hereroa tugwelliae L. Bolus in S. Afr.
Gard. 253. 1928; Notes, 50, 143; pl. 20.
1928; S. Afr. Gard. 121. 1930; Succulenta, 109. 1930. - Labarre (red.), Mesembr. 124; fig. 26. 1931.
Juttadinteria tugwelliae Schwantes in
D. Kakt. Ges. 184. 1925/26.
Mesembr. tugwelliae L. Bolus in Ann.
Bol. Herb. 1: 129. 1915; Succulenta,
5, 1925.

A stemless or tufted succulent perennial. Leaves opposite, somewhat con-
nated at the base, soft, form very
irregular and dissimilar, shaped by im-
pressions of the shorter leaf to the larger
one, overtopping (positive and negative),
smooth, whitish, green to light-green,
obtuse, somewhat spreading, obliquely
and clavately trigonous, strongly keeled,
about 5 cm long, 2 cm broad and 1,5 cm
thick. Flowers 1-3, pedicillate, peduncle
not longer than the leaves and not exserted from between the leaves, pedicels
very short, 0,2-1 cm long, with bracts,
0,3-0,6 cm long, yolk-yellow, about 3 cm
in diam. Sepals 5, subequal, 2 longer
than the rest, with keeled and terete project. Petals 2-seriate, wedge-shaped to
linear, free, obtuse. Stamens numerous,
in an erect column, exposed to view to
their base, yellow. Ovary inferior; glands
5 single ones; placentas on the outer
walls of the chambers; stigmas 5, filiform. Capsule shortly and broadly
obconic, flattish, with raised sutures on
the top, 5-locular, valves of expanded
capsule inflexed and separating only
slightly from one another, not spreading;
keels small, slightly prominent, about
half as long as the valve and without
marginal wings; loculi acutely roofed
with slightly stiff loculi-roofs, with acute
depressions between the ridges and the
opening nearly closed by a large tubercle;
seeds, several in each loculus, ovoid,
pointed at one end.

Monotypic genus from Prince Albert.
(Type: *B. cana* (Haw.) N. E. Brown.)

M.Page. VI. 1920.
(fresh from the veld).
Krige, Prince Albert.
June 1920
Bijlia cana   95

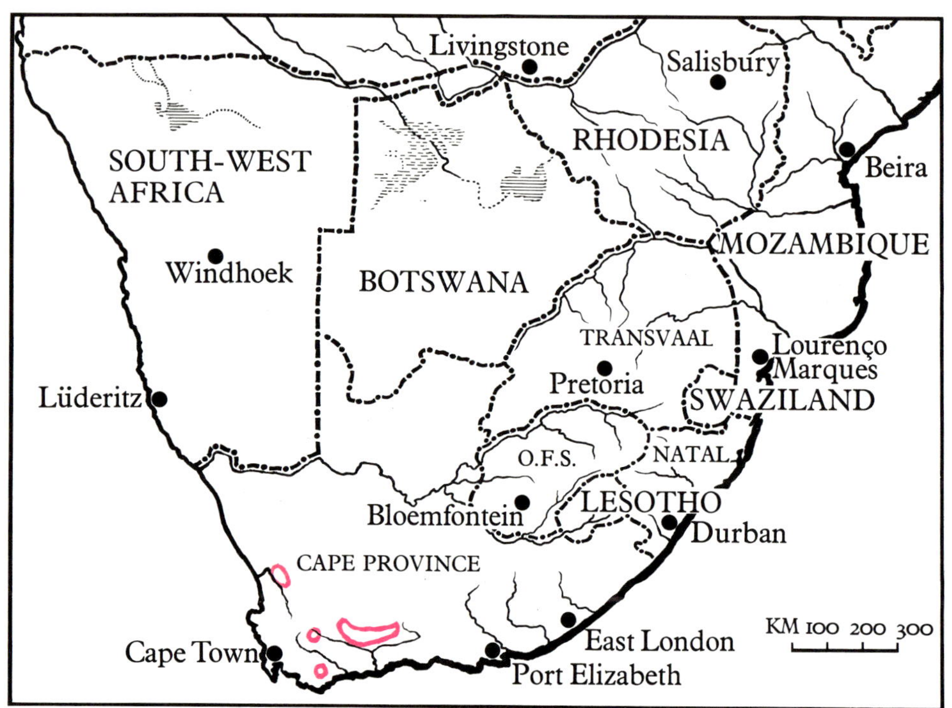

Braunsia

Braunsia Schwantes
Named in honour of Dr H. Brauns
of Willowmore.

Schwantes, Gartenwelt, 32: 644. 1928. -
v. Pöllnitz, Aufteilung, 28. 1933. - Jacobsen, Verzeichnis, 24. 1938. - Jacobsen,
Herre, Volk, Mesembr. 86. 1950. -
Phillips, Genera, 319. 1951. - Jacobsen,
Handbuch, 1191. 1955; Handbook,
1000. 1960. - Bolus, J. S. Afr. Bot. 31:
172. 1965. - Jacobsen, Lexikon, 367,
t.147/4. 1970.

Echinus L. Bol., Fl. Pl. S. Afr. 7: pl.
266. 1927. - N. E. Brown, Kew. Bull. 56.
1929; Gard. Chron. 67: 71 (in key).
1930. - v. Pöllnitz, Aufteilung, 41. 1933.
- Pax, Natürl. Pfl. 214. 1934. - Jacobsen, Succ. Pl. 173, 1935; Verzeichnis,
77. 1938. - Goossens, Blomplante, 145
(in key) 1940. - Jacobsen, Herre, Volk,
Mesembr. 60 (in key), 97. 1950. - Phillips,
Genera, 320. 1951. - Jacobsen, Handbuch, 1354. 1955. - Schwantes, Fl.
Stones, 96, 338; pl. 26b. 1957. - Jacobsen, Handbook, 952 (in system of
Schwantes), 961 (in key of Bolus), 968
(in key of Herre & Volk), 1135. 1960.

Dwarf perennials, erect or prostrate,
woody. Leaves opposite, always nearly
equal, pair of leaves $\frac{1}{4}$ to $\frac{1}{2}$ of its length connate, crescent, 3-angled, keeled, not pa-
pillate (seen without pocket lens), velvet-
like, rarely with prominently scattered
dots, mostly smooth, about 2,5 cm long,
1,5 cm broad and thick, always with a
white cartilaginous margin. Flowers
solitary or in cymes, often pedicillated,
rarely sessile, about 4 cm in diam. Sepals
5, 2 exteriors much compressed, interiors with large membranous margins.
Petals 4-5-seriate, acuminate, white or
pink. Staminodes nearly not appressed to
the stamens, sometimes already from
about the middle recurved, equal to the
stamens, filaments, exteriors below obscure, interiors numerous, slightly
ciliate-papillate, anthers small, golden-
yellow. Ovary slightly elevated; glands inconspicuous, minutely crenulate; stigmas 5, acuminate. Capsule 5-locular,
rarely 4-, 6-, or 7-locular, valves without
wings, keels minute lacerate, aristate,
loculi-roofs well-developed or reduced
to a limb, tubercles absent; seeds short,
echinate.

Species: 4 recorded from the Cape
Province: Laingsburg, Matjiesfontein,
Touwsriver, Prince Albert, Little
Karroo, Bredasdorp. (Type: *B. geminata*
(Haw.) L. Bol.)

Braunsia apiculata 97

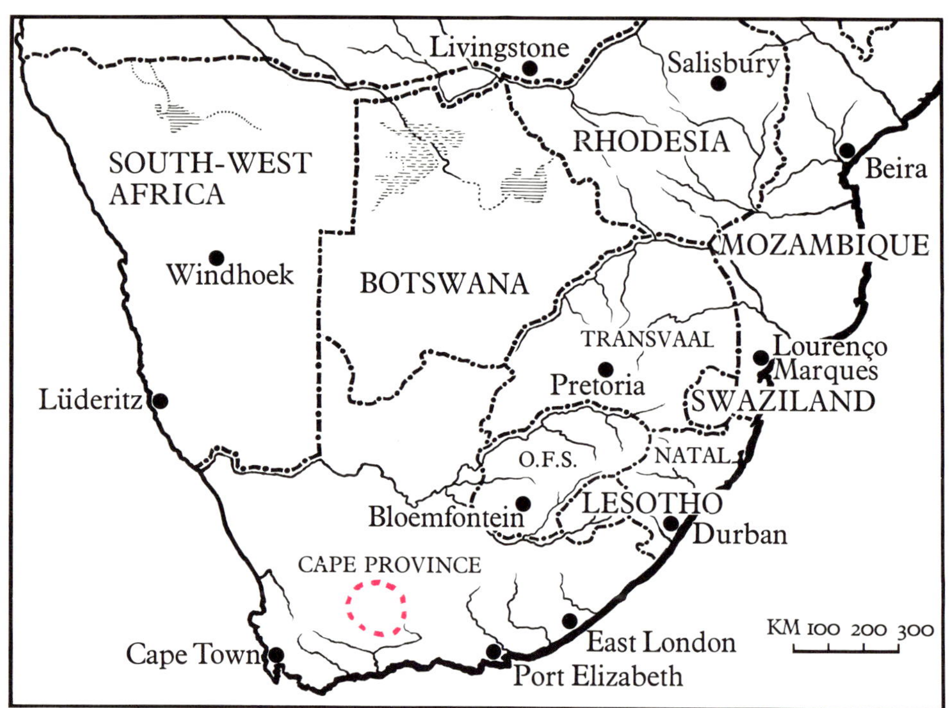

Calamophyllum

Calamophyllum Schwantes
From the Greek *calamus* = reed and
phyllon = leaf.

Schwantes, Z. Sukk. 15, 28, 29, 106 (in key). 1927/28. - N. E. Brown, J. Bot. 66: 322. 1928. - v. Pöllnitz, Aufteilung, 29. 1933. - Jacobsen, Sukk. 96. 1933. - Pax, Natürl. Pfl. 211. 1934. - Jacobsen, Succ. Pl. 133. 1935; Verzeichnis, 25. 1938. - Jacobsen, Herre, Volk, Mesembr. 53 (in key), 86. 1950. - Phillips, Genera, 319. 1951. - Jacobsen, Handbuch, 1193; fig. 1021. 1955. - Schwantes, Fl. Stones, 338. 1957. - Jacobsen, Handbook, 952 (in system of Schwantes), 956 (in key of Bolus), 965 (in key of Herre & Volk) 1001; fig. 1220. 1960; Lexikon, 368, t. 148/3. 1970.

Mesembrianthemum, group *Teretifolia* Haw. in Berger, Mesembr. & Crassulaceae. 224. 1908.

Clumped small plants with often invisible internodes, freely branching. Leaves opposite, crowded, nearly terete at the base, upper surface fairly flat towards the apex, lower surface roundish-keeled or subcylindrical or nearly terete, very much curved to one side and widely spreading, often dotted, 5-10 cm long and about 0,8-1,0 cm broad and thick. Flowers solitary, terminal, short pedicellated, with 2 bracts, sometimes surpassing the flowers, red, open in the afternoon, about 3 cm in diam. Calyx unequally 4-lobed. Petals several series, sometimes emarginated or obtuse, red. Stamens many, white or red, with red anthers. Ovary ? Placentas parietal. Stigmas 10-12, diverging. Capsule 10-12-locular, otherwise unknown. Distribution unknown.

Species: 3 ? Imperfectly known. (Type: *C. teretifolium* (Haw.) Schwant.)

R · DARROLL '71.

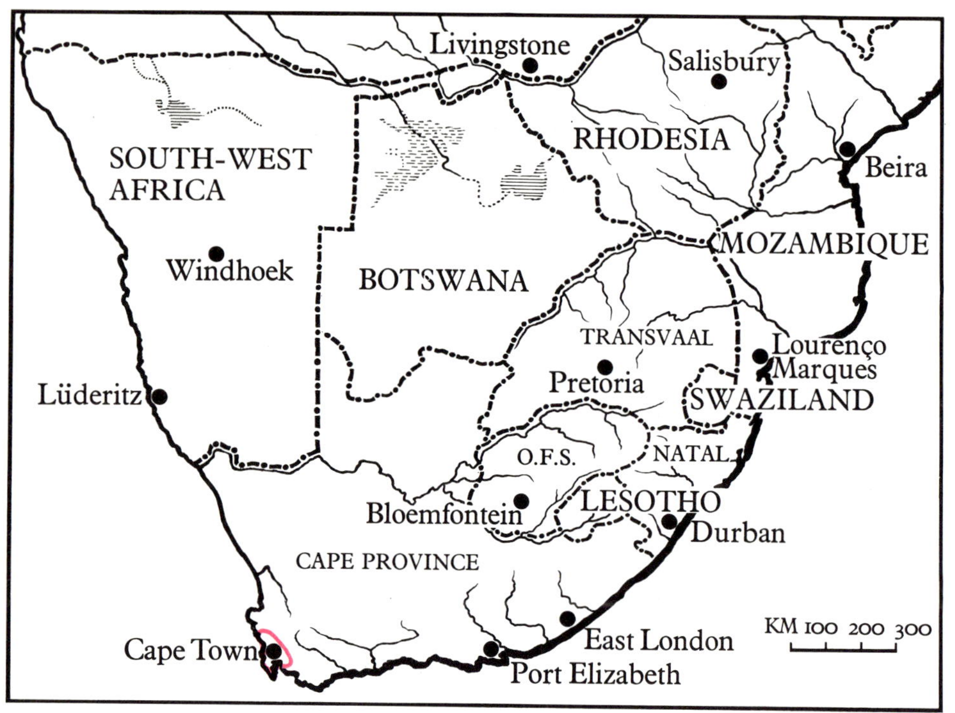

Carpanthea

Carpanthea N. E. Brown
From the Greek *karpos* = fruit and *anthe* = flower.

N. E. Brown, Gard. Chron. 78: 412 (in key). 1925. - Phillips, Genera, 245. 1926. - N. E. Brown, J. Bot. 66: 323. 1928. - Bolus, Notes, 15; pl. IV, fig. C. 1928. - v. Pöllnitz, Aufteilung, 29. 1933. - Pax, Natürl. Pfl. 211. 1934. - Jacobsen, Sukk. 96. 1933; Succ. Pl. 133. 1935; Verzeichnis, 26. 1938. - Goossens, Blomplante, 141 (in key). 1940. - Jacobsen, Herre, Volk, Mesembr. 53 (in key), 86. 1950. - Adamson, Salter, Fl. Cape Pen. 377. 1950. - *Phillips, Genera, 298. 1951.* - Jacobsen, Handbuch, 1195; fig. 1023. 1955. - Schwantes, Fl. Stones, 50, 341; pl. 10 (capsule), 11b (flower); fig. 10 & 11 (capsule) 1957. - Jacobsen, Handbook, 954 (in system of Schwantes), 956 (in key of Bolus), 965 (in key of Herre & Volk) 1002; fig. 1221, 1222. 1960; Lexikon, 368, t. 148/2. 1970.

An annual succulent herb. Leaves opposite, radical or at the lower part of the flowering branches, leaf-blade shorter than the petiole, somewhat connate at the base, spathulate or lanceolate, obtuse, narrower towards the caniculated petiole, 3-nerved, along the margins ciliolate, covered with scattered villous, white hairs, about 3,5-10 cm long and 1-2,5 cm broad. Flowers 1-3, terminal, pedicels 3-10 cm long, the lateral ones with 2 fairly large, lanceolate, sheathed bracts, open during the afternoon, golden-yellow, about 4-7 cm in diam., prostrate in fruit. Sepals 5, down to the top of the ovary, unequal, 2 lobes leafy, as long as or longer than the petals, the shorter ones broad-ovate and with broad, dry membranes. Petals very numerous, in several series, slender, free or nearly so, narrow-linear, pointed, silk-like, outside often somewhat reddish. Stamens very numerous, at first inflexed to the centre over the stigmas, afterwards spreading outward and exposing the stigmas, outer stamens hair-like and without anthers; staminodes present. Ovary inferior, broader than deep; placentas on the floor of the chambers; style none; stigmas 12-20 filiform, finely pointed. Capsule much broader than deep, 12-20-locular, valves narrow, with a pair of contiguous expanding keels ending in awns, without marginal wings, loculi not roofed but with a very narrow space for the seeds to escape; seeds small, roundish, not winged.

A South-Western Cape genus with 2 species. (Type: *C. pomeridiana* (L.) N. E. Br.)

Carpanthea pomeridiana 101

Carpobrotus

Carpobrotus N. E. Brown
From the Greek *karpos* = fruit and
brota = edible things (because the fruits
can be eaten).

N. E. Brown, Gard. Chron. 78: 433 (in key). 1925. Phillips, Genera, 249. 1926. - Bolus, Fl. Pl. S. Afr. 7: pl. 247. 1927; Notes, 10, pl. II E (fruit of *C. fourcadei*). 1928. - v. Pöllnitz, Aufteilung, 29. 1933. - Jacobsen, Sukk. 96; fig. 89. 1933. - Pax, Natürl. Pfl. 211. 1934. - Jacobsen, Succ. Pl. 134; fig. 112. 1935; Verzeichnis, 26. 1938. - Goossens, Blomplante, 143 (in key). 1940. - Jacobsen, Herre, Volk. Mesembr. 66 (in key), 86; fig. 18. 1950. - Adamson, Salter, Fl. Cape Pen. 389. 1950. - *Phillips, Genera, 298, 1951.* - Jacobsen, Handbuch, 1195. fig. 1024. 1955. - Schwantes, Fl. Stones, 95, 342; fig. 25 (fruit). 1957. - Jacobsen, Handbook, 954 (in system of Schwantes), 957 (in key of Bolus), 970 (in key of Herre & Volk), 1002; fig. 1223. 1960; Lexikon, 368, t.149/1. 1970.

Perennial succulents with trailing stems. Leaves opposite, slightly united at the base, sharply 3-angled. Flowers large, terminal, pedicel 5-7 cm long, white, yellow, red, largest of the whole family, up to 15 cm in diam. Sepals 5, down to the top of the ovary. Petals numerous, free. Stamens numerous erect. Ovary inferior; placentas on the outer walls of the chamber; style none; stigmas 10-16, radiating, plumose. Fruit fleshy or pulpy, indehiscent and without valves, 10-16-locular, often edible; seeds obovoid, slightly compressed, on long funicles.

Species: 23 in South Africa from Clanwilliam through the Cape Province to Natal. - Chile 1, California 1, Australia 4 (together with New Zealand and Tasmania), mostly along the coast. (Type: *C. edulis* (L.) N. E. Br.)

M.M. Page. IX. 1924.

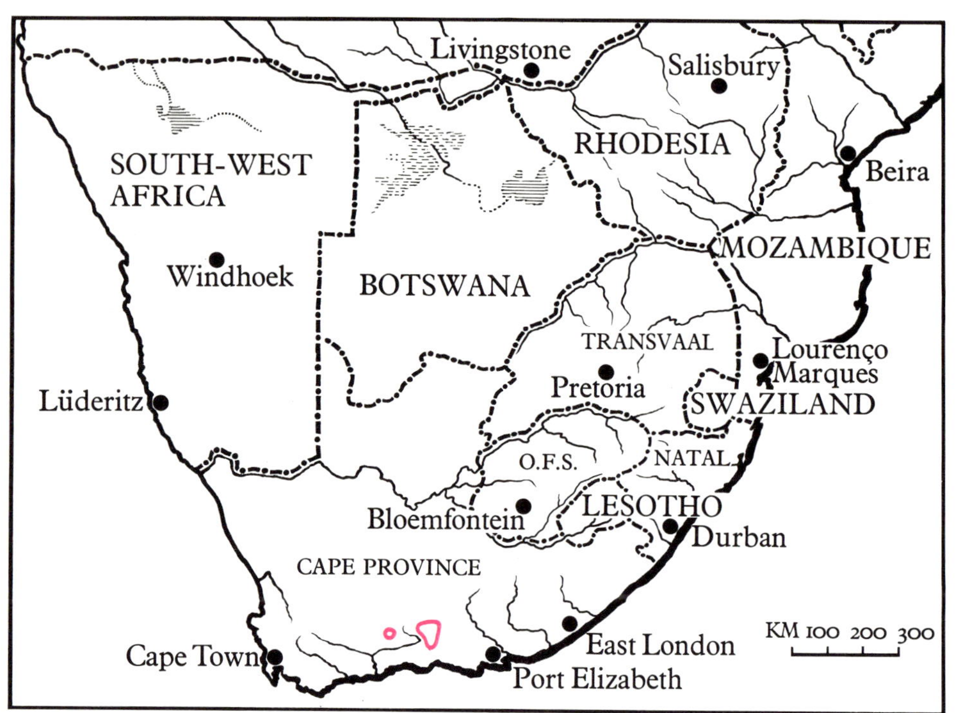

Carruanthus

Carruanthus Schwantes
From *carru* = karoo and the Greek
anthos = flower.

Schwantes, Z. Sukk. 181. 1926. - N. E.
Brown, J. Bot. 66: 325. 1928. - v. Pöllnitz, Aufteilung, 30. 1933. - Jacobsen,
Sukk. 97; fig. 90. 1933. - Pax, Natürl.
Pfl. 212. 1934. - Jacobsen, Succ. Pl. 134;
fig. 113. 1935; Verzeichnis, 28. 1938. -
Jacobsen, Herre, Volk, Mesembr. 73 (in
key) 87. 1950. - Phillips, Genera, 319.
1951. - Jacobsen, Handbuch, 1199; fig.
1025. 1955. - Schwantes, Fl. Stones,
94, 338. 1957 - Jacobsen, Handbook, 952
(in system of Schwantes), 959 (in key
of Bolus), 973 (in key of Herre & Volk),
1006; fig. 1224. 1960; Lexikon, 369,
t. 149/3 & 4. 1970.

Tischleria Schwantes Sukkulentenkunde 4: 78. 1951. - Jacobsen, Handbuch, 1698; fig. 1331. 1955. - Schwantes,
Fl. Stones, 95; pl. 26a. 1957. - Jacobsen,
Handbook, 952 (in system of Schwantes)
952 (in key of Bolus), 973 (in key of
Herre & Volk), 1427; fig. 1603. 1960.

Tufted succulent perennial with short
branchlets and with fleshy, almost rape-
shaped roots and invisible internodes.
Leaves opposite, crowded, slightly con-
nated at the base, lanceolate, 3-angled,
upper-side flat or somewhat concave,
lower side towards the apex with
broadened, often chin-like produced
keel, not dotted, erect, ascending or
spreading, often incurved, grey-green to
shiny green, smooth, along the margins
and towards the apex more or less finely
toothed, about 6 cm long and 1,8 cm
broad. Flowers solitary, or in dichotomous cymes, of which the axillary
flowers are often suppressed, peduncle 2-
angled, shorter than the leaves, pedicel
about 10 cm long, slender, terete,
thickened towards the point, with 1-2
pairs of small, leafy sheathed bracts,
yellow, outside reddish, open during afternoon, about 4-5 cm in diam. Sepals
5, unequal, partly keeled. Petals several
series, linear-lanceolate, radiate and
incurved. Stamens many, erect, yellow.
Ovary conical; glands separated from
each other; placentas parietal; stigmas 5,
filiform or subulate, erect, recurved,
somewhat shorter than the stamens. Capsule 5-locular, keels ending in wings,
loculi-roofs reduced to a limb, tubercle
reduced, sometimes absent; seeds pear-
shaped or round or trigonous with
rounded margins, about 1 mm long.

Species: 2, Cape Province, Willowmore. (Type: *C. caninus* (Haw.)
Schwant.)

Carruanthus sp. 105

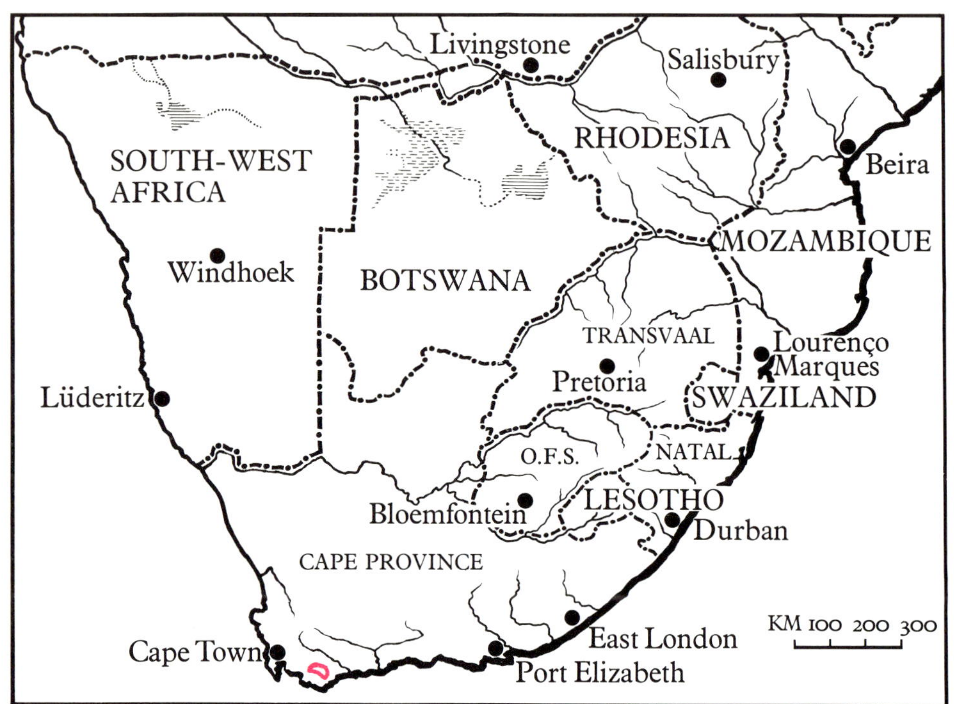

Caryotophora

Caryotophora Leistner
From the Greek *karyotos* = nut and *phorein* = carrying (since the fruits look like nuts).

Leistner in Bolus, Notes, 289; pl. 93, 94. 1958. - Jacobsen, Handbook, 952 (in system of Schwantes), 956 (in key of Bolus), 1007; fig. 1225, 1226. 1960; Lexikon, 369, t. 149/2. 1970.

Perennial herb, only slightly succulent and glabrous; stem very short, 1,2 cm long; main root horizontal, 5-20 cm below the surface of the sandy soil, suckering freely, the flowering branches annual, decumbent, 315 cm long. Leaves alternate or sub-opposite, flat, ascending, decurrent, clasping at the base, entire, lanceolate-spathulate, obtuse or subacute, leaves on stem 6-16 cm long, 1-3 cm broad; leaves on branches 4-6 cm long, 1-2 cm broad. Flowers opening during the day, terminal, 4-6 cm in diam., white, peduncle 3-10 cm long, 0,3-0,5 cm in diam., white, peduncle 3-10 cm long, 0,3-0,5 cm in diam. Sepals 5, unequal, 2 exterior ones triangular at the base, entire, becoming linear at the tip, 1,2-2 cm long, third lobe with membranous margins, 1-1,5 cm long. Petals numerous, white, linear narrowing towards the tip, 2-3 cm long, 0,17-2 cm broad. Stamens and staminodes many, incurved, filaments 0,7-0,8 cm long, papillose near the base. Ovary inferior, receptacle subturbinate, slightly convex on top, 0,4-0,8 cm long, 1-1,6 cm in diam.; glands annular; placentas axile; stigmas subulate, 3-4, 0,5-0,7 cm long. Capsule a schizocarp with nut-like, woody mericarps, fruit depressed globose to lenticular 1,4-2,4 cm in diam. 0,8-1,5 cm high, consisting of 3-4 nut-like, bilocular woody mericarps which contain a single ovule in each locule; seed reniform, verrucose, reddish-brown, often sterile, 0,18-0,2 cm long.

Species: 1 (monotypic genus), Cape Province: Brandfontein, distr. Bredasdorp. (Type: *C. skiatophytoides* Leistner.)

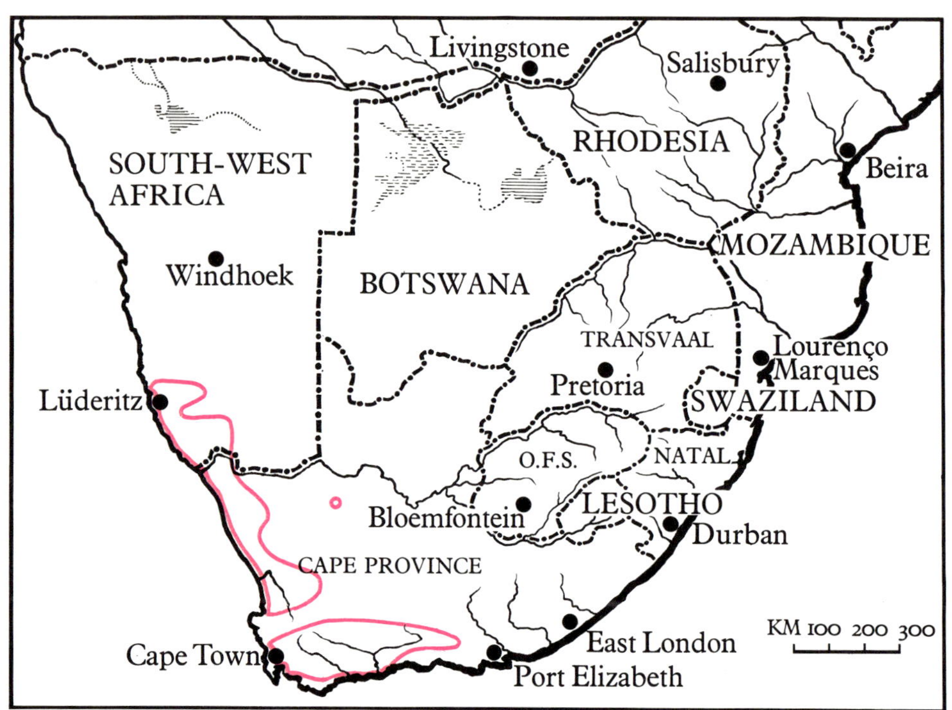

Cephalophyllum

Cephalophyllum N. E. Brown
From the Greek *kephale* = head and
phyllon = leaf.

N. E. Brown, Gard. Chron. 78: 433 (in
key). 1925. - Phillips, Genera, 247.
1926. - Bolus, Fl. Pl. S. Afr. 7: pl. 257.
1927; Notes, 116; fig. 38-40. 1928. -
Rendle, J. Bot. 66: 325. 1928; 67: 17.
1929. - v. Pöllnitz, Aufteilung, 30. 1933.
- Jacobsen, Sukk. 98; fig. 91. 1933. -
Pax, Natürl. Pfl. 212. 1934. - Jacobsen,
Succ. Pl. 135; fig. 114. 1935; Verzeichnis,
28.1938. - Goossens, Blomplante, 147
(in key). 1940 - Jacobsen, Herre, Volk,
Mesembr. 67 (in key), 87; fig. 28a (kelk).
1950. - Adamson, Salter, Fl. Cape Pen.
389. 1950. - *Phillips, Genera, 298. 1951.* -
Jacobsen, Handbuch, 1199; fig. 1026-
1028. 1955. - Schwantes, Fl. Stones, 105.
338; pl. 30a, 92b (capsule) 1957. -
Jacobsen, Handbook, 952 (in system of
Schwantes), 964 (in key of Bolus), 971 (in
key of Herre & Volk), 1008; fig. 1227-
1231. 1960. - Bolus, J. S. Afr. Bot. 28:
15 (key of 7 sub-groups). 1962. - Jacobsen, Lexikon, 370, t. 150/1, 2, 3. 1970.
- Friedrich in Merxmüller, Prodromus,
20. 1970.

Dwarf succulent perennial; main stem
short or very short, producing decumbent or prostrate branches with distinct
internodes between some of the leaf-
pairs. Leaves opposite, crowded into a
tuft on the main stem and ends of the
branches, elongated, subterete or 3-
angled, with more or less convex sides,
minutely dotted; not connate, erect,
ascending or spreading, rarely club-
shaped, about 8 cm long and 0,8 cm in
diam., the leaves may be so crowded, that
they look like a ball or head, hence the
name of the genus. Flowers usually
3 together, often also 2 or solitary, terminal, on long bractless pedicels, mostly
with 2 bracts, all colours, about 5 cm in
diam. Sepals 5, down to its union with
the ovary. Petals numerous, free, linear.
Stamens numerous, erect, not collected
into a column. Ovary inferior, with the
top raised into a crater-like rim, en-
circling the base of the stigmas, merely
depressed at the centre; glandular wall
undivided; placentas on the floor of
the chambers; style none; stigmas 15-20,
plumose. Capsule 8-20-locular, valve-
wings well developed, keels parallel;
loculi-roofs present; tubercles often
large, sometimes hidden under the
loculi-roof; seeds ovate or pear-shaped,
small, brown to light-brown.

Species: 63 recorded from South-
West Africa, Cape Province: Richters-
veld, Namaqualand, Bushmanland,
South-Western Cape, Little Karroo and
Karroo. (Type: *C. tricolorum* (Haw.)
N. E. Br.)

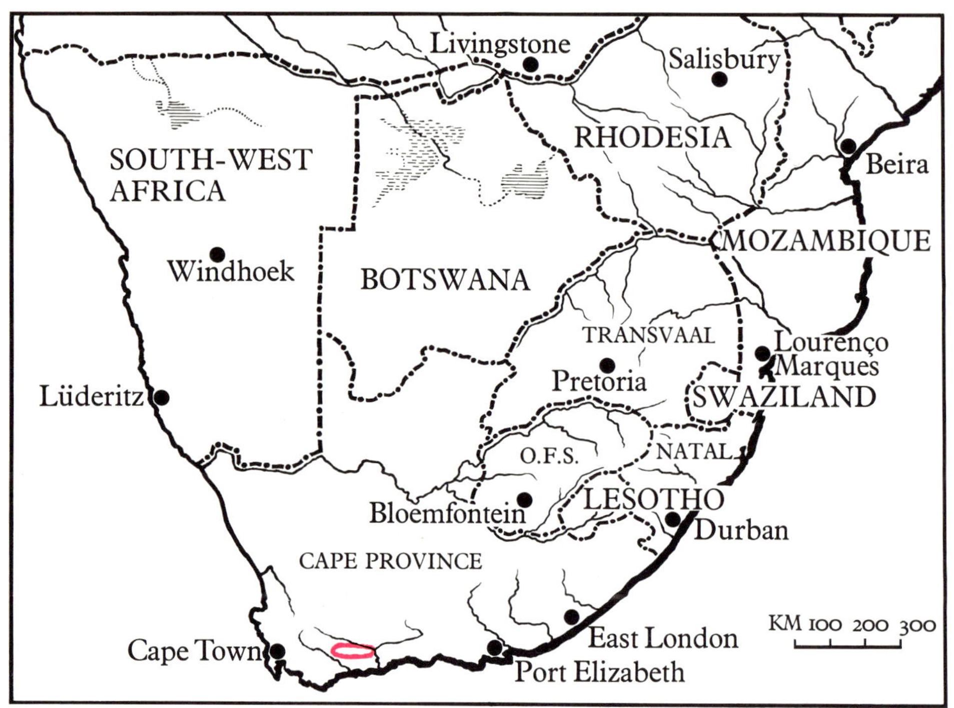

Cerochlamys

Cerochlamys N. E. Brown
From the Greek *keros* = wax and *chlamys* = mantle, on account of the coating of wax on the leaves.

N. E. Brown, J. Bot. 66: 171. 1928. - v. Pöllnitz, Aufteilung, 30. 1933. - Pax, Natürl. Pfl. 212. 1934. Jacobsen, Verzeichnis, 33. 1938. - Goossens, Blomplante, 141 (in key). 1940. - Jacobsen, Herre, Volk, Mesembr. 53 (in key), 88. 1950. - Bolus, Notes, 227. 1950. - *Phillips, Genera, 299. 1951.* - Jacobsen, Handbuch, 1211; fig. 1029. 1955. - Schwantes, Fl. Stones, 94, 338; pl. 256(b). 1957. - Jacobsen, Handbook, 952 (in system of Schwantes), 956 (in key of Bolus), 965 (in key of Herre & Volk), 1017; fig. 1232. 1960. - Lexikon, 374, t. 150/4, 1970.

A stemless succulent perennial or, with age, becoming tufted and with short branching stems. Leaves opposite, 1-3 pairs to a growth, shortly united at the base, trigonously clavate with the keel very oblique and so much turned to one side that one side is flat and makes nearly a right-angle with the flat surface or upper side of the leaf, while the other side is convex, very firm in substance, covered with a waxy secretion, dotless, about 4 cm long. Flowers 1-3, terminal, pedicel very short, bracts almost hidden among the leaves, pink, about 3 cm in diam. Sepals subequally 5-lobed down to their union with the ovary, some of the lobes with narrow membranous edges. Petals numerous, free 2-3-seriate, linear. Stamens numerous, all apparently connivent in a cone, filaments not bearded; staminodes filiform, surrounding the stamens. Ovary inferior, obconic, flattish on the top; glands 5, dark green; placentas on the outer walls of the chambers; stigmas 5, minute, papilla-like, acute. Capsule very shortly obconic, flattish or slightly convex, with raised sutures on the top, 5-locular; valves erect when expanded; keels with free awn-like and incurved tips; loculi acutely roofed with slightly stiff loculus-wings; tubercle absent or obscure but with a dense mass of funicles at the opening in empty loculi; seeds ovoid, pointed at one end, microscopically tuberculate.

Monotypic genus from the Little Karroo. (Type: *C. pachyphylla* (L. Bol.) L. Bolus.)

Cerochlamys pachyphylla III

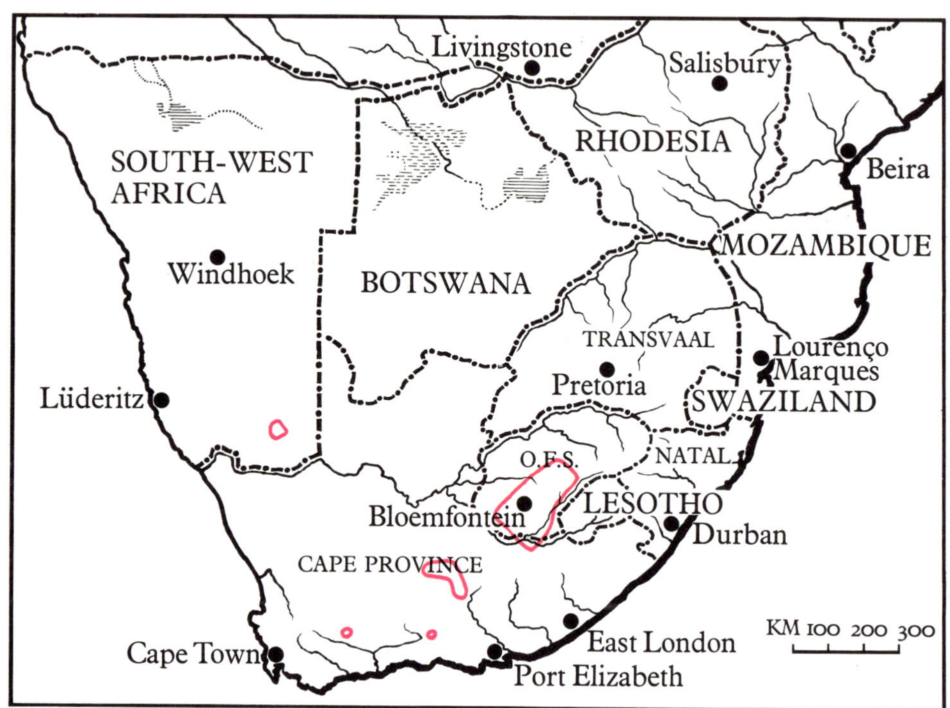

Chasmatophyllum

Chasmatophyllum Dinter et Schwantes
From the Greek *chasma* = open mouth
and *phyllon* = leaf.

Dinter et Schwantes, Z. Sukk. 15 (in key), 17, 30. 1927. - Schwantes, Z. Sukk. 176 (in key). 1926. - Bolus, Notes, 96; fig. 21 E 8 (capsule). 1928. - Berger, Gartenflora, 44. 1928. - N. E. Brown, J. Bot. 67: 18. 1929. - v. Pöllnitz, Aufteilung, 31. 1933. - Jacobsen, Sukk. 99. 1933. - Pax, Natürl. Pfl. 212. 1934. - Jacobsen, Succ. Pl. 136; fig. 115. 1935; Verzeichnis, 33. 1938. - Jacobsen, Herre, Volk, Mesembr. 53 (in key), 88. 1950. - Phillips, Genera, 319. 1951. - Jacobsen, Handbuch, 1212; fig. 1030. 1955. - Schwantes, Fl. Stones, 137, 339. 1957. - *Jacobsen, Handbook*, 953 (in system of Schwantes), 956 (in key of Bolus), 965 (in key of Herre & Volk), *1017*; fig. 1233. *1960*; Lexikon, 374, t.150/5. 1970. - Friedrich in Merxmüller, Prodromus, 23. 1970.

Dwarf perennial shrublets, in the beginning erect, later prostrate, short shrubbily branched, very slender, branchlets 6-8 leaved, with flowers 3-5 cm high. Leaves opposite, sheathed at the base, more or less erect, oblong-spathulate, with an obtuse keel, more or less semi-terete in diam., often along the margins, apex and keel with 1-2 obtuse teeth, on both sides covered with whitish prominent warts, about 2 cm long and 0,5 cm broad. Flowers solitary, terminal, pedicillated, without bracts, golden-yellow, open during afternoon, about 2,5 cm in diam. Sepals 5, unequal, 3-angled. Petals 2-3-seriate, rather broad-lanceolate, yellow, reddish on the apex. Stamens mostly slightly conically collected, yellow. Ovary flat, rarely somewhat conical; glands 5, distinct; placentas parietal; stigmas 5, filiform, pinnate, as long or longer than the stamens. Capsule 5-locular, with broad valve-wings; loculi-roofs present but the seeds only partly covered. (*Rhinephyllum* without loculi-roofs, main difference!); without tubercles but with C. maninum with distinct bifid tubercle; seeds obtuse-rounded, obscurely hooked, somewhat rough, whitish-yellow, about $\frac{3}{4}$ mm long.

Species: 6, recorded from South-West Africa (Karas Mts.), Cape Province: Karroo and the Free State. (Type: *C. musculinum* (Haw.) Dint. et Schwant.)

M.M. Page. XII 1923.

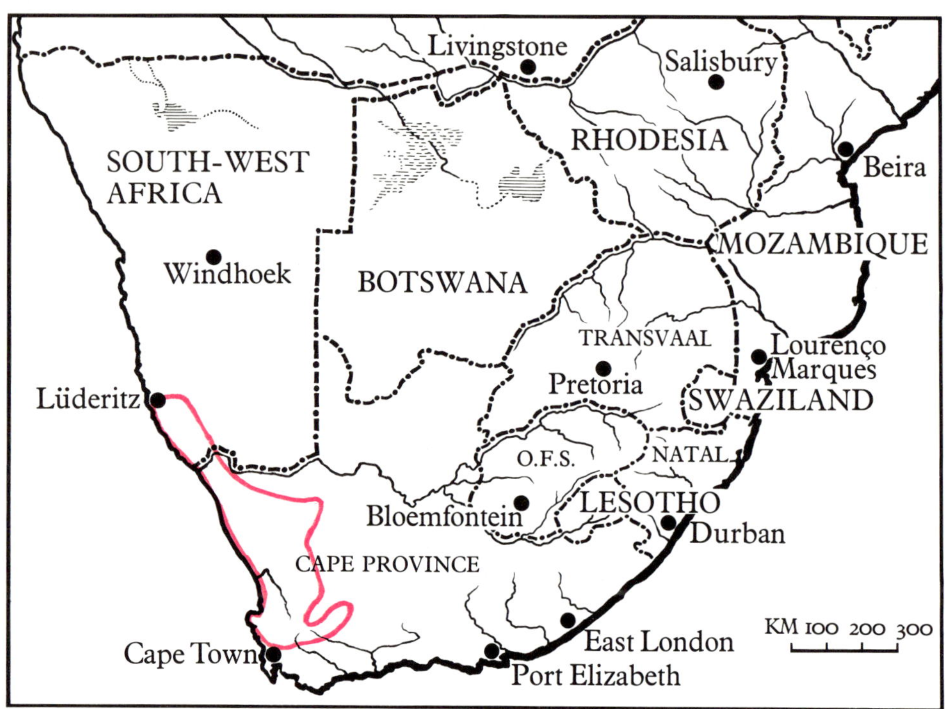

Cheiridopsis

Cheiridopsis N. E. Brown
From the Greek *cheiris* = sleeve and
opsis = like.

N. E. Brown, Gard. Chron. 78: 433 (in key). 1925; 79: 406 with fig. 1926. - Phillips, Genera, 244. 1926. - Tischer, Z. Sukk. 3: 348. 1928. - Bolus, Notes, 73; fig. 5, 6, 7. 1928. - N. E. Brown, J. Bot. 67: 19. 1929. - Labarre (red.), Mesembr. 128; fig. 28-36, 94-96. 1931. - v. Pöllnitz, Aufteilung, 31. 1933. - Jacobsen, Sukk. 99; fig. 93-96. 1933. - Pax, Natürl. Pfl. 212. 1934. - Jacobsen, Succ. Pl. 137; fig. 116-121. 1936; Verzeichnis, 33. 1938. - Goossens, Blomplante, 141 (in key), 1940. - Jacobsen, Herre, Volk, Mesembr. 53 (in key), 88. 1950. - *Phillips, Genera, 299. 1951.* - Jacobsen, Handbuch, 1213; fig. 1031-1044. 1955. - Schwantes, Fl. Stones, 109, 338; pl. 2 (coloured), 32, 33. 1957. - Jacobsen, Handbook, 952 (in system of Schwantes), 956 (in key of Bolus), 965 (in key of Herre & Volk), 1019; fig. 1234-1247. 1960. - Bolus, J. S. Afr. Bot. 27: 263 (key of 6 sub-groups). 1961. - Jacobsen, Lexikon, 374, t.151/1 2, 3, 4, 152/1. 1970.

Deilanthe N. E. Brown, Gard. Chron. 88: 278. 1930; 89: 259 (in key). 1931. - v. Pöllnitz, Aufteilung, 36. 1933. - Pax, Natürl. Pfl. 214. 1934. - Jacobsen, Succ. Pl. 124, 163; fig. 223. 1935; Verzeichnis, 61. 1938. - Phillips, Genera, 302, 1951. Bolus, Notes, 136. 1958. - Friedrich in Merxmüller, Prodromus, 24. 1970.

114

Dwarf succulent perennials forming clumps. Leaves opposite, in nature only 1-2 pairs to a growth but in cultivation, up to 3 pairs are sometimes present at the same time, the alternating pairs often dissimilar in size, form or degree of union at the base, one pair being shortly united at the base, and the next pair united from one-third to three-fourths or, in few species, for nearly all their length, withering, and the basal part forming during the resting period a truncate sheath surrounding the next pair of leaves, or in a few species, completely enclosing the new pair which usually have their flat faces closely applied to one another so that the pair somewhat resembles the beak of a bird, green, glaucous-green or white, often conspicuously dotted, sometimes without dots, terete with wing-like produced keel and then often hatchet-shaped, 3-angled, boat-shaped, rhombic to spathulate, roundish or ovate, more or less oblong-triquetrous or linear-lanceolate, firm or soft, often with 1-2 small teeth along the keel or the apex, about 10 cm long and 1,5 cm broad. Flower solitary, terminal, mostly pedicelled with bracts on the base and sometimes also in the middle of the pedicels, mostly yellow, orange, white, rarely violet-red, of about 1,3 cm -10 cm in diam. Sepals 4-5 down to its union with the ovary. Petals numerous, free, several series. Stamens numerous, erect or more or less connivent. Ovary partly superior, becoming inferior in fruit; disc annular; placentas on the outer wall or floor of the chambers; stigmas 8-19, sessile, more or less plumose, without a style. Capsule 8-19-loculi, valves narrow, widely spreading or reflexed when wetted; keels with membranous awn-like or linear points, usually without marginal wings, but sometimes with narrow marginal wings at the basal half; with loculi-roofs and the opening nearly closed by a large tubercle; seeds pear-shaped smooth or finely tuberculate, 3-angled with obtuse angles, whitish to brownish.

Species: 99, recorded from South-West Africa, Richtersveld, Namaqualand and the South-Western Cape. (Type: *C. tuberculata* (Mill.) N. E. Br.)

Cheiridopsis tuberculata   115

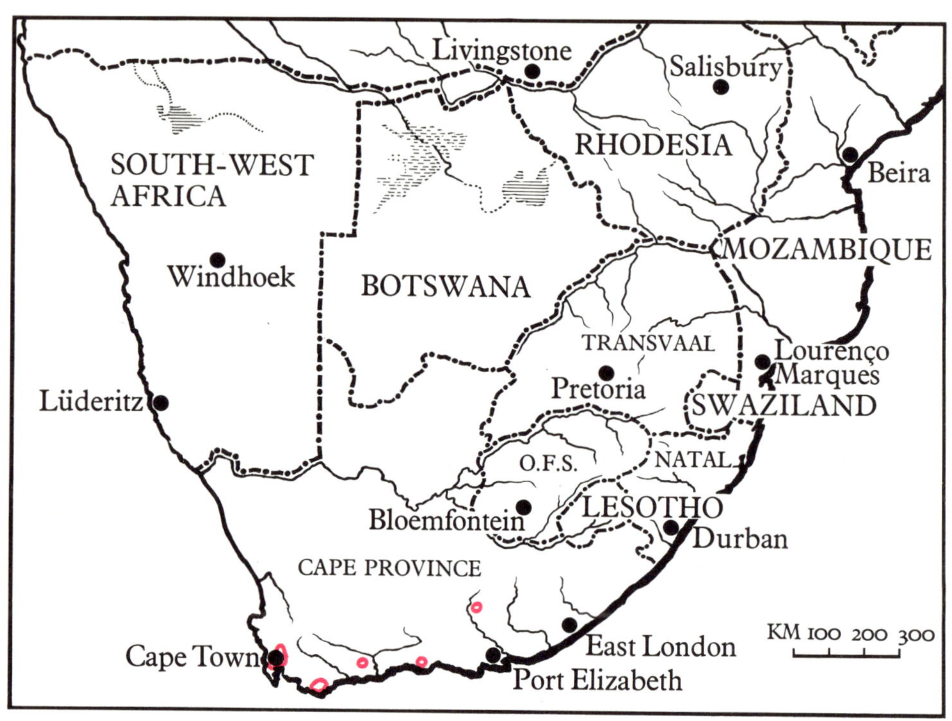

Conicosia

Conicosia N. E. Brown
From the Greek *konikos* = cone-shaped, from the form of the capsule.

N. E. Brown, Gard. Chron. 78: 433 (in key). 1925. - Phillips, Genera, 247. 1926. - Bolus, Notes, 10; pl. II, F. 1928. - Schwantes, M. Inst. Allg. Bot. Hamburg, 8: 164. 1929. - N. E. Brown, Gard. Chron. 90: 13; fig. 138. 1931. - v. Pöllnitz, Aufteilung, 32. 1933. - Jacobsen, Sukk. 104; fig. 97. 1933. - Pax, Natürl. Pfl. 212. 1934. - Jacobsen, Succ. Pl. 143; fig. 122. 1935; Verzeichnis, 40. 1938. - Goossens, Blomplante, 143 (in key). 1940. - Jacobsen, Herre, Volk, Mesembr. 53 (in key), 89. 1950. - Adamson, Salter, Fl. Cape Pen. 376. 1950. - *Phillips, Genera, 300, 1951.* - Jacobsen, Handbuch, 1234; fig. 1045. 1955. - Schwantes, Fl. Stones, 59, 341; fig. 15 (capsule); pl. 13a. 1957. - Jacobsen, Handbook, 954 (in system of Schwantes), 956 (in key of Bolus), 965 (in key of Herre & Volk), 1034; fig. 1248, 1249. 1960; Lexikon, 380, t.152/2. 1970.

Perennials or in 1 or 2 species, only lasting 1 or 2 years, either with fibrous roots and a single main stem terminating in a perennial dense tuft of leaves with flowering branches produced from the basal part of the tuft, or with a long fleshy tuberous rootstock emitting from its top a central leaf-tuft and prostrate flowering branches or flowering branches only, with opposite or sometimes some alternate leaves scattered along them and a small tuft at the ends, deciduous after fruiting. Leaves in the large tufts crowded and appearing alternate but in reality opposite in their origin, slightly stem-clasping at the base, erect or ascending except when old, long and narrow, either sharply 3-angled or subterete, slightly channelled down the face, firm, light or dark dotted, those of one pair equal, to about 25 cm long. Flowers solitary, terminal, with long pedicels without bracts, yellow, often with an unpleasant odour, of various size, to about 8-13 cm in diam. Sepals 5, subequally or unequally, with broad bases (the 3 inner with membranous margins), subulate or terete at the tips. Petals very numerous, free, in several series, linear, often ciliate at the basal part. Stamens numerous, at first incurved around or over the stigmas, afterwards spreading and exposing the stigmas; filaments bearded at the base; staminodes numerous, filiform or hair-like, at first more or less incurved. Ovary inferior or partly superior, with a flattish top rising at the centre into a short cone; placentas on the outer walls of the chambers, ovules several in each chamber; stigmas 10-20, filiform, sessile, without a style. Capsule 10-20 locular, dry $\frac{1}{2}$ superior, with the conical top separating into 10-20 narrow valves or segments which stand more or less erect, not spreading when wetted, without expanding keels, each valve with a pair of thin submembranous wing-like loculus-partitions on the inner face; seeds rather large for the group, subglobose, or somewhat lenticular, often slightly keeled at the margin, not winged, smooth.

Species: 10, Namaqualand and the South-Western Cape to Graaff-Reinet. (Type: *C. pugioniformis* (L.) N. E. Brown.) Common name: varkwortel.

M.Page. XI. 1920.

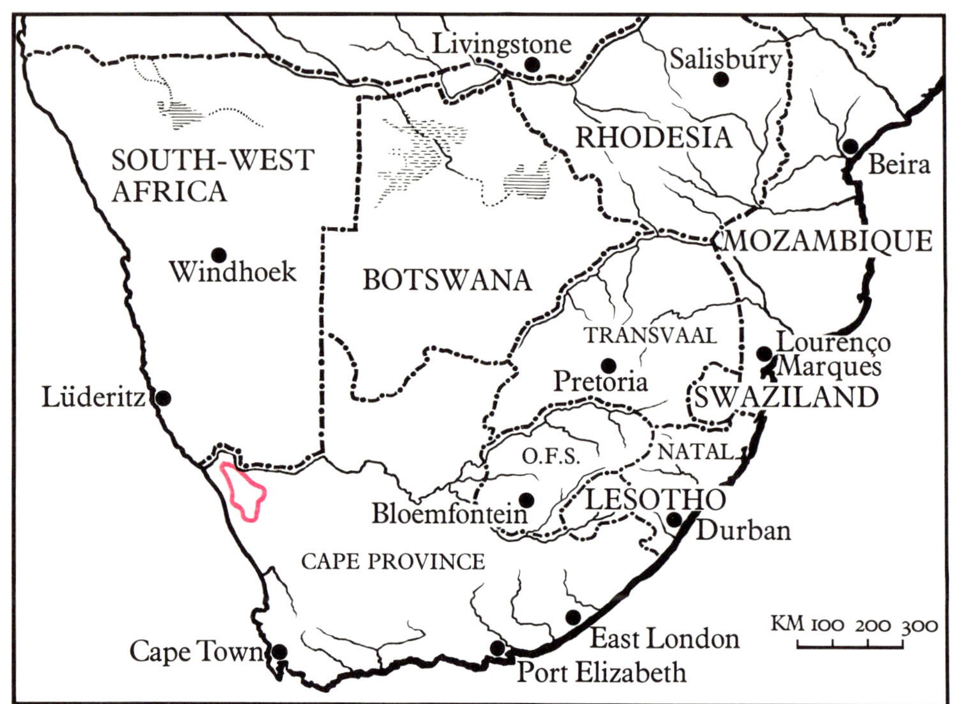

Conophyllum

Conophyllum Schwantes
From the Greek *konos* = cone and
phyllon = leaf.

Schwantes, Z. Sukk. 321. 1928. - v. Pöllnitz, Aufteilung, 32. 1933. - Jacobsen,
Sukk. 105; fig. 98. 1933; Succ. Pl. 144;
fig. 123. 1935; Verzeichnis, 41. 1938. -
Jacobsen, Herre, Volk, Mesembr. 51 (in
key), 89. 1950. - Phillips, Genera, 319.
1951. - Bolus, Notes, 237; pl. 89 (with distribution map); pl. 61-80. 1954. - Jacobsen, Handbuch, 1237; fig. 1046, 1047.
1955. - *Schwantes, Fl. Stones, 115, 341;
pl. 34 b, 35, 36. 1957.* - Jacobsen, Handbook, 953 (in system of Schwantes),
956 (in key of Bolus), 965 (in key of
Herre & Volk), 1036; fig. 1250, 1251.
1960.

Small, middle-sized and high shrubs
(12-70 cm), often with thick stems sometimes articulated; each branchlet
developing yearly only 2 pairs of leaves.
Leaves: 1st pair only connate at the
base, upper side flat, lower side keeled or
rounded, about 5-6 cm long to about
2 cm broad and 1 cm thick, papillose;
2nd pair of leaves nearly completely or
to one third of its length connate to a
conicolor terete body, only one part of the
apex free, about 3-5 cm long and 1 cm
thick. During the resting period another
body like the 1st pair of leaves will be
formed into this body, which sucks the
older one out, so that only a dried,
membranous sheath will be left, which
protects the young growth during summer and falls off after the young pair
of leaves have grown out; 3rd pair
of leaves is the last one of a flowering
branch separated from the "cone-leaf"
by a more or less elongated internode.
Flowers at first solitary, later on in cymes,
terminal, pedicillated, with and without
bracts, sometimes bearing connate
leaves in its axils, white, yellow to light
violet-red, to about 12 cm in diam. Sepals
5, nearly equal, sometimes 2 longer
than the others, free to the ovary. Petals
numerous, free, in about 3-4-series,
linear, stamens numerous, erect, column-
like collected. Ovary flat, with 5-7
sutures; glands annular; placentas parietal; stigmas 5-7, filiform or subulate.
Capsule 5-7-locular, very short, broad,
obconical; valve-wings as long as or
longer than the valves; expanding keels
very broad, diverging from the base,
with incurved points, with membranous
wings; loculi-roofs stiff, without tubercle; seeds ovate to pear-shaped, tuberculated.

Species: 25, recorded from Richters-
veld, Kleinzee, Kommaggas, Steinkopf,
Cape Province. (Type: *C. marlothianum*
Schwant.)

Conophyllum dissitum 119

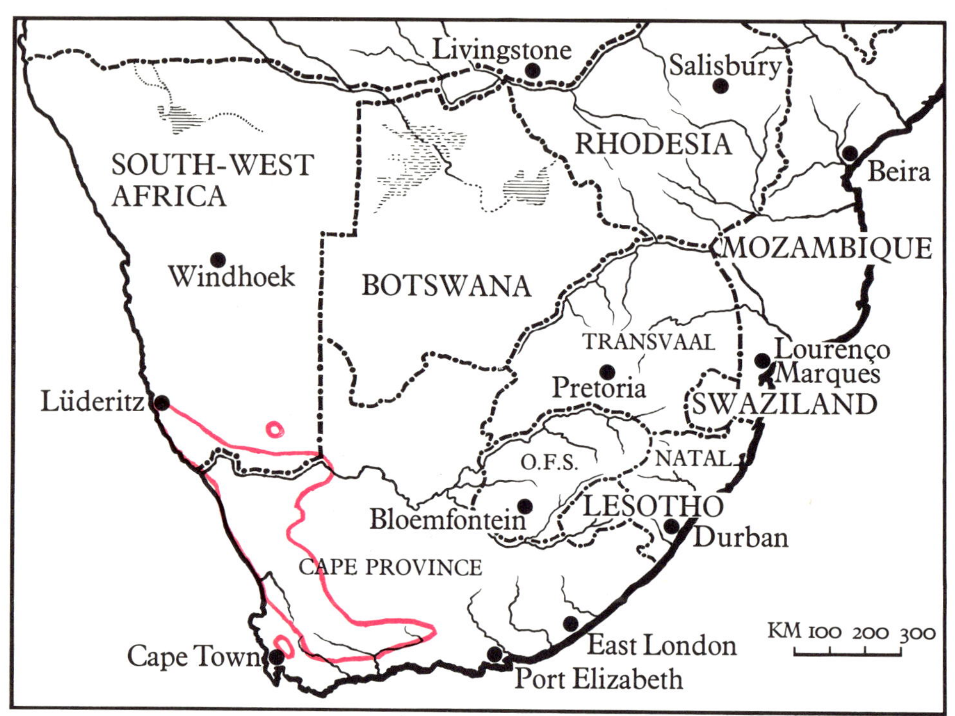

Conophytum

Conophytum N. E. Brown
From the Greek *konos* = cone and *phyton* = plant.

N. E. Brown, Gard. Chron. 71: 198 with fig. 1922; 72: 8 with fig. 1922; 78: 413 (in key), 450 with fig. 1925. - Phillips, Genera, 241. 1926. - Schwantes, Z. Sukk. 2: 137. 1925. - Tischer, Z. Sukk. 222. 1928; Succulenta, 10: 198. 1928. - Bolus, Notes, 17. pl. 1-12 (coloured). 1928. - Labarre (red.), Mesembr. (73), 137 with fig. 1931. - v. Pöllnitz, Aufteilung, 32. 1933. - Jacobsen, Sukk. 107. fig. 99-109. 1933. - Pax, Natürl. Pfl. 213. 1934. - Jacobsen, Succ. Pl. 146; fig. 124-134. 1935; Verzeichnis, 42. 1938. - Goossens, Blomplante, 147 (in key). 1940. - Jacobsen, Herre, Volk, Mesembr. 79 (in key), 89; fig. 24. 1950. - *Phillips, Genera, 300. 1951.* - Jacobsen, Handbuch, 1243; fig. 1049-1067. 1955. - Schwantes, Fl. Stones, 247; pl. 7 & 8 (coloured); pl. 68-87, fig. 32 (diagram of the development of the plant bodies); fig. 33 (forms of plant body of *C. halenbergense*). 1957. - Jacobsen, Handbook, 953 (in system of Schwantes), 959 (in key of Bolus), 975 (in key of Herre & Volk), 1041; fig. 1252-1311. 1960; Lexikon, 381, t.153/1, 3, 4, 5, 154/1, 2, 3, 4, 155/1, 2, 3, 4, 156/1-6, 157/1-6. 1970. - Friedrich in Merxmüller, Prodromus, 27. 1970.

Very small succulent plants, stemless or developing stems with age, with several or numerous growths in a clump; each growth formed of 2 leaves fused into one fleshy body, globose, obconical, ovoid, subcylindric, or oblong in shape, convex, flat, depressed, notched or two-lobed at the top, with a small orifice resembling a closed mouth at the centre of the top or between the lobes which with the biloba group may be wing or somewhat hatchet-like, or somewhat contorted or acuminate or long-acuminate, rarely hairy or papillate, mostly smooth, often dotted, with or without a window which may be small or dot-like, there may be drawings on the lobes or not. Flowers solitary, growing up from the interior of the growth through the central orifice, more or less pedicillated, with bracts exserted or not without them, all colours, open during day-time, some species also open during the night, up to 3 cm in diam. Calyx with a distinct slender membranous tube above the ovary. Sepals 4-7 at the top more or less included in or partly or entirely exserted from the body of the growth. Corolla with a distinct slender tube as long as or longer than the calyx-tube. Petals numerous or occasionally few, spreading or recurved, in 1 to several series, the inner series at the mouth of the tube sometimes much smaller than the others and differently coloured. Stamens few or many, erect, not collected into a column, included in the corolla-tube or partly exserted from it. Ovary flat or convex at the top, with a marginal crenulate glandular ring; style long or short (rarely almost absent); stigmas 4-7, filiform; placentas parietal. Capsule 4-7-locular, small, keels often ending in wings, with or without loculus-wings covering the seeds; seeds ovate, navelled, very small, smooth.

Described species about 290, recorded from South West Africa (12), Namaqualand (160), Karroo and Little Karroo (40) and Clanwilliam and Vanrhynsdorp districts. Some are described twice or thrice and more. In the course of time it will probably be the largest group of all. Dr A. Tischer at Heidelberg (Germany) has finished a monograph of this very difficult group. (Type: *C. minutum* (Haw.) N. E. Br.)

Conophytum bilobum

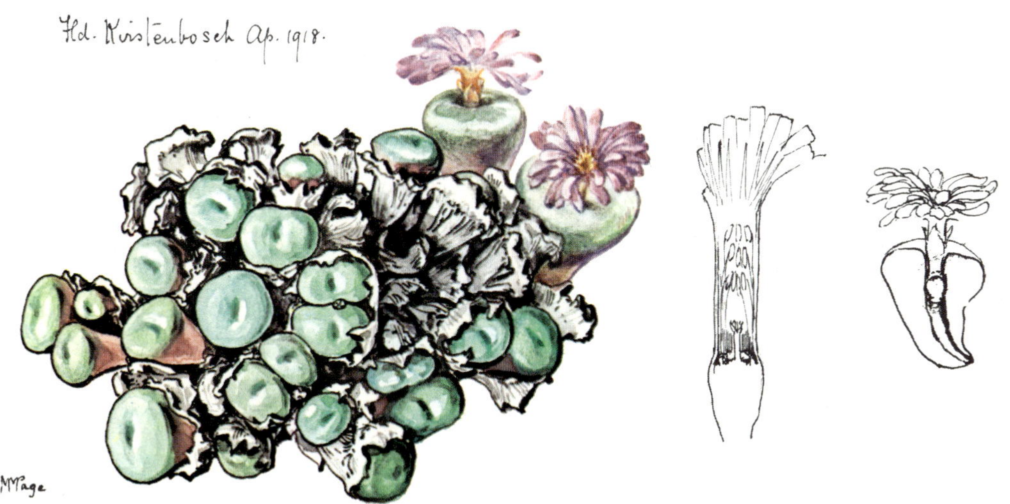

Conophytum minutum   121

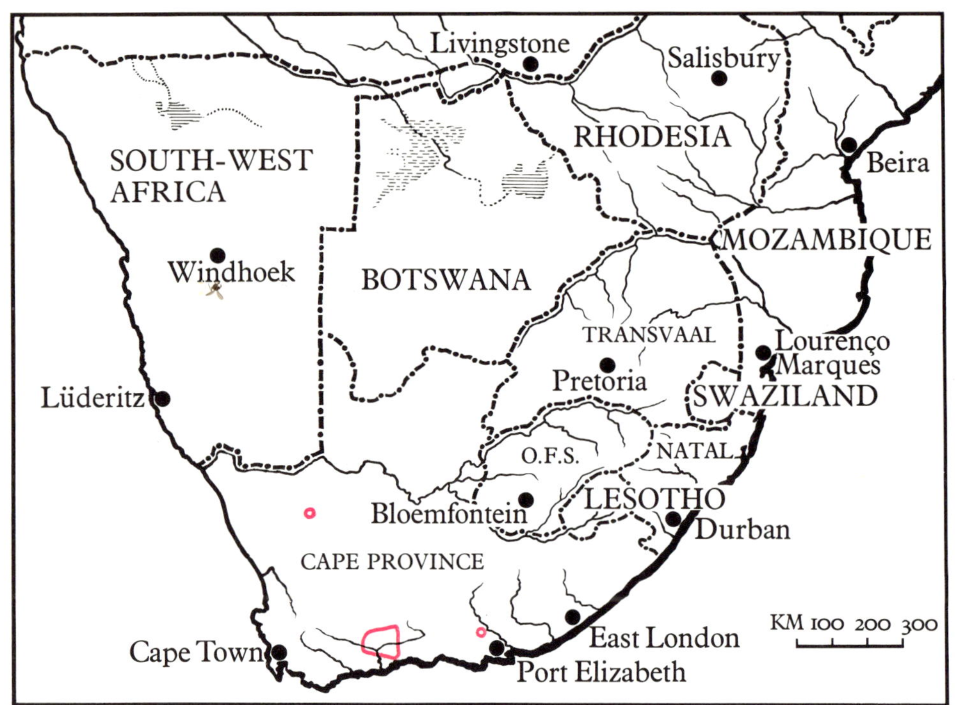

Cylindrophyllum

Cylindrophyllum Schwantes
From the Greek *kylindros* = cylinder
and *phyllon* = leaf.

Schwantes, Z. Sukk. 3: 15 (in key), 28. 1927/28. - v. Pöllnitz, Aufteilung, 36. 1933. - Jacobsen, Sukk. 119. 1933. - Pax, Natürl. Pfl. 220. 1934. - Jacobsen, Succ. Pl. 161. 1935; Verzeichnis, 61. 1938. - Jacobsen, Herre, Volk, Mesembr. 67 (in key), 93. 1950. - Phillips, Genera, 319. 1951. - Jacobsen, Handbuch, 1302; fig. 1088. 1955. - Schwantes, Fl. Stones, 104. 338. 1957. - Jacobsen, Handbook, 952 (in system of Schwantes), 962 (in key of Bolus), 971 (in key of Herre & Volk), 1091; fig. 1312. 1960; Lexikon, 402. t.158/1. 1970.

Tufted succulent perennials with short branchlets, leaves crowded, only about 5 cm high. Leaves opposite, connate at the base, cylindrical or 3-angled with much rounded edges to terete, obtuse, grey-green, finely dotted, crowded, about 7 cm long and 0,8 cm in diam. Flowers solitary, terminal, short stalked, without bracts i.e. there is no difference with the leaves, large, yellow or red, open at noon, about 7 cm in diam. Sepals 5, unequal, 3 lobes larger and keeled. Petals in several series, short. Stamens in several series, together with the surrounding staminodes incurved, conically collected, orange. Ovary conically elongated; glands annular; placentas parietal; stigmas 5-8, rather short, subulate, conically collected. Capsule 5-8-locular, valve-wings narrow, long, awn-like pointed (with *C. tugwelliae* valve-wings absent); keels toothed, parallel; loculi-roofs present; tubercle small; seeds dark-brown, tesselate, without navel, round.

Species: 5 recorded from the Jansenville, Ladismith, Laingsburg and Prince Albert districts. (Type: *C. calamiforme* (L.) Schwant.)

Cylindrophyllum tugwelliae 123

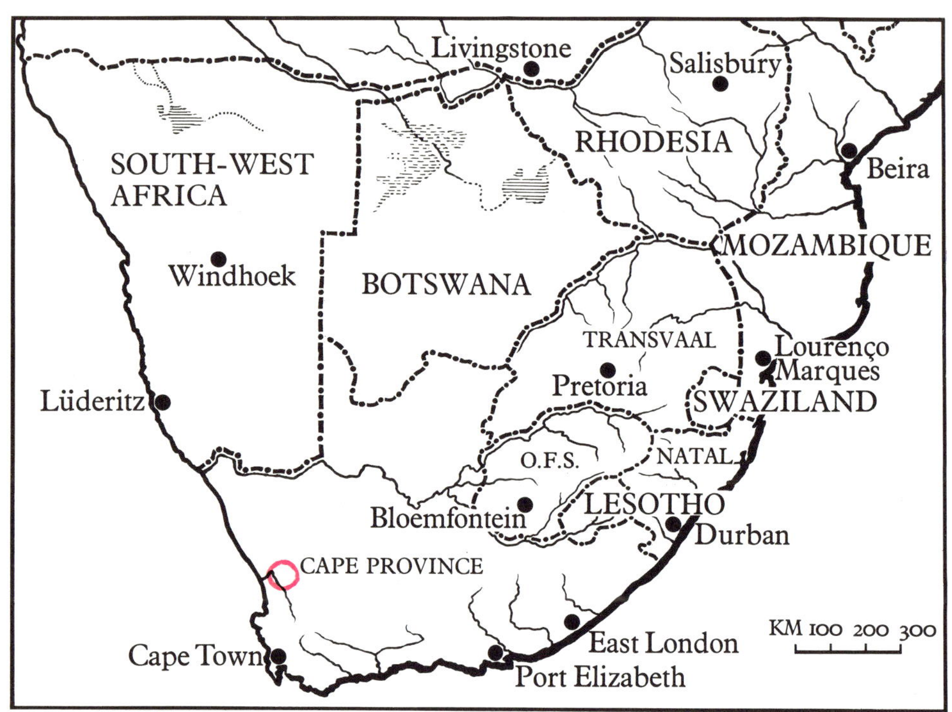

Dactylopsis

Dactylopsis N. E. Brown
From the Greek *dactylos* = finger and
opsis = like.

N. E. Brown, Gard. Chron. 78: 413 (in key). 1925. - Phillips, Genera, 244. 1926. - N. E. Brown, Gard. Chron. 83: 338 with fig. 1928. - Labarre (red.), Mesembr. 202; fig. 92, 93. 1931. - v. Pöllnitz, Aufteilung, 36. 1933. - Jacobsen, Sukk. 120; fig. 112. 1933; Succ. Pl. 161; fig. 137. 1935; Verzeichnis, 61. 1938. - Goossens, Blomplante, 142 (in key). 1940. - Jacobsen, Herre, Volk, Mesembr. 54 (in key), 93. 1950. - *Phillips, Genera, 301. 1951.* - Jacobsen, Handbuch, 1308; fig. 1089. 1955. - Schwantes, Fl. Stones, 25, 35; pl. 5a & b. 1957. - Jacobsen, Handbook, 951 (in system of Schwantes), 956 (in key of Bolus), 965 (in key of Herre & Volk), 1092; fig. 1313. 1960; Lexikon, 403, t.158/4. 1970.

Mesembr. digitiforme Thunb. Prodr. 88. 1775. - Haworth, Rev. 86. 1821. - De Candolle, Prodromus syst. 418. 1828. - Sonder, Fl. cap. 2; 405. 18. 1894.

Gibbaeum digitiforme (Thunb.) N. E. Brown, Fedde, Rep. 36: 15. 1934.

Mesembr. digitatum Ait. Hort. Kew. II: 181. 1810. - Berger, Mesembr. & Portul. 229; fig. 47. 1908. - Marloth, Trans. Roy. Soc. S. Afr. 2: 407. 1910/12. - Fl. Pl. S. Afr. 2: 405, 1922. 3: 102, 1923.

A dwarf succulent perennial, forming clumps, very pulpy. Leaves 2 or 3 to each growth, alternate, terete, very stout, with large tubular sheaths clasping one another closely and concealing the short stem, soft and pulpy, dotless, withering completely and disappearing when at rest; the white membranous rests include the young leaf and protect it; also during this period there are still flowers or remains of it. Flowers terminal, solitary, sessile, small, about 2 cm in diam., white, open for about 3 weeks, day and night. Calyx produced above its union with the ovary into a short tube. Sepals 5, equal. Petals numerous, in several series. The petaloid staminodes are, in contrast to all the other genera, dry and stiff. The inner staminodes concealing the stamens, linear, stiff, united into a tube at the base and arising in the angle where the calyx unites with the ovary. Stamens numerous, erect, 4-seriate, arising from the tube of the corolla and not exserted from it. Ovary $\frac{1}{2}$ superior; placentas axil; ovules numerous in each chamber. Capsule 5-locular, $\frac{1}{2}$ superior, subglobose, with the 5 sutures raised into grooved ridges on the top; valves deltoid, with the expanding keels not contiguous at the lower part but distant from each other and these forming the loculus-partitions, more or less diverging at the apical part and there bearing broad membranous marginal wings or flaps that are united in pairs between each pair of valves; loculi open, without loculus-wings or tubercles, central axis ending in 5 short hard rays arranged in a small star; seeds very small, many in each loculus, compressed, ellipsoid, somewhat pointed at one end, smooth.

There are 2 species recorded from the Knersvlakte near Vanrhynsdorp. (Type: *D. digitata* (Ait.) N. E. Br.)

Bet. von Rhynsdorp. and
Clanwilliam. digitata. ⅔⅞ Frames. 17-7-28

Clanwilliam. [illegible].
19-2-1929.

Dactylopsis digitata   125

Delosperma

Delosperma N. E. Brown
From the Greek *delos* = visible and
sperma = seed.

N. E. Brown, Gard. Chron. 78: 412 (in key), 433 (in key). 1925. - N. E. Brown, Burtt-Davy, Fl. Pl. Tvl. & Swazil. 1: 157. 1926. - Schwantes, Möllers Gärtnerztg. 42: 234, 245, 258. 1927. - Bolus, Notes, 105; fig. 29. 1928. - v. Pöllnitz, Aufteilung, 36. 1933. - Jacobsen, Sukk. 120; fig. 113. 1933. - Pax, Natürl. Pfl. 214. 1934. - Jacobsen, Succ. Pl. 163; fig. 138, 139. 1935; Verzeichnis, 62. 1938. - Goossens, Blomplante, 144 (in key). 1940. - Jacobsen, Herre, Volk, Mesembr. 62 (in key), 93. 1950. - Adamson, Salter, Fl. Cape Pen. 379. 1950. - Phillips, Genera, 320. 1951. - Jacobsen, Handbuch, 1304; 1090-1095. 1955. - Schwantes, Fl. Stones, 83, 338; pl. 21a & b. 1957. - L. F. Willems, Morfologiese en Taksonomiese Studie van Enkele Delosperma Spesies. M.Sc. thesis, Univ. Stellenbosch. 1957. - Jacobsen, Handbook, 952 (in system of Schwantes), 956 (in key of Bolus), 969 (in key of Herre & Volk), 1093; fig. 1314-1319. 1960. - *Bolus, J. S. Afr. Bot. 32: 209 (with naming of 5 sub-groups). 1966.* - Jacobsen, Lexikon, 403, t.158/2 & 3, 159/1-4. 1970. - Friedrich in Merxmüller, Prodromus, 34. 1970.
Ectotropis N. E. Brown, Gard. Chron.

81: 12. 1927. - Bolus, Ann. Bol. Herb. 98. 1927. - v. Pöllnitz, Aufteilung, 41. 1933. - Pax, Natürl. Pfl. 214. 1934. - Jacobsen, Verzeichnis, 78. 1938. - Jacobsen, Herre, Volk, Mesembr. 61 (in key), 97. 1950. - Phillips, Genera, 303. 1951. - Jacobsen, Handbuch, 1355. 1955. - Schwantes, Fl. Stones, 332, 342. 1957. - Jacobsen, Handbook, 957 (in key of Bolus), 968 (in key of Herre & Volk); 1136. 1960; Lexikon, 420. 1970.

Erect or decumbent or prostrate creeping shrubs, as also herbs, rootstock sometimes tuberous. Leaves opposite, sessile, slightly or not connate, more or less finely papillate, often hook-like bent at the apex, broadly triangulate to cylindrical or ovate or sometimes flat and soft, sometimes 3 or more times as long as broad, erect or dilated, to 7 cm long and 2,5 cm broad. Flowers solitary or with few to many flowered cymes, pedicels long or short, bracts distinct or leaf-like, often 2,5-8 cm in diam., all colours. Sepals 4-8, unequal, often equal the longer ones sometimes horn- or tail-like. Petals few series, linear, obtuse or emarginate. Staminodes often present. Stamens erect, sometimes conical-collected, whitish, sometimes hairy to the base, mostly surrounded by staminodes. Ovary often slightly convex; glands separated, partially crenulated, often obscure, rarely annular, sometimes connate with each other by false glands; placentas parietal; stigmas 5, subulate or caudate, acute, papillate. Capsule 5-locular, keels mostly with membranous marginal wings; loculi-roofs mostly absent, rarely, reduced to a limb; without tubercle; seeds roundish, pale brown, sometimes areolate, 0,05-0,15 cm in diam.

Species: 140, recorded from South West Africa, Cape Province, Natal, Transvaal, Rhodesia. East Africa: Massai-Highlands; Abyssinia, Arabia. (Type: *D. echinatum* (Ait.) Schwantes.)

Delosperma tradescantioides   127

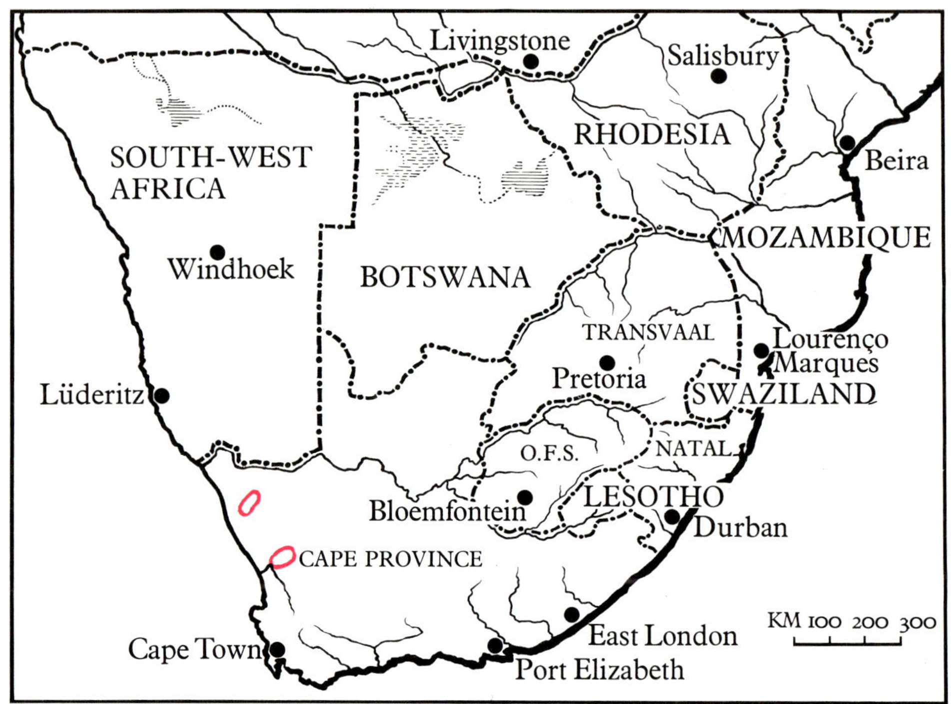

Dicrocaulon

Dicrocaulon N. E. Brown
From the Greek *dikros* = fork and *kaulon* = stem, stalk.

N. E. Brown, J. Bot. 66: 141. 1928; Gard. Chron. 87: 71 (in key). 1930. - v. Pöllnitz, Aufteilung, 37. 1933. Pax, Natürl. Pfl. 214. 1934. - Jacobsen, Verzeichnis, 68. 1938. - Goossens, Blomplante, 143 (in key). 1940. -'Jacobsen, Herre, Volk, Mesembr. 64 (in key), 95. 1950. - Phillips, Genera, 302. 1951. - Jacobsen, Handbuch, 1326. 1955; Handbook, 952 (in system of Schwantes), 960 (in key of Bolus), 970 (in key of Herre & Volk); 1109, fig. 1320, 1321. 1960; Lexikon, 410, t. 160/1. 1970.

Low shrubs with very thin forked branchlets with two kinds of leaves: 1st pair of leaves, small, more or less elliptic or roundish, with or without small free points; 2nd pair of leaves oblong-terete, only connate at the base, here also inflated, like all the leaves papillate. Flowers terminal, solitary, white. Sepals 4-5, unequal. Petals very numerous, free, filiform, with the inner ones merging into staminodes. Stamens erect. Ovary inferior; placentas on the walls of the chambers; stigmas 4-7, minute, subulate, acute. Capsule inferior, 4-7-locular, valves prominently keeled, contiguous or parallel, with broad membranous margins; loculus-wings membranous; tubercle absent; seeds light-brown, globose, with navel, shiny.

Species: 6, recorded from the Cape Province: Namaqualand, Kommaggas (Riethuis), Vanrhynsdorp (Knersvlakte), Piketberg. (Type: *D. pearsonii* N. E. Br.)

24-2-1928
Frames.
198
28
North of
Van Rhyns-
dorp
B.O.Carter
bud
Plan. of bud

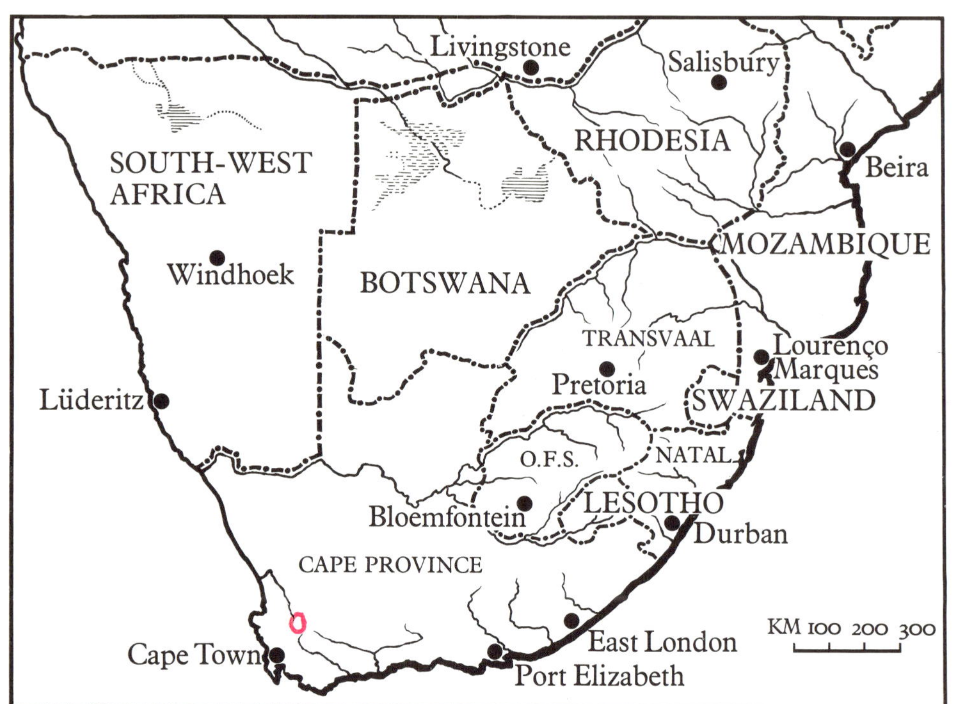

Didymaotus

Didymaotus N. E. Brown
From the Greek *didymos* = double and
aotus = flower.

N. E. Brown, Gard. Chron. 78: 433 (in key) 1925; *80: 149 with fig. 1926. - Bolus, Notes, pl. 16 (coloured) 1928.* - Labarre (red.), Mesembr. 204, fig. 94, 94a. 1931. - v. Pöllnitz, Aufteilung, 38. 1933. - Jacobsen, Sukk. 122; fig. 115. 1933. - Pax, Natürl. Pfl. 214. 1934. - Jacobsen, Succ. Pl. 165; fig. 410. 1935; Verzeichnis, 68. 1938. - Goossens, Blomplante, 146 (in key), 1940. - Jacobsen, Herre, Volk, Mesembr. 79 (in key), 95. 1950. - Phillips, Genera, 320. 1951. - Jacobsen, Handbuch, 1327; fig. 1097. 1955. - Schwantes, Fl. Stones, 170, 340. 1957-. Jacobsen, Handbook, 953 (in system of Schwantes), 960 (in key of Bolus), 975 (in key of Herre & Volk), 1109; fig. 1322. 1960; Lexikon, 410, t.160/2. 1970.

Perennial forming clumps with mostly 2 and very rarely with several (up to 12) pairs of leaves. Leaves opposite, connate and forming a body, thick, ovate, broader than long, to about double as long as broad, roughly dotted, in the sun red of colour, thus mimicking very well the ironstones of its surroundings, upper side 3-angled, lower side keeled and the keel chin-like produced, bodies to 2,5 cm long. Flowers axillary from a sprout which develops the same time as the oldest pairs of leaves are formed and consists only of 2 bracts each side of the body; the more the main leaves wither, the more the bracts develop till flowering one day; in meantime the next pair of leaves already develop, together with their bracts. Flowers whitish to light or dark violet-red, centre always darker in colour, about 4 cm in diam. Sepals 6, rounded, keeled at the back, 2 larger than the others, pairs apparently equal. Petals in several series, free at the base, more or less pink, towards the points darker. Stamens erect, hairy at the base, filaments violet, anthers yellow. Ovary in the middle convex; glands distinct, annular, granulated, dark brownish-green; placentas parietal; stigmas 6, shorter than the stamens, free, forming a fine point, inside somewhat plumose, subulate, stouter at the base, 1-1,5 cm long. Capsule 6-locular, with loculi-roofs, tubercle absent; seeds bulb-shaped, light-brown, with a navel.

Monotypic genus, Ceres Karroo, along the road to Calvinia. (Type: *D. lapidiformis* (Marloth) N. E. Br.)

Didymaotus lapidiformis 131

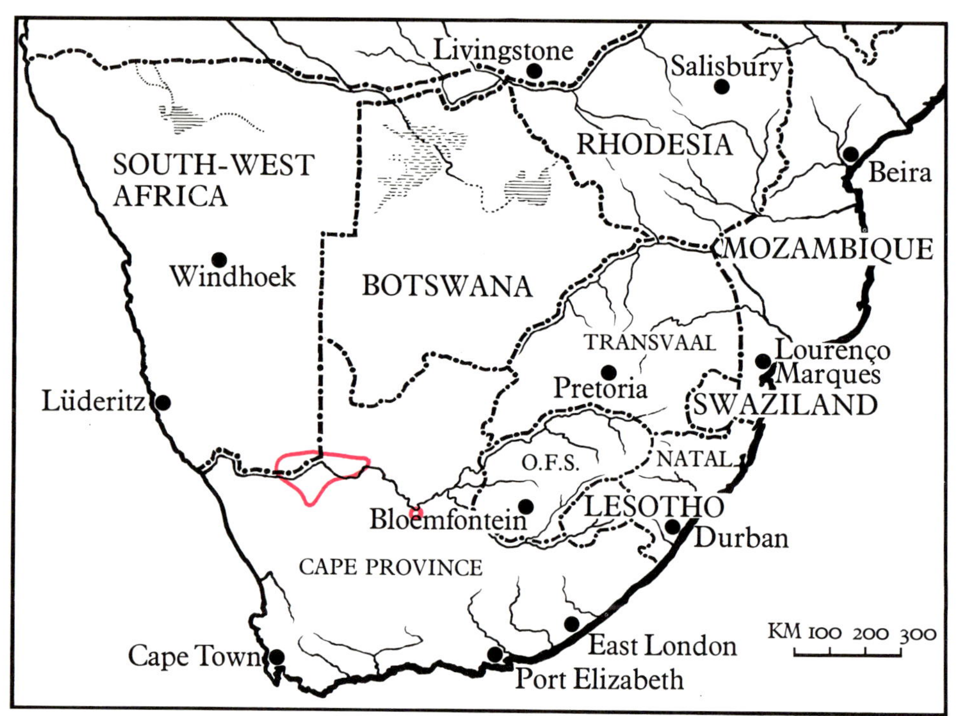

Dinteranthus

Dinteranthus Schwantes
Named in honour of the late Prof.
K. Dinter, and the Greek *anthos* = flower.

Schwantes, Z. Sukk. 2: 184, 264; fig. 266, 267. 1926; Möllers Gärtnerztg. 42: 233. 1927. - N. E. Brown, Gard. Chron. 82: 85. 1927; 84: 472. 1928. - Labarre (red.), Mesembr. 206; fig. 8, 96, 97. 1931. - v. Pöllnitz, Aufteilung, 38. 1933. - Jacobsen, Sukk. 122; fig. 116, 117. 1933. - Pax, Natürl. Pfl. 214. 1934. - Jacobsen, Succ. Pl. 165; fig. 141-143. 1935. - Bolus, Notes, 71, 75 (key of species); pl. 10, 11, 12. 1937. - Jacobsen, Verzeichnis, 69. 1938. - Jacobsen, Herre, Volk, Mesembr. 77, 78 (in key), 95; fig. 13. 1950. - *Phillips, Genera, 302. 1951.* - Jacobsen, Handbuch, 1328; fig. 1098-1102. 1955. - Schwantes, Fl. Stones, 178, 340; pl. 53b. 1957. - Jacobsen, Handbook, 953 (in system of Schwantes), 958 (in key of Bolus), 975 (in key of Herre & Volk), 1110 (with the Bolus key of species); fig. 1323-1328. 1960; Lexikon, 410, t.160/3-6. 1970. - Friedrich in Merxmüller Prodromus, 35. 1970.

Stemless perennials, forming clumps; each growth with one pair (or when making new growths 2 pairs) of very thick fleshy leaves which are united at the base for one third to two thirds of their length and convex on the face, causing them to be united higher up at the middle part than at the edges, not dotted or with few small, dark-green dots, smooth or with a very minutely granulated surface, the pair of leaves about equal in size, about 3 cm long and 2 cm thick; one species with lithops-like leaves nearly as long as broad, forming a body, with a narrow fissure between the 2 leaves, dove-grey often yellowish or reddish coloured, not dotted, granulated rough (pocket lens!) about 4-5 cm in diam., without a window. Flower solitary, terminal, appearing sessile, pedicelled, bractless, pedicel usually pubescent, flat, acutely 2-edged, entirely concealed between the united part of the leaves, visible in the fruit; opening late afternoon, yellow to dark-yellow, about 3-5 cm in diam. Sepals (6)7-rarely 8-10, unequally, pubescent, some of the lobes having membranous edges. Petals numerous, free or somewhat connate at the base, spreading horizontally. Stamens numerous, erect, collected, exposed to view from their base, lower part incumbent on ovary, upper erect papillate near the base or glabrous; staminodes none. Ovary inferior, convex or rarely conically elevated; glands crenulate; placentas on the outer walls of the chambers; stigmas 7-9-11-15, rather slender, free, papillate at the apex or plumose. Capsule 6-7 more rarely up to 15-locular, shortly and broadly obconic, flattish with slightly raised sutures on top, pedicel winged; valves broadly winged, expanding-keels closely contiguous into a central keel with broad membranous margins; loculi-roofs reduced to a limb, very narrow or absent; without a tubercle; seeds very numerous in each loculus, extremely small, minutely rough, white or brownish.

Species: 6, recorded from the southern part of South West Africa; Cape Province: Prieska, Keimoes, Pella, Pofadder and Bushmanland. (Type: *D. microspermus* (Dinter et Derenberg) Schwantes.)

B.O.Carter.

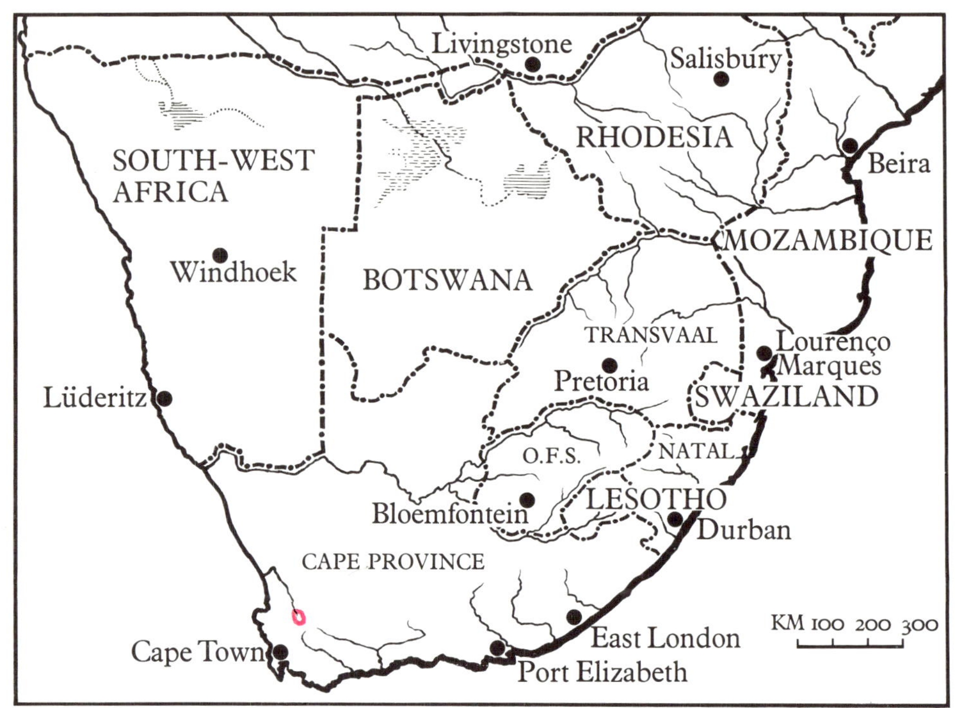

Diplosoma

Diplosoma Schwantes
From the Greek *diplos* = double and
soma = body.

Schwantes, Z. Sukk. 179. 1926. - N. E. Brown, Gard. Chron. 83: 251. 1928. - Bolus, Notes, 69; fig. 1. 1928. - v. Pöllnitz, Aufteilung, 38. 1933. - Pax, Natürl. Pfl. 214. 1934. - Jacobsen, Succ. Pl. 167. 1935; Verzeichnis, 69. 1938. - Goossens, Blomplante, 145 (in key). 1940. - Jacobsen, Herre, Volk, Mesembr. 72 (in key), 95; fig. 30 (capsule). 1950. - *Phillips, Genera, 303. 1951.* - Jacobsen, Handbuch, 1331; fig. 1103, 1104. 1955. - Schwantes, Fl. Stones, 122, 341; pl. 40c. 1957. - Jacobsen, Handbook, 954 (in system of Schwantes), 960 (in key of Bolus), 973 (in key of Herre & Volk), 1113; fig. 1189 (capsule), 1329, 1330. 1960; Lexikon, 411, t.161/1, 1970.

Small succulent perennials, forming clumps, stemless and deciduous, without evident rootstock but with numerous short fibrous roots from a dense cluster of sessile growths; each growth formed of 2 opposite leaves which are horizontally directed and shortly and very obliquely united at the base and along the basal part of one margin, with the free parts diverging, soft, pulpy, with some hyaline dots, withering completely away annually, about 2,5 cm long and 0,5 cm broad. Flowers solitary, terminal, sessile between the bases of the leaves; bracts none, violet-red to white, opened at noon, about 2,5 cm in diam. Sepals (5-6)-7, very unequal, roundish, dishlike exserted above the ovary. Petals 2-3-seriate, numerous, linear. Stamens numerous, erect, hairy to the base, yellowish-white, surrounded by staminodes. Ovary inferior, flat or pear-shaped at the top; glands annular, dark-green, slightly crenulate; placentas parietal; stigmas (mostly) erect, filiform. Capsule 6-7-locular, very flat; expanding keels very broad, dark-brown, either ending in an awn or in a wing; loculi-roofs present; without tubercle; seeds pear-shaped, light-brown, small.

Species: 2, recorded from the Piketberg district, in limestone. (Type: *D. retroversum* (Kensit) Schwantes.)

Disphyma

Disphyma N. E. Brown
From the Greek *dis* = two, double, and
phyma = tubercle.

N. E. Brown, Gard. Chron. 78: 433 (in key). 1925. - *Bolus, Fl. Pl. S. Afr. 7: pl. 276. 1927;* Notes, 96, 123. 1928. - v. Pöllnitz, Aufteilung, 39. 1933. - Jacobsen, Sukk. 123. 1933. - Pax, Natürl. Pfl. 214. 1934. - Jacobsen, Succ. Pl. 167. 1935; Verzeichnis, 70. 1938. - Goossens, Blomplante, 145 (in key). 1940. - Jacobsen, Herre, Volk, Mesembr. 63 (in key), 95. 1950. - *Adamson, Salter, Fl. Cape Pen. 389. 1950.* - Phillips, Genera, 320. 1951. - Jacobsen, Handbuch, 1322. 1955. - Schwantes, Fl. Stones, 338. 1957. - Jacobsen, Handbook, 952 (in system of Schwantes), 961 (in key of Bolus), 969 (in key of Herre & Volk), 1114; fig. 1331. 1960; Lexikon, 411, t.161/2. 1970.

Perennial shrubs, prostrate, creeping and rooting at the nodes, internodes 2-3 cm long; in each of the axils short branchlets with many leaves. Leaves along the main-stem and on top of the branchlets in rosettes, opposite, about all of the same size, terete to 3-angled, erect, ascending or spreading, firm, hyaline dotted, especially against translucent light, 3- to more times as long as broad, bright green, later reddish or yellowish, to about 3 cm long and 0,6 cm in diam. Flowers terminal, solitary, or 2-3, pedicel to about 3 cm long without bracts, open at noon, white to violet-red, various in size, about 4 cm in diam. Sepals 5, often unequal, 3 lobes sometimes with membranous margin, sometimes very narrow, fairly acute. Petals numerous, about 2-series, in the inner one very few, free, linear, obtuse, entire, at the base often narrow, stamens numerous, erect, anteriors after a short time reflexed, posteriors papillate at the base, anthers and pollen light-yellow, staminodes none. Ovary convex, glands deeply crenulate; placentas parietal; stigmas 5, subulate, ending in a long awn, densely plumose, to about 0,4 cm long. Capsule 5-locular, light yellow, somewhat sponge-like, when open valves very reflexed, keels from the beginning diverging, with a few teeth, with large, somewhat oblong wings, as long as the valves, loculi-roofs present, tubercle pale, bifid (*D. australe* sometimes without tubercle), seeds nearly ovate, nearly smooth.

Species: 3, Cape Province: from Cape Town to the Swartkopsriver on sandy, slightly brackish soil, sometimes together with Salicornia; West and South Australia, Victoria, New South Wales, Tasmania and New Zealand along the coast (*D. australe*). (Type: *D. crassifolium* (L.) L. Bolus.)

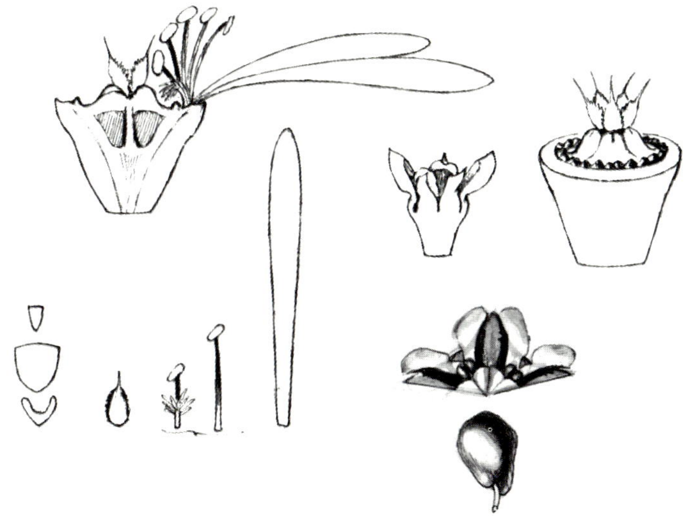

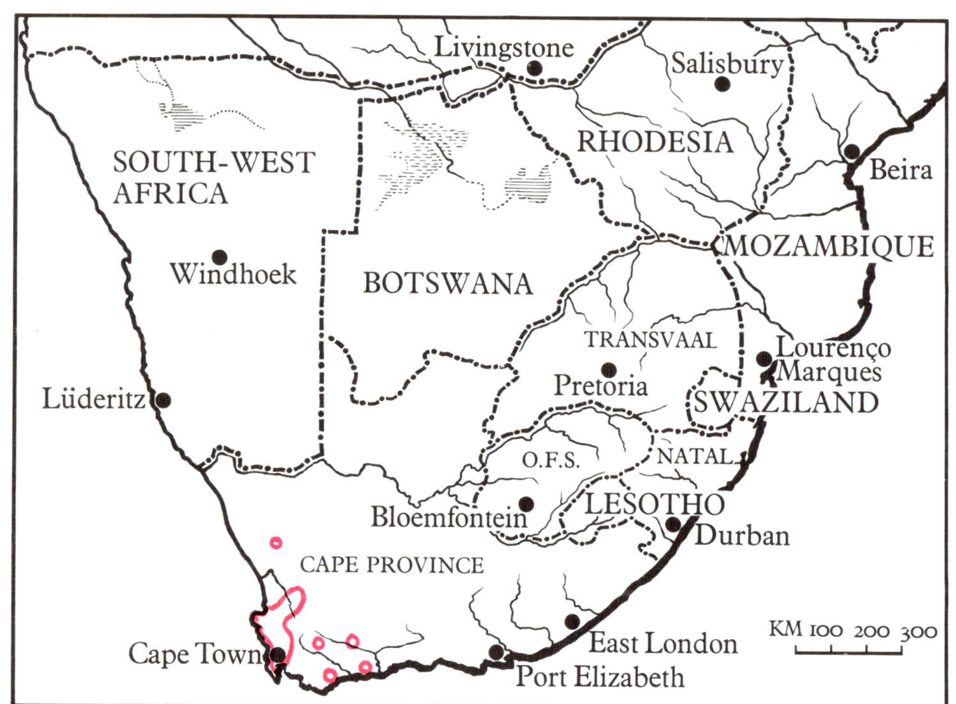

Dorotheanthus

Dorotheanthus Schwantes
Named in honour of Mrs Dorothea
Schwantes, mother of Prof. G.
Schwantes.

Schwantes, Möllers Gärtnerztg. 42: 283. 1927. - N. E. Brown, Möllers Gärtnerztg. 43: 400. 1928. - N. E. Brown, Kew Bull. 56. 1929. - v. Pöllnitz, Aufteilung, 39. 1933. - Jacobsen, Sukk. 124; fig. 118. 1933. - Pax, Natürl. Pfl. 214. 1934. - Jacobsen, Succ. Pl. 168; fig. 145. 1935; Verzeichnis, 70. 1938. - Goossens, Blomplante, 149 (in key). 1940. - Jacobsen, Herre, Volk, Mesembr. 59 (in key), 95; fig. 17. 1950. - *Adamson, Salter. Fl. Cape Pen. 377. 1950.* - Phillips, Genera, 320. 1951. - Jacobsen, Handbuch, 1333; fig. 1105, 1106. 1955. - Schwantes, Fl. Stones, 44, 340; pl. 9b; fig. 7 (capsule), 8 (flower), 9 (capsule). 1957. - Jacobsen, Handbook, 953 (in system of Schwantes), 959 (in key of Bolus), 968 (in key of Herre & Volk), 1114 (with key of species); fig. 1332, 1333, 1334. 1960. Lexikon, 411, t.161/3 & 4, 1970.

Cleretum N. E. Brown. Gard. Chron. 78: 412 (in key). 1925; 82: 228. 1927.

Stigmatocarpum L. Bolus. S. Afr., Gard. & Country Life, 17: 327. 1927; Notes, 67, 89; fig. 1B, fig. 16 C. 1928.

Annual herbs, about 10 cm high, with short branchlets, decumbent, sometimes red in colour, papillate. Leaves forming rosettes, later on opposite or alternate, sometimes connate at the base, very narrow to the base, petiole-like, entire, soft, narrow linear or lanceolate-spathulate, obtuse, about 8 cm long and 0,8 cm broad. Flowers 5-partite, terminal or axillary, pedicels about 6 cm long, without bracts, white with red points or reddish with darker centre, about 4 cm in diam. Sepals 5, sometimes leaf-like and longer than the petals, with pendulous, tap-like crystal clear papillae; cylindrical, unequal. Petals 2-seriate, free, narrow lanceolate. Stamens brown or reddish, the inner ones knee-jointed above the ovary, sometimes hairy at the base. Ovary flat, concave in the midst without ribs and tubercles; glands annular, branched; placentas parietal; stigmas 5, thin, subulate, papillate along the inside, persistent on the fruit. Capsule 5-locular, flat with a stout column in the midst; expanding keels flat, 3-angular, diverging from the base, with narrow wings ending in points; loculi-roofs present; tubercle absent; seeds pear-shaped to globose, smooth.

Species: 10, South-Western Cape and Namaqualand. (Type: *D. gramineus* (Haw.) Schwant.)

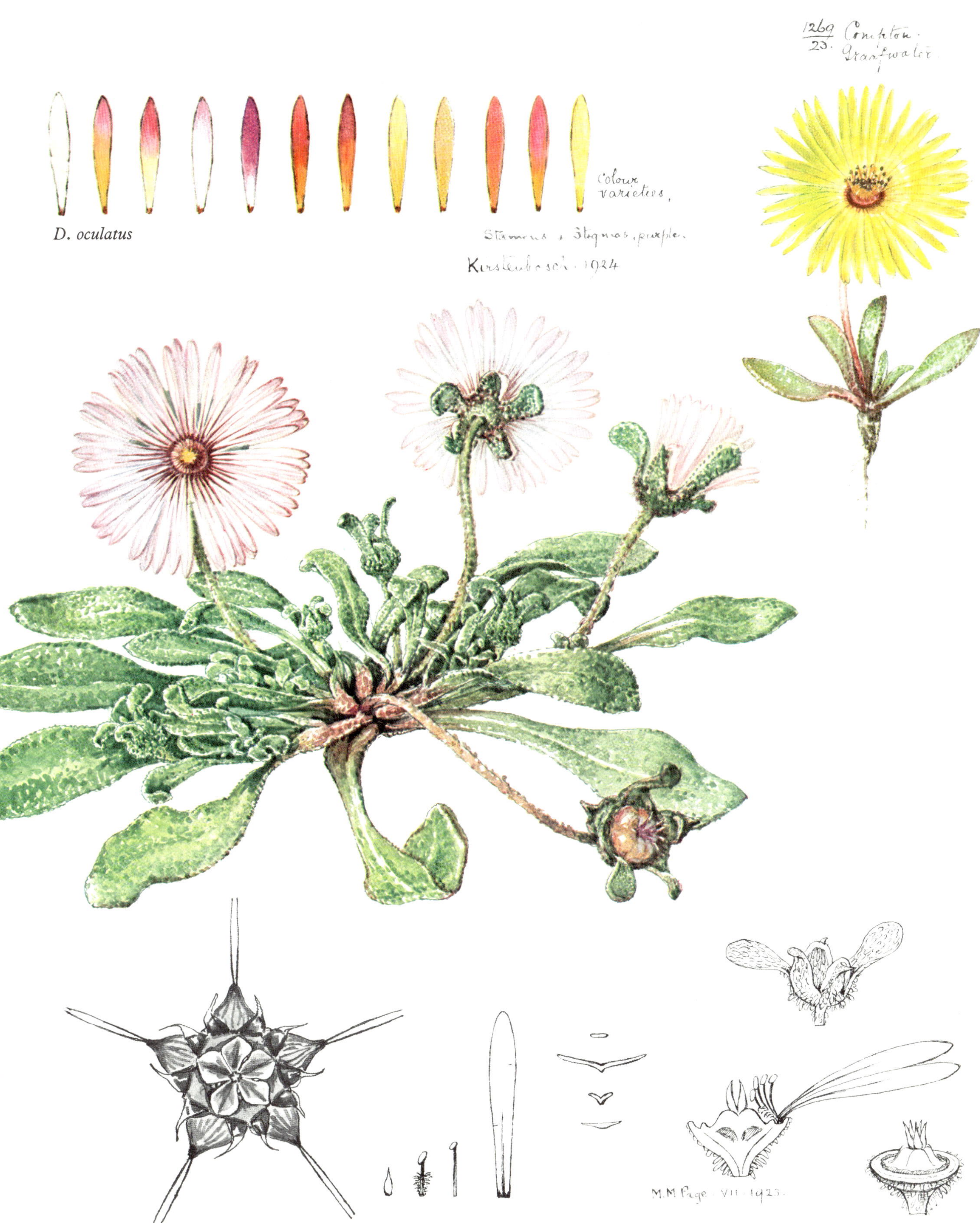

Dorotheanthus bellidiformis 139

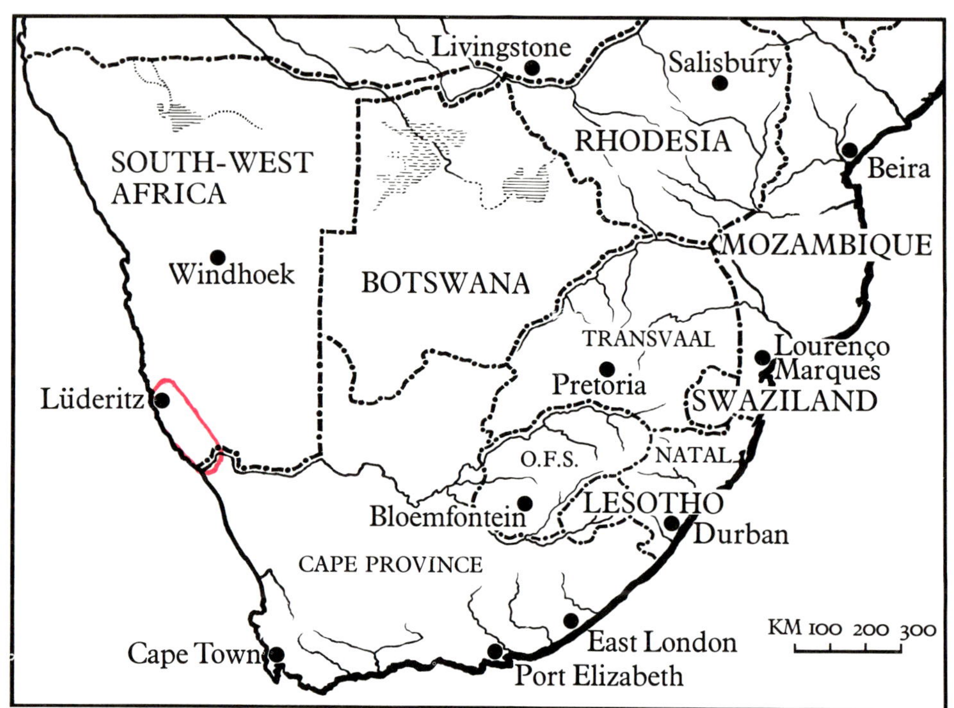

Dracophilus

Dracophilus Schwantes
From the Greek *draco* = dragon and *philos* = friend. (One species grows on the Drachenfels near Pomona, South-West Africa.)

Schwantes, Möllers Gärtnerztg. 42: 187. 1927. - Schwantes, Z. Sukk. 2: 183 (sub-genus). 1926. - Labarre (red.), Mesembr. 209; fig. 98, 99. 1931. - v. Pöllnitz, Aufteilung, 39. 1933. - *Bolus, Notes, 182 (with key of species). 1939.* - Jacobsen, Herre, Volk, Mesembr. 77 (in key), 95. 1950. - Phillips, Genera, 320. 1951. - Jacobsen, Handbuch, 1335; fig. 1107. 1955. - Schwantes, Fl. Stones, 136. 399; pl. 43b. 1957. Jacobsen, Handbook, 953 (in system of Schwantes), 960 (in key of Bolus), 974 (in key of Herre & Volk), 1118 (with key of species); fig. 1335. 1960; Lexikon, 413, t.161/5. 1970. - Friedrich in Merxmüller, Prodromus, 62. (Juttadinteria) 1970.

Perennial, small succulents forming clumps. Leaves opposite, 3-angled, margins usually with distant teeth or projections, very fleshy and soft, thick, often keeled, young leaves dotted, older ones not dotted, fine, rough, whitish to bluish or grey-green, about 5 cm long, 1,5 cm broad and thick. Flowers terminal, solitary, pedicels about 2,5 cm long with 2 bracts, jointed high above, open in the afternoon, white-violet, about 3 cm in diam. Sepals 5, nearly equal, obtuse. Petals 2-3-seriate, free, nearly as long as the sepals, with reflexed points, 7-12 mm long. Stamens conically collected, anthers yellow, filaments white, hairy, at the base; staminodes none. Ovary more or less concave, centre abruptly raised; glands annular, crenulated; placentas parietal; stigmas 8-14, shorter, equal or slightly larger than the stamens, subulate, recurved, yellowish-green. Capsule 8-14-locular, expanding keels diverging, toothed, ending in an awn, without wings, with well developed loculi-roofs; seeds pearl-like, verrucose, with navel, brown.

Species: 4, Lüderitz, S.W.A. and Richtersveld. (Type: *D. delaetianus* (Dint.) Dtr. et Schw.)

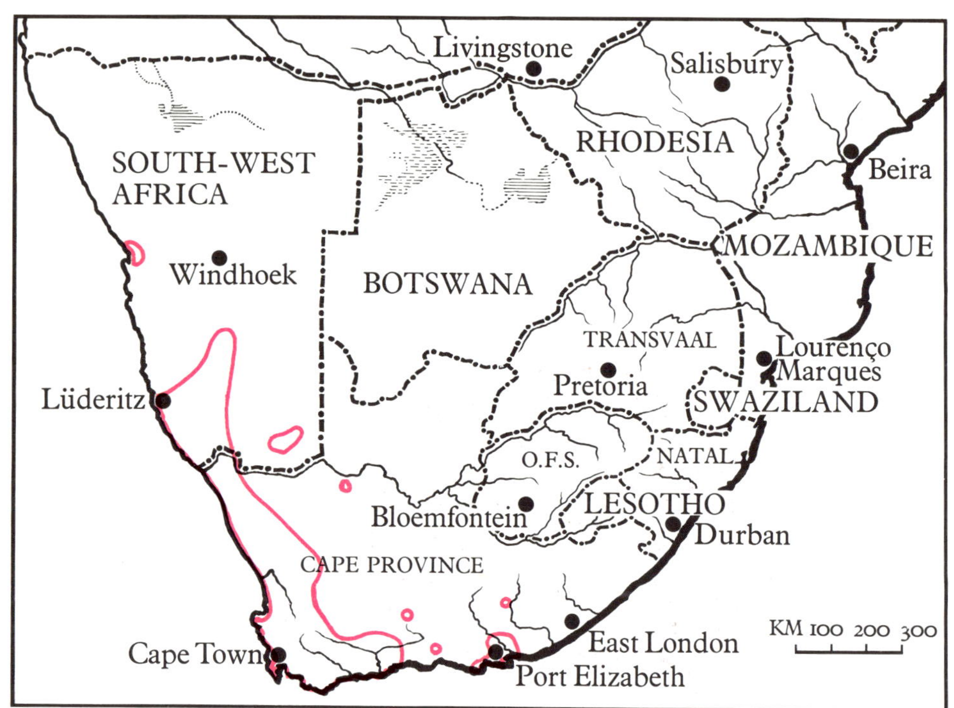

Drosanthemum

Drosanthemum Schwantes
From the Greek *drosos* = dew and *anthemon* = flower.

Schwantes, Z. Sukk. 3: 14 (in key), 29, 106. 1927. - Bolus, Notes, 99; fig. 24. 1928. - N. E. Brown, Gard. Chron. 87: 14. 1930. - v. Pöllnitz, Aufteilung, 40. 1933. - Jacobsen, Sukk. 126; fig. 120. 1933. - Pax, Natürl. Pfl. 214. 1934. - Jacobsen, Succ. Pl. 170; fig. 147, 148. 1935; Verzeichnis, 71. 1938. - Goossens, Blomplante, 144 (in key). 1940. - Jacobsen, Herre, Volk, Mesembr. 62 (in key), 95. 1950. - Adamson, Salter, Fl. Cape Pen. 378. 1950. - Phillips, Genera, 320. 1951. - Jacobsen, Handbuch, 1336; fig. 1108, 1109, 1110. 1955. - Schwantes, Fl. Stones, 83, 338; pl. 20a. 1957. - Jacobsen, Handbook, 952 (in system of Schwantes), 959 (in key of Bolus), 969 (in key of Herre & Volk), 1120; fig. 1336, 1337, 1338. 1960. - Herre et Friedrich, Mitt. Bot. Staatssammlung München, Verbreitungskarten, 46. 1961. - Bolus, J. S. Afr. Bot. 30: 36 (key of groups). 1964. - Jacobsen, Lexikon, 413, t.161/6. 1970. - Friedrich in Merxmüller, Prodromus, 37. 1970.

Small or middle-sized shrubs with erect or decumbent to climbing and rooting or prostrate branches with distinct internodes. Leaves opposite, always equal, with glittering papillae, compressed, 3-angled to terete, without fringe of bristles at the apex, to about 2,5 cm long and 0,4 cm in diam., the papillae may be blister-like or acute on both points. Flowers solitary or in cymes, mostly pedicelled but sometimes sessile, on short side-branchlets, all colours with the exception of blue, to about 3 cm in diam. Sepals 5-rarely 6-lobed, with papillae. Petals 1-2-series, reflexed. Stamens conically collected or erect, often surrounded by staminodes. Ovary flat or convex above. Glands in several parts; placentas parietal. Stigmas 4-6, often long filiform to lanceolate and plumose, rarely shortly 3-angled, often as long as or longer than the stamens, very rarely shorter but then not plumose. Capsule 4-6-locular, with valve-wings; expanding keels contiguous, parallel or diverging; loculi-roofs present, tubercle very small or absent. Seeds fairly small, light-brown, fairly rough, with ribs and tubercles.

Species: about 95, in S.W.A. 7 and in the Western Cape Province and Namaqualand about 88. (Type: *D. hispidum* (L.) Schwant.)

Drosanthemum speciosum 143

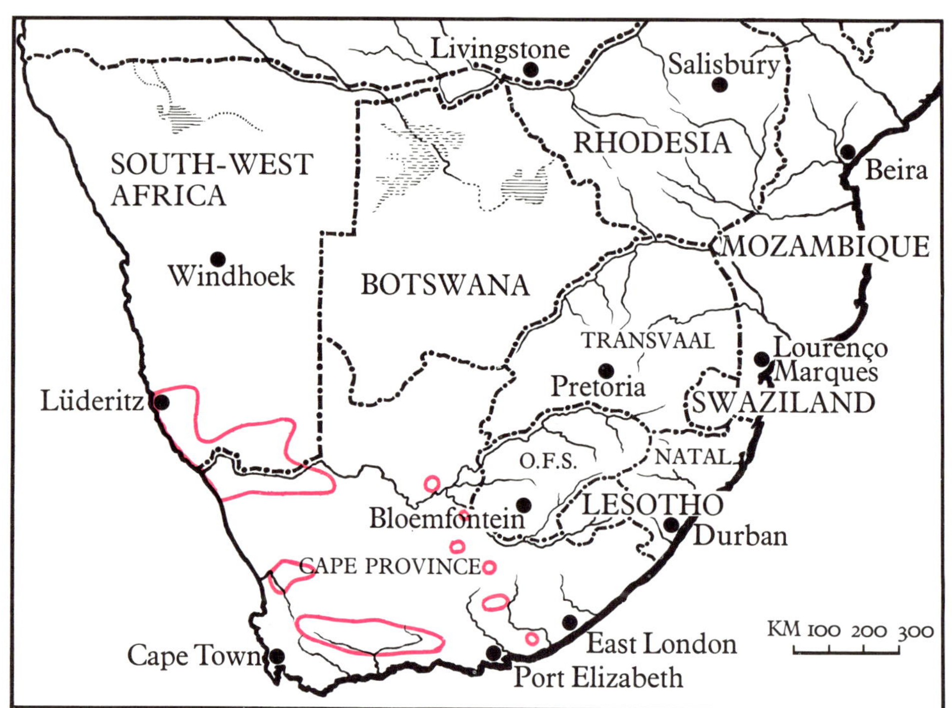

Eberlanzia

Eberlanzia Schwantes
Named in honour of the late
Mr F. Eberlanz, Lüderitz, S.W.A.

Schwantes, Z. Sukk. 2: 189. 1926. - N. E. Brown, Gard. Chron. 87: 32. 1930. - v. Pöllnitz, Aufteilung, 40. 1933. - Jacobsen, Sukk. 127. 1933; Succ. Pl. 172. 1935; Verzeichnis, 76. 1938. - Jacobsen, Herre, Volk, Mesembr. 63 (in key) 96. 1950. - Phillips, Genera, 320. 1951. - Jacobsen, Handbuch, 1351; fig. 1111. 1955. - Schwantes, Fl. Stones, 82, 338; pl. 18a. 1957. - Jacobsen, Handbook, 952 (in system of Schwantes), 961 (in key of Bolus), 969 (in key of Herre & Volk), 1131; fig. 1339. 1960; Lexikon, 419, t.162/2. 1970. - Friedrich in Merxmüller Prodromus, 42. 1970.

Erect shrubs and undershrubs with woody roots and with distinct internodes. Leaves opposite, connate at the base, obscure 3-angled with the edges often very much rounded, long to rather short and thick, obtuse with a very small mucro, rarely flat, papillate, grey-green to blue-green with fine dark dots, to about 2 cm long and 0,6 cm in diam. Flowers in branched cymes, often spinescent and persistent (with some exceptions), pedicels with 2 leafy bracts, not easily distinguishd from the leaves, white to light and dark-violet, about 1 to 2 cm in diam. Sepals 5, nearly equal, 3-angled, obtuse or acute. Petals 1-series, linear or absent, with only staminodes taking their place, hairy at the base or in the middle like the stamens. Ovary nearly flat or somewhat raised; glands annular, dark-green, crenulated; placentas parietal. Stigmas 5, erect, subulate, acute, free. Capsule 5-locular, valve-wings nearly completely connate with the valves, so that it looks like absent; expanding keels very much diverging; loculi-roofs present, tubercles present, sometimes, as with *E. sedoides*, hidden under the loculi-roofs. Seeds middle-sized, more or less edged to roundish to pear-shaped, sometimes rough or nearly smooth, yellowish or yellowish-brown.

Species: 28, S.W.A., Cape Province, Transvaal (Johannesburg). (Type: *E. clausa* (Dint.) Schwant.)

Eberlanzia spinosa   145

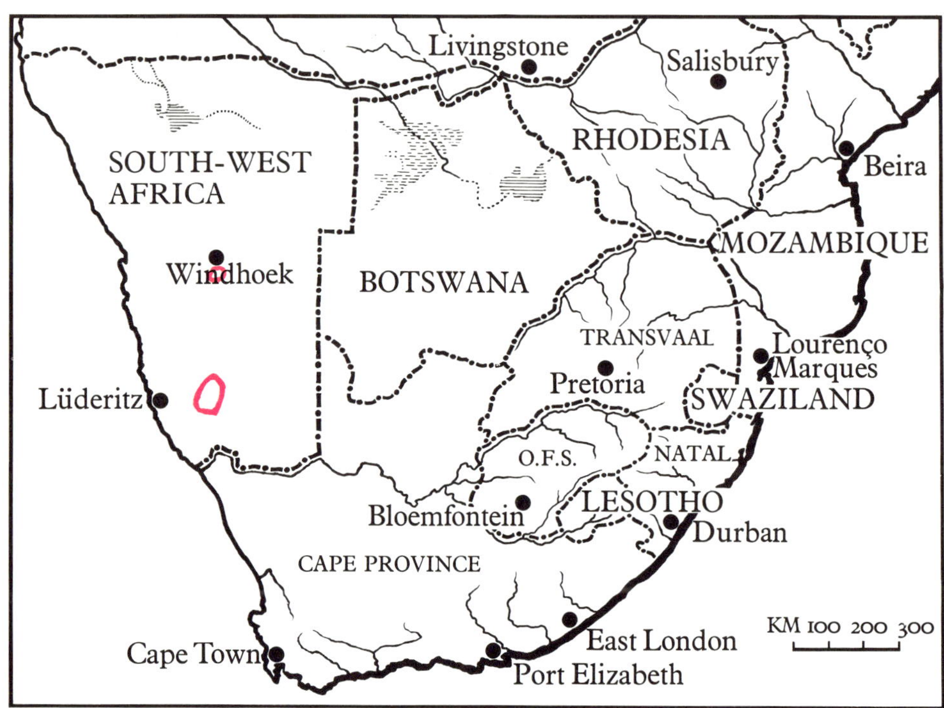

Ebracteola

Ebracteola Dinter et Schwantes
From the Latin *bracteola* = bract,
ebracteola = without bracts.

Dinter et Schwantes, Z. Sukk. 3: 15 (in key), 24. 1927. - v. Pöllnitz, Aufteilung, 40. 1933. - Jacobsen, Sukk. 127; fig. 121. 1933. - Pax, Natürl. Pfl. 220. 1934. - Jacobsen, Succ. Pl. 172; fig. 149. 1935; Verzeichnis, 77. 1938. - Jacobsen, Herre, Volk, Mesembr. 70 (in key), 97. 1950. Phillips, Genera, 320. 1951. - Jacobsen, Handbuch, 1353; fig. 1112. 1955. - Schwantes, Fl. Stones, 93, 338. 1957. - Jacobsen, Handbook, 952 (in system of Schwantes), 962 (in key of Bolus), 972 (in key of Herre & Volk), 1134; fig. 1341. 1960; Lexikon, 420. t.162/3. 1970. - Friedrich in Merxmüller Prodromus, 44. 1970.

Small tufted plants with very thickened rootstock and short branchlets, so that the internodes are nearly always invisible. Leaves opposite, connate at the base, compressed, elongated 3-angled-prismatic to nearly terete, light bluish-green, often densely dotted, about 3-4 cm long and 1 cm in diam. Flowers terminal, solitary, shortly pedicelled, without bracts, white to red, to about 2,5 cm in diam. Sepals 5, free to the ovary. Petals red or white. Stamens broad, tape-like. Glands not separated. Placentas parietal. Stigmas 5, subulate, more or less plumose. Capsule 5-locular with narrow to very narrow wings along the valves; expanding keels broad, sheet-like thin, very much toothed; loculi-roofs stiff; tubercle small, often crenulate as in Hereroa. Seeds roundish, smooth, whitish.

Species: 2, South West Africa, Auas Mts. nr. Windhoek. (Type: *E. montis-moltkei* (Dint.) Dint.et Schwant.)

146

M. M. Page. 2. 1924.

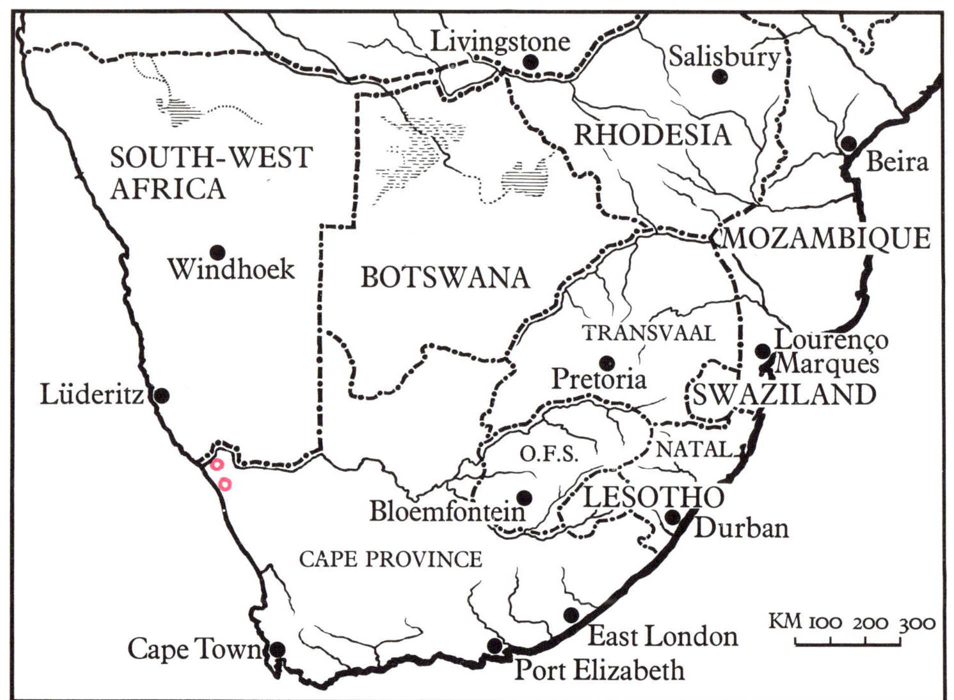

Enarganthe

Enarganthe N. E. Brown
From the Greek *enarges* = shining and *anthe* = flower.

N. E. Brown, Gard. Chron. 87: 71 (in key), 151. 1930. - v. Pöllnitz, Aufteilung, 41. 1933. - Pax, Natürl. Pfl. 214. 1934. - Jacobsen, Verzeichnis, 78. 1938. - Goossens, Blomplante, 144 (in key), 1940. - Jacobsen, Herre, Volk, Mesembr. 65/66 (in key), 97. 1950. - *Phillips, Genera, 303. 1951.* - Jacobsen, Handbuch, 1355. 1955. - Schwantes, Fl. Stones, 332, 342. 1957. - Jacobsen, Handbook, 963 (in key of Bolus), 970 (in key of Herre & Volk), 1136; fig. 1342. 1960. - Jacobsen, Lexikon, 420. t.162/1. 1970.

Mesembr. octonaria L. Bolus in Fl. Pl. S. Afr. 7: pl. 271. 1927.

Erect shrub with woody branches and distinct internodes. Leaves opposite, not united at the base, stoutly trigonous, not papulose, seen from the side somewhat club-shaped, obtuse, to 3,5 cm long, 1,5 cm in diam. Flowers solitary or threenate, then the lateral ones developing very slowly, nearly sessile or on pedicel about 0,8 cm long, bud compressed, opening at noon, violet-red with obscure stripe at the centre about 6-7 cm in diam. Sepals 4, unequal, 2 much larger and leafy, to about 2 cm long, 2 others with broad membranous margin about 1,2 cm long. Petals numerous, in about 2 series, free, obtuse. Stamens, anteriors erect, 1-series, posteriors densely crowded, filaments bearded below, violet-red; staminodes 0. Ovary inferior, convex at the top, 8-chambered; glands in a crenulate ring; placentas parietal. Stigmas 8, short subulate, stout, acuminate about 2,5 mm long. Capsule 8-locular, fairly large and woody, with loculi-roofs and tubercle; expanding-keels ending in short spines. Seeds brown, pear-shaped.

Species: 1, Namaqualand: Richtersveld (betw. Aughrabies Mts. & Brakfontein). (Type: *E. octonaria* (L. Bol.) N. E. Brown.)

Fruit added 3.1930.
7-1930
B.O.Carter. 5 1927.
Returned from S.A. Plants april 1929
octonarium,
L. Bolus,

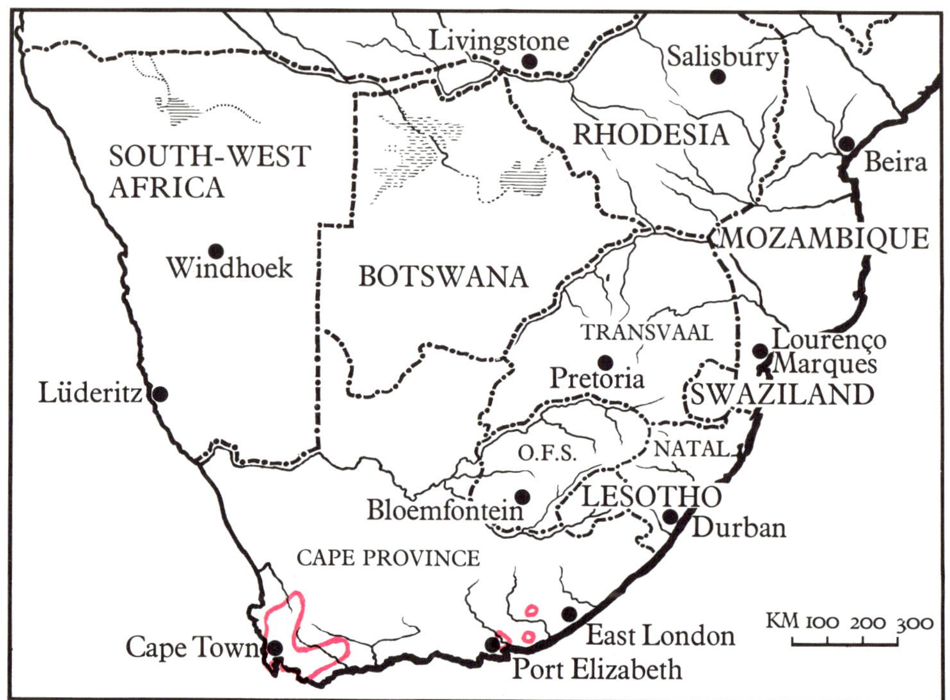

Erepsia

Erepsia N. E. Brown
From the Greek *erepsis* = the cover, because the stamens are covered by the staminodes.

N. E. Brown, Gard. Chron. 78: 433 (in key). 1925. - Phillips, Genera, 248. 1926. - Bolus, Notes, 8, pl. 1E. 1928. - v. Pöllnitz, Aufteilung, 41. 1933. - Jacobsen, Sukk. 128. 1933. - Pax, Natürl. Pfl. 214. 1934. - Jacobsen, Succ. Pl. 174. 1935; Verzeichnis, 78. 1938. - Goossens, Blomplante, 145 (in key). 1940. - Bolus, Notes, 228. 1950. - Jacobsen, Herre, Volk, Mesembr. 60 (in key), 97. 1950. - Adamson, Salter, Fl. Cape Pen. 385. 1950. - *Phillips, Genera, 300, 303. 1951.* - Jacobsen, Handbuch, 1355; fig. 1113. 1955. - Schwantes, Fl. Stones, 69, 339; fig. 18. 1957. - Jacobsen, Handbook, 952 (in system of Schwantes), 959 (in key of Bolus), 968 (in key of Herre & Volk), 1136; fig. 1343. 1960; Lexikon, 420, 4.162/4. 1970.

Circandra N. E. Brown in Gard. Chron. 87: 126. 1930. - Bolus, Notes, 228, 1950.

Perennial succulent shrublets, erect or decumbent, entirely glabrous, often with 2-edged branches and distinct internodes. Leaves opposite, very slightly or scarcely connate at the base, acutely 3-angled, laterally compressed and dilated and scabrid or serrulate at the upper part of the keel or with cartilaginous keel, often hyaline dotted, margins often reddish to about 3 cm long and 0,8 cm in diam. Flowers solitary, or 2- or 3-nate, later axillary, without bracts, pedicels to about 5 cm long, violet-red or white, about 4 cm in diam. The flowers of *E. inclaudens* remain always open. Sepals 5, tubular connate above the ovary, unequal, acuminate, 2-keeled, 3 with a membranous margin. Petals free, linear-spathulate or linear-cuneiform, obtuse, many series, sometimes posteriors papillate towards the base, arising from the top of the calyx tube. Stamens many, arising from the inner surface or at the top of the calyx tube, all inflexed (not erect); staminodes numerous, some inflexed and more or less concealing the stamens and arising with the petals; filaments white, bearded, anthers and pollen yellow. Ovary inferior, flat or slightly convex at the top. Glands scarcely visible. Placentas on the outer walls of the chambers. Stigmas 5-6, rarely 4 or 7, very small, stout, obtuse, roundish to almost globular. Capsule 5-6, rarely 4 or 7-locular, slightly convex, woody, sharply 2-edged; valve-wings nearly absent or very narrow; expanding keels mostly parallel, diverging towards the apex, ending in broad awn-pointed membranes; loculi roofed with stiff wings but without tubercles at the openings. Seeds obovate, minutely dotted (*E. viridis*).

Species: 38, Cape Province: from Clanwilliam and right round the coast to Uitenhage and Graaff-Reinet, also recorded in the Ladismith district. (Type: *E. inclaudens* (Haw.) N. E. Brown.)

Erepsia bracteata   151

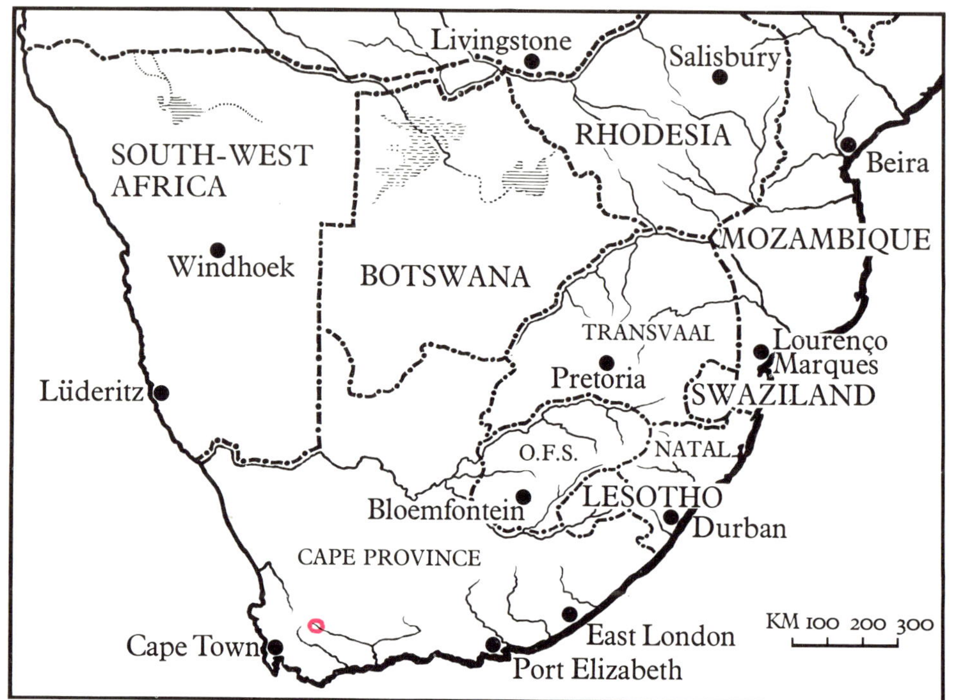

Esterhuysenia

Esterhuysenia L. Bol. gen. nov.
Named in honour of Miss E. Esterhuysen.

Bolus, Journ. S. Afr. Bot. Vol. XXXIII, Part IV, Oct., p. 308/9. 1967. - Jacobsen, Lexikon, 422. 1970.

Plants smooth, 20-33 cm high, 42 cm in diam.; trailing roots up to 50 cm long and 7 mm in diam.; stem at base 8 mm in diam.; densely branched, branches 0,5-4 mm in diam.; internodes often 2-10 mm, rarely 2 cm long; herbaceous parts minutely tuberculate; leaves ascendent or spreading or sometimes subfalcate, above linear, plane, towards the end slightly pointed, subconvex at broadest part, lower side roundish, obscurely keeled, keel reduced to a line, apiculate, glaucous-green or when older reddish, 8-10 mm long with vagina 1,5 mm long, generally 1,5-2,5 mm in diam.; flowers solitary at daytime, often 1,7 cm in diam.; peduncle 6-9 mm long; receptacle obconical, 3-4 mm long, 4-5 mm in diam.; sepals 5, acute or subacute, more or less marginate, 2,5-3 mm long, at base 1,5-2 mm wide or 4-5 mm long, at base 2-3 mm wide; staminodes very thin, towards the point recurved, pink, surpassing the stamens; filaments ca. 4-seriate, pink, the inner ones up to the base papillate; anthers and pollen yellowish; disc distinctly crenulate; ovarium towards its middle slightly elevated; stigmas 5, fairly slender, papillate at the base, caudate above, ca. 2,5 mm long; capsule obconical, often 3 mm long and up to 4 mm in diam., expanded 10 mm in diam., keels thick, marginate, parallel at base, at the points far divergent, in the middle touching the valves, without tubercles but with well developed cell wings.

This genus is closely allied to *Lampranthus* and *Ruschia* but the capsule lacks the wings of the former and the tubercles of the latter.

One species: *E. alpina* L. Bol. Cape Province: distr. Worcester: Hex River Mountains, Milner Peak, East Side on ledges 5 500-6 000 ft. Dec. 1948, leg. Miss E. Esterhuysen. Also: district Worcester, Fonteintjiesberg, "Growing from crevices and depressions in massive rock formations on plateau leading to beacon, ca. 6 000 ft.; small compact low shrub."

Esterhuysenia alpina   153

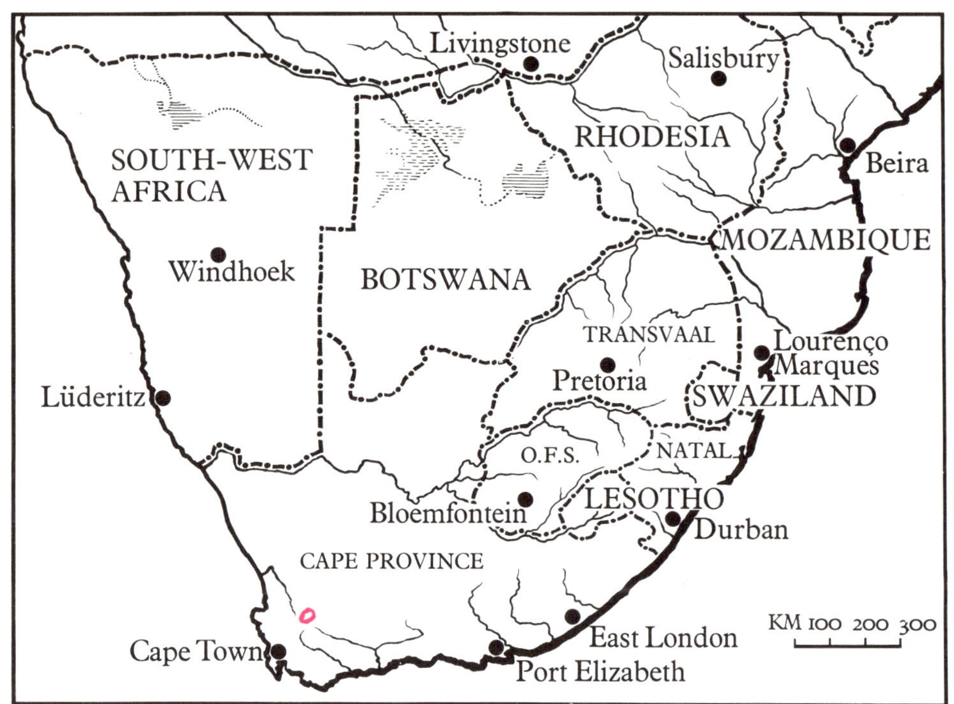

Eurystigma

Eurystigma L. Bolus
From the Greek *eurys* = broad and
stigma, on account of the broad stigma.

Bolus, Notes 2: 179. 1930. - v. Pöllnitz,
Aufteilung, 41. 1933. - Pax, Natürl. pfl.
220. 1934. - Jacobsen, Verzeichnis, 81.
1938. - Jacobsen, Herre, Volk, Mesembr.
55 (in key), 97. 1950. - Phillips, Genera,
320. 1951. - Jacobsen, Handbuch,
1361. 1955. - Schwantes, Fl. Stones,
19. 1957. - Jacobsen, Handbook, 951
(in system of Schwantes), 957 (in key of
Bolus), 966 (in key of Herre & Volk)
1141. 1960; Lexikon, 422. 1970.
Cryophytum, Bolus, Notes, 2: 151.
1930.

Annual, erect, spreading herb. All the
green parts very smooth and covered with
glittering papillae, thus with a greasy
lustre, green or yellowish-green, often
reddish coloured, without distinct
papillae. Branches erect, thickened towards the apex, hence club-shaped.
Leaves a few, the lower 2 or 3 pairs opposite, the upper ones alternate, terete
or semi-terete, lax during the flowering-
time, to about 6,5 cm long. Flowers 3-5
on each plant in cluster-like cymes, for
25 days, open at noon and in afternoon,
about 6 cm in diam. Sepals with the
club-shaped pedicel 2 cm long, at the
apex 1,4 cm in diam. with a 2 mm long
tube. Sepals 5, unequal, anteriors to
about 1,3 cm broad, ending in leafy
points, to about 2,2 cm long; posteriors
to about 1,3 cm long. Stamens and
staminodes in several series, slender,
anthers and pollen golden-yellow. Ovary
rarely 0,1 cm raised, obscurely 5-angled;
glands deep crenulated, placentas axile.
Stigmas 5, thick, broad, subulate or
rather filiform, acuminate, yellow,
3-4 mm long. Capsule 5-locular, unknown.
Species: 1, Ceres Karroo. (Type:
Eurystigma clavata (L. Bolus) L. Bolus.)

Eurystigma clavata 155

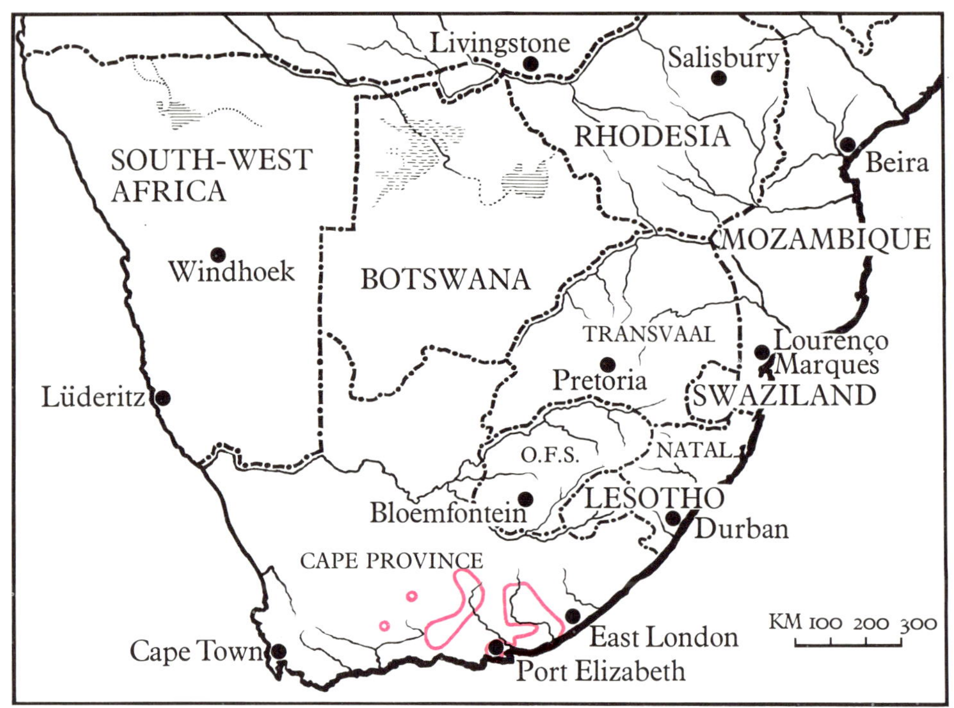

Faucaria

Faucaria Schwantes
From the Latin *faux* = jaw; *faucaria* =
a collection of jaws.

Schwantes, Z. Sukk. 2: 176. 1925/26;
3: 105. 1927. - Bolus, Fl. Pl. S. Afr. 7:
pl. 267. 1927. - Labarre (red.), Mesembr.
fig. 100-102. 1931. - v. Pöllnitz, Aufteilung, 41. 1933. - Jacobsen, Sukk. 129;
fig. 122. 1933. - Pax, Natürl. Pfl. 214.
1934. - Jacobsen, Succ. Pl. 174; fig. 150.
1935; Verzeichnis, 81, 1938. - *Bolus,
Notes, 109; pl. 16 (leaves and fruits),
pl. 17. 1939.* - Jacobsen, Herre, Volk,
Mesembr. 71 (in key), 97. 1950. -
Phillips, Genera, 320. 1951. - Jacobsen,
Handbuch, 1362; fig. 1114-1119. -
Schwantes, Fl. Stones, 129, 340; pl. 4
(coloured), pl. 41a; fig. 41, 42 (capsule).
1957. - Jacobsen, Handbook, 953 (in
system of Schwantes), 959 (in key of
Bolus), 972 (in key of Herre & Volk),
1141; fig. 1344-1353. 1960; Lexikon, 422,
t.163/1-6, 1970.

Dwarf, rosette-forming plants with
fleshy, rapiform roots, during youth
without a stem, later on with only a short
one, and also with short branchlets.
Leaves opposite, 4-6 together on one
short stem, very much crowded, somewhat connate at the base and often more
or less terete, towards the apex keeled
and 3-angled or oblong-lanceolate or
shortly broad 3-angled with all possible
stages between these leaf-forms; lower
side often chin-like produced, margins
and upper surface with cartilaginous
or often with stout, sometimes in awn-
tipped teeth, very rarely without it
(e.g. *F. ryneveldiae*), light- or dark-green,
often with white, scattered warts, to
about 4,5 cm long and at the broadest
place nearly 2 cm broad. Flowers short
pedicelled or sessile, without bracts,
solitary, open during afternoon, yellow
to orange, outside and when withering
reddish-copper coloured, 3-5 cm in diam.
Sepals 5, free to the ovary, unequal,
often with membranous margins and
keeled. Petals 2-3-series, linear-lanceolate, obtuse or emarginate, narrowed to
the base, towards the apex recurved,
yellow, golden or orange, rarely lemon-
yellow, outside reddish to copper
coloured. Stamens numerous, erect,

diffuse, yellowish-white, as long as or
longer or shorter than the stigmas. Ovary
long, more or less spherical to conical
above. Glands 5, separate, small glands,
so characteristically coloured that it is
worth while to examine a faded flower
(Schwantes l.c.); placentas parietal.
Stigmas 5-6, filiform, as long as, longer
or shorter than the stamens; style none.
Capsule 5-6-locular, deep barrel-like,
loculi overgrown. Quite peculiar!
Schwantes l.c. gives more details about
it; expanding keels lateral on the valves,
touching the neighbouring valve, ending
in a vertical spreading winged awn;
seeds can only escape through the orifice
between the keels; ordinary loculi-roofs
and tubercles none. Seeds reddish-
brown, with stalked, minute warts,
about 1,5 mm long.

Species: 33 from the Eastern Cape
Province and the Karroo. (Type:
F. tigrina (Haw.) Schwant.)

M.M.Page. V. 1923.

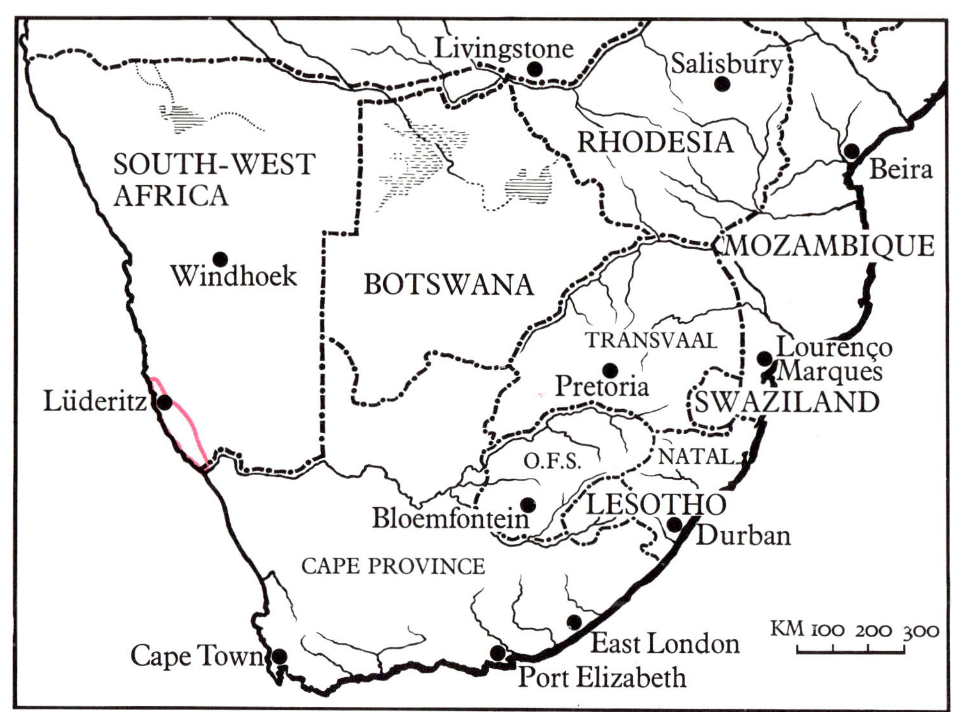

Fenestraria

Fenestraria N. E. Brown
From the Latin *fenestra* = window,
fenestraria = a number of windows.

N. E. Brown, Gard. Chron. 78: 433
(in key), 1925. - Phillips, Genera, 246.
1926. - Bolus, Notes, 54. 1928. - Labarre (red.), Mesembr. 36, 213. fig. 103,
104. 1931. - v. Pöllnitz, Aufteilung,
42. 1933. - Jacobsen, Sukk. 131; fig. 123.
1933. - Pax, Natürl. Pfl. 215. 1934. -
Jacobsen, Succ. Pl. 176, fig. 151. 1935;
Verzeichnis, 83. 1938. - Goossens, Blomplante, 144 (in key). 1940. - Jacobsen,
Herre, Volk, Mesembr. 75 (in key), 98;
fig. 9. 1950. - *Phillips, Genera, 304.
1951.* - Jacobsen, Handbuch, 1369; fig.
1120, 1121. 1955. - Schwantes, Fl.
Stones, 106, 331; pl. 30b. 1957. - Jacobsen, Handbook, 952 (in system of
Schwantes), 962 (in key of Bolus), 974
(in key of Herre & Volk), 1146; fig. 1354,
1355. 1960; Lexikon, 424, t.164/2.
1970. - Friedrich in Merxmüller Prodromus, 46. 1970.

Very small stemless succulent perennials. Leaves clustered, erect, club-
shaped, truncate at the apex and resembling windows. Flowers solitary, terminal, pedunculate, bracteate, white or
yellow, about 8 cm in diam. Sepals 5
free down to the junction with the ovary.
Petals many, 1-series, free, linear.
Stamens numerous, erect, lax; staminodes o. Ovary nearly superior; placentas
parietal. Stigmas 10-11 or 13-16, radi-
ately spreading, very plumose. Capsule
10-11 or 13-16-locular, expanding
keels parallel, at the points diverging and
with some teeth, as also with small wings,
with a firm axile column; locular-wings
well developed, with tubercles. Seeds
whitish with brown nipple, somewhat
conical.

Species: 2, Lüderitz, S.W.A. Richtersveld: Grootderm, Alexander Bay.
(Type: *F. aurantiaca* N. E. Brown.)

1368
34 Pomona
2x
R.O.Barter

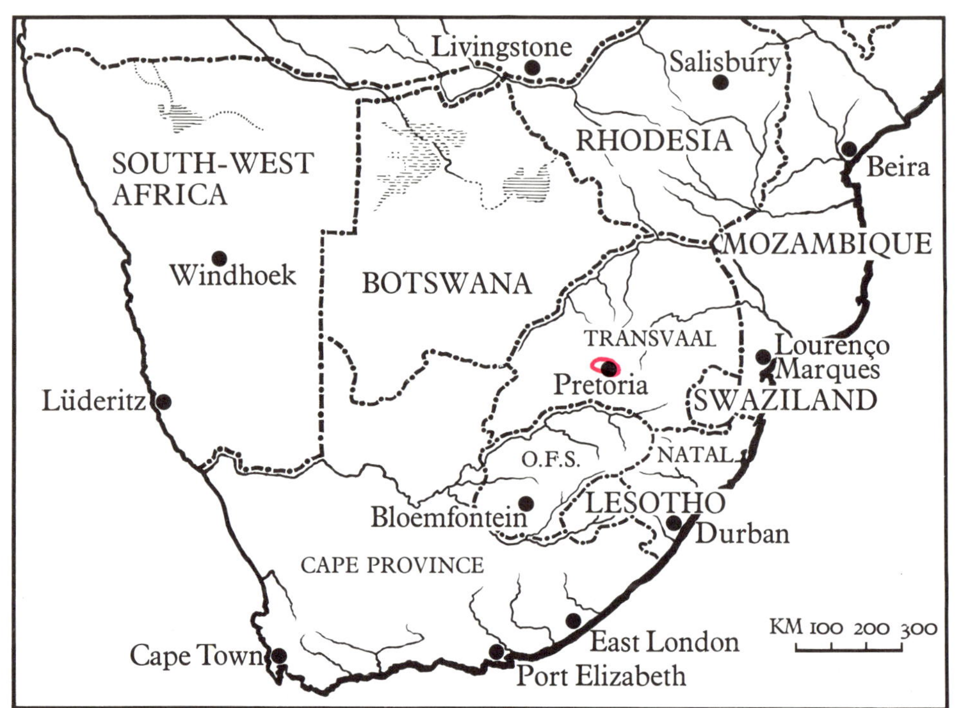

Frithia

Frithia N. E. Brown
Named in honour of the late
Mr Frank Frith.

N. E. Brown, Gard. Chron. 78: 433 (in key), 1925; Burtt Davy, Fl. Pl. & Ferns, 1: 162. 1926. - Phillips, Genera, 247. 1926. - Bolus, Fl. Pl. S. Afr. 7: pl. 275. 1927. Labarre (red.), Mesembr. 215. fig. 205. 1931. - v. Pöllnitz, Aufteilung, 42. 1933. - Jacobsen, Sukk. 132. 1938. - Goossens, Blomplante, 143 (in key). 1940. - Jacobsen, Herre, Volk, Mesembr. 71 (in key), 98. 1951. - Phillips, Genera, 304. 1951. - Jacobsen, Handbuch, 1371; fig. 1122. 1955. - Schwantes, Fl. Stones, 108, 340; fig. 26; pl. 31a. 1957. - Jacobsen, Handbook, 953 (in system of Schwantes), 957 (in key of Bolus), 972 (in key of Herre & Volk), 1147; fig. 1356. 1960; Lexikon, 425, t.164/1. 1970.

Very small stemless perennial herb. Leaves 5-7 to a plant or growth, alternate, in a tuft, erect, clavate, subterete, truncate and without chlorophyll at the apex. Flowers sessile or subsessile among the leaves, light violet or white, to 2,5 cm in diam. Sepals 5, with a short green tube above the ovary. Petals numerous, in several series, linear, posterior ones smaller, shortly tubular connate at the base, near the mouth one ring of dark-yellow, narrow-linear staminodes. Stamens about 3-series, erect, inserted into the tube and probably concealed by the posterior petals or staminodes. Ovary inferior, placentas on the outer wall of the chambers; style 0; stigmas 5, minute, stout, acute. Capsule 5-locular, when open, margins of the valves recurved; expanding keels parallel to the points diverging, with very small wings; without loculi-roofs and tubercles 0. Seeds small, with fine tubercles.

Species: 1, Pretoria, Magalies Mts. and Witbank. (Type: *F. pulchra* N. E. Brown.)

Frithia pulchra 161

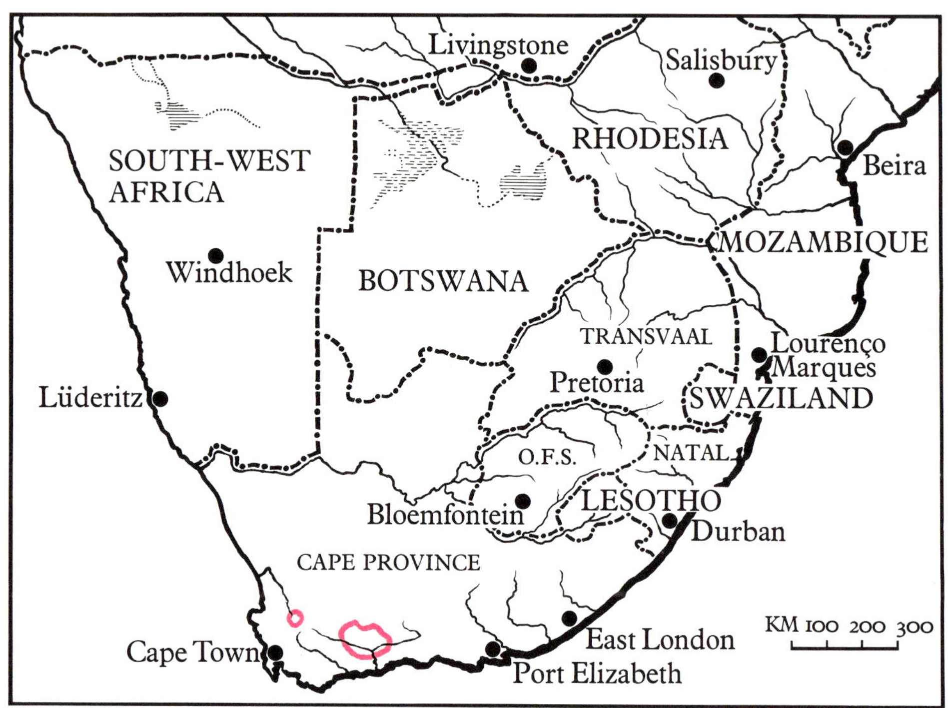

Gibbaeum

Gibbaeum (Haw.) N. E. Brown
From the Latin *gibba* = hump.

N. E. Brown, Gard. Chron. 71: 129, 151 with fig. 1922; 78: 413, 433 (in key). 1925; 79: 172 with fig. 1926. - Haworth, Rev. Pl. Succ. 104. 1821. - Phillips, Genera, 242. 1926. - Tischer, Z. Sukk. 3: 227. 1927. - Bolus, Notes, 45; pl. 17 (coloured). 1928. - Labarre (red.), Mesembr. 31, 216; fig. 106-112. 1931. - v. Pöllnitz, Aufteilung, 43. 1933. - Jacobsen, Sukk. 132; fig. 124-129. 1933. - Pax, Natürl. Pfl. 215. 1934. - Jacobsen, Succ. Pl. 178; fig. 153-159. 1935; Verz., 83. 1938.- Goossens, Blompl., 143 and 147 (in key). 1940. - Jacobsen, Herre, Volk, Mesembr. 80 (in key), 98; fig. 3. 1950. - *Phillips, Genera, 304. 1951.* - Nel, Gibbaeum Handbook, 1953. - Jacobsen, Handbuch, 137; fig. 1123-1146. 1955. - Schwantes, Fl. Stones, 162. 340; fig. 27a; pl. 50, 51. 1957. - Jacobsen, Handbook, 953 (in system of Schwantes), 961 (in key of Bolus), 976 (in key of Herre & Volk), 1148; fig. 1358-1380. 1960; Lexikon, 425, t.164/3-6, 165, 166/1-6. 1970.
Argeta N. E. Br., Gard. Chron. 82: 113; fig. 52. 1927; J. Bot. 66: 265. 1928.- Tischer, Monatsschr. Kakt. Ges. 1: 234. 1929. - Labarre (red.), Mesembr. 258, fig. 18. 1931. - Jacobsen, Sukk. 90, fig. 82. 1933; Succ. Pl. 125; fig. 104. 1935.- Goossens, Blom., 146 (in key). 1940.
Mentocalyx N. E. Brown, Gard. Chron.

81: 251, fig. 128. 1927. - Labarre (red.) Mesembr. 258; fig. 139, 140. 1931. - Goossens, Blompl., 146 (in key). 1940.
Rimaria N. E. Brown, Gard. Chron. 78: 413 (in key). 1925; 79: 85, 134; fig. 81. 1927. - Tischer, Monatsschr. Kakt. Ges. 1: 226. 1929. - Jacobsen, Sukk. 173; fig. 179. 1933; Succ. Pl. 240; fig. 227. 1935. - Goossens, Blompl., 147 (in key). 1940.

Clumps with erect, decumbent or prostrate branches, with tap- or fibrous roots, rarely with fleshy roots (*G. schwantesii* & *G. velutinum*), internodes distinct or invisible. Leaves of one pair very unequal and dissimilar, free or only connate at the base, or much connate, forming a body, then ball-like or ovate or conical or terete or semiterete with all stages between these forms, fissure nearly closed or more or less open, often pubescent and velvet-like, often glabrous, rarely with a window at the apex (*G. cryptopodium*), sometimes (*G. schwantesii* and *G. velutinum*) with contiguous pairs of leaves, keeled towards the apex when young like a bird's beak, rarely toothed along the margins, green, dark-green, whitish, covered with white membranous sheaths or not, very variable in size, 1,8-8 cm long, up to 3 cm broad and 2 cm thick.

Flowers, solitary, terminal, pedicillate; bracts none, white, light to dark-violet, 2-8 cm in diam. Sepals 5-9, unequal, free to the ovary, 2 lateral ones longer and often keeled, also with membranous margin. Petals 1-3-series, free, often flaccid, linear. Stamens numerous, erect, rarely column-like collected (*G. pentrese*), often bearded at the base, rarely not, but then lax as with *G. schwantesii* and *G. velutinum*; with or without flaccid staminodes (*G. heathii* without). Ovary often raised above the body, rarely e.g. *G. petrense* narrow and flat above. Glands 6, large, separate; placentas parietal; stigmas 6-7 (8-9 *G. heathii*) often stout, subulate, acute, plumose, style none. Capsule 6-7-locular (only *G. heathii* 8-9) short-conical or obconical, often convex, with prominent sutures; valves deltoid reflex spreading when expanded; expanding keels parallel or at first so and later diverging, or diverging, rarely toothed, ending in large broad wings; with loculi-roofs, without tubercles. Seeds, many in a loculus, ovoid, pointed at one end, smooth.

Species: 21, C.P.: Ladismith, Laingsburg, Calitzdorp, Oudtshoorn, McGregor, Swellendam, Ceres Karroo (South). (Type: *G. pubescens* (Haw.) N. E. Br.)

M.M.Page . X . 1924 .

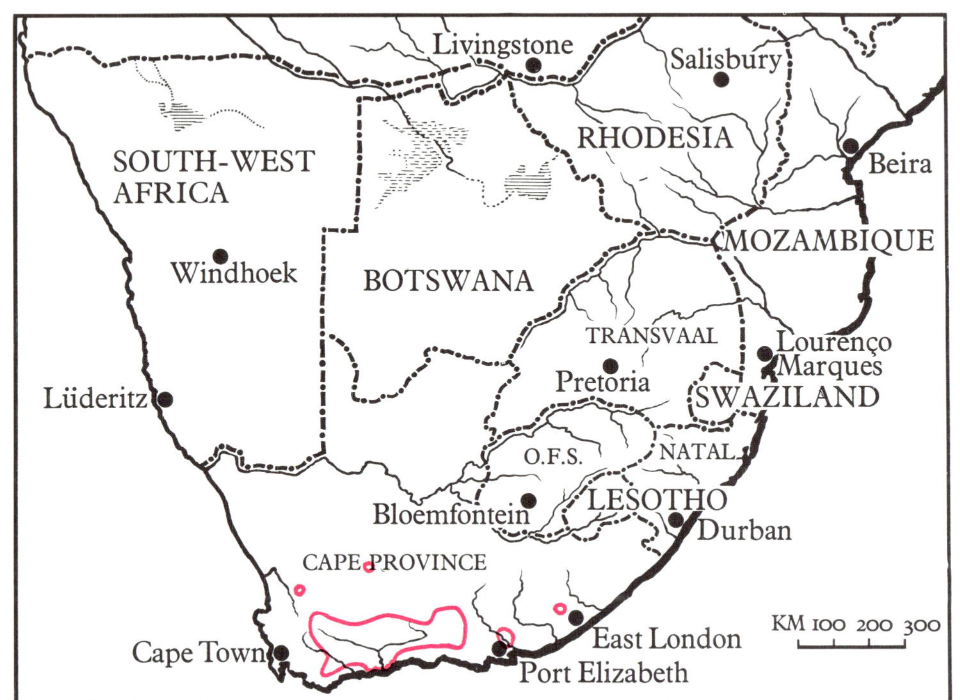

Glottiphyllum

Glottiphyllum (Haw.) N. E. Brown
From the Greek *glota* = tongue and
phyllon = leaf.

N. E. Brown, Gard. Chron. 70: 311.
1921; 71: 9 with fig. 1922; 78: 433 (in
key). 1925. - Haworth, Rev. Pl. Succ. 103.
1821. - Phillips, Genera, 243. 1926. -
N. E. Brown, Gard. Chron. 82: 290
with fig. 1927; 83: 13 with fig. 1928. -
Bolus, Notes, 49; pl. 19 (coloured). 1928.
- Tischer, Monatsschr. Kakt. Ges. 1:
105. 1929. - Labarre (red.), Mesembr.
224; fig. 113-118. 1931. - v. Pöllnitz,
Aufteilung, 43. 1933. - Jacobsen, Sukk.
135; fig. 130-132. 1933. - Pax, Natürl.
Pfl. 215. 1934. - Jacobsen, Succ. Pl. 182;
fig. 160-163. 1935; Verzeichnis, 86.
1938. - Goossens, Blomplante, 146
(in key). 1940. - Jacobsen, Herre, Volk,
Mesembr. 76 (in key), 98; fig. 15.
1950. - *Phillips, Genera, 305, 1951.* -
Jacobsen, Handbuch, 1387; fig. 1147-
1163. 1955. - Schwantes, Fl. Stones,
97, 340; pl. 27, 28, 29. 1957. - Jacobsen,
Handbook, 953 (in system of Schwantes), 962, 964 (in key of Bolus), 974 (in
key of Herre & Volk), 1158; fig. 1381-
1398. 1960; Lexikon, 427, t.167/1 & 2.
1970.

Very small succulent perennials, branching close to the ground, each branch
with 4 or more leaves. Leaves opposite,
crowded or very closely placed, usually
3 to several times as long as broad, but
in few species occasionally not much
longer than broad under natural conditions, those of each pair subequal
or unequal in size and usually with the
terminal part of one leaf different in shape
from that of the leaf opposite it, either
arranged in 2 rows or the alternating
pairs crossing one another, thick, soft
and pulpy in substance, green, rarely
whitish green or brownish tinted, without dots or, in some species, pellucidly
dotted. Flowers solitary, axillary, without bracts, sessile or pedicelled, bright
yellow, only one species white and comes
from Steytlerville, species as yet un-
named, to about 7 cm in diam. Lobes 4,
down to its union with the ovary, often
angular. Petals numerous, free or
slightly united at the base, cuneately
linear. Stamens numerous, erect; staminodes 0. Ovary partly superior, convex or
flattish on the top; placentas parietal.
Stigmas 8-20, half erect or reflexed,
shortly plumose, style none. Capsule
convex or flattish-convex on the top,
8-20-locular, flat above or convex or
pear-shaped, without marginal wings; expanding keels diverging, ending in awns
or with membranous points; loculi
roofed with rigid loculus wings and their
opening nearly closed by a large tubercle. Seeds small, ovoid, with a small
nipple at one end.

Species: 58, Little Karroo and Karroo
up to Victoria East. (Type: *G. lingui-
forme* (L.) N. E. Br.)

Glottiphyllum scalpratum 165

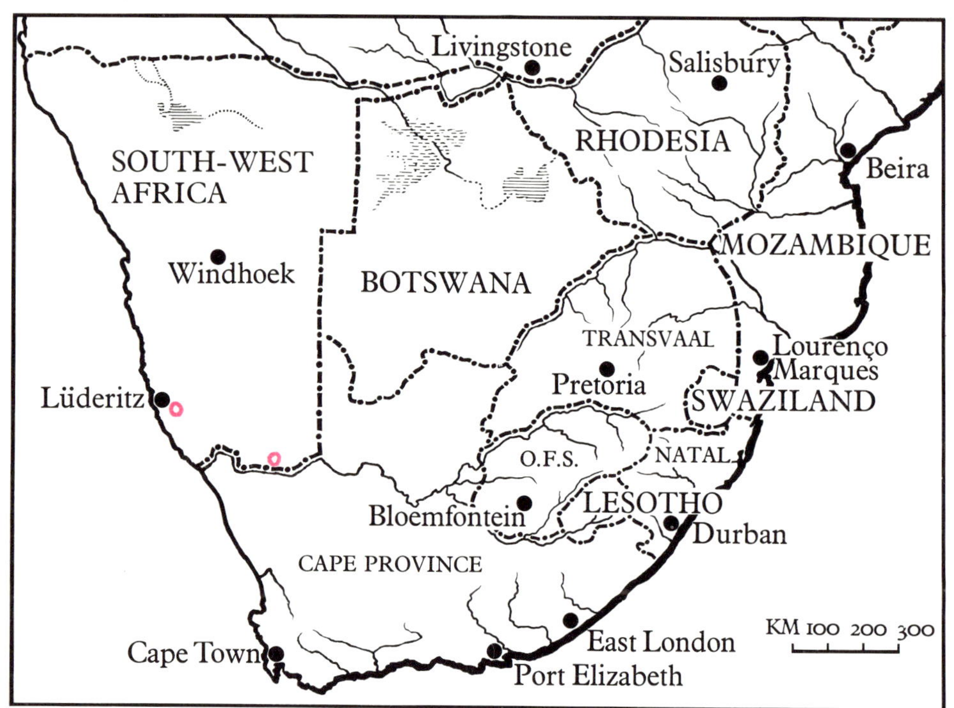

Halenbergia

Halenbergia Dinter
Named after Halenberg near Lüderitz
in South West Africa.

Dinter, Dtsch. Kakt. Ges., 158. 1937. -
Schimper-Faber, Pflanzengeographie, 2,
fig. 1039, 1935. - Jacobsen, Verzeich-
nis, 90. 1938. - Jacobsen, Herre, Volk,
Mesembr., 55 (in key), 99. 1950. - Phil-
lips, Genera, 320. 1951. - Jacobsen,
Handbuch, 1403. 1955. - Schwantes, Fl.
Stones, 20, pl. 3. 1957. - Jacobsen,
Handbook, 951 (in system of Schwantes),
957 (in key of Bolus), 966 (in key of
Herre & Volk), 1170. 1960. - Jacobsen,
Lexikon, 430, t.168/1. 1970. - Friedrich
in Merxmüller, Prodromus, 85 (*Me-
sembryanthemum hypertrophicum*) 1970.

Schwantes l.c. under *Hydrodea* gives
the following details about it: "*Hydrodea
hypertrophica* (Dint.) Schwant. also belongs to this genus although Dinter
has established a distinct genus for it,
Halenbergia, on account of the petals
being slit into three to five long strips.
Except for the difference in the form of
the petals, however, this species is so
similar to the other *Hydrodeas* that it is
hardly desirable to separate them. The
large, pure white flowers are so close
that they overlap and the cushions
formed by this species look, according
to Dinter, like linen spread out to bleach
on the yellow patches of desert scree
near Halenberg in South West Africa."

Through the kindness of the late
Mr. F. Eberlanz of Lüderitz S.W.A., I
received in 1942/43 some young plants
of it in the Internment Camp of Andalusia, C.P., where they developed and
flowered. The drawing made of them
will surely give a good idea how the plants
look. Therefore they have been published here.

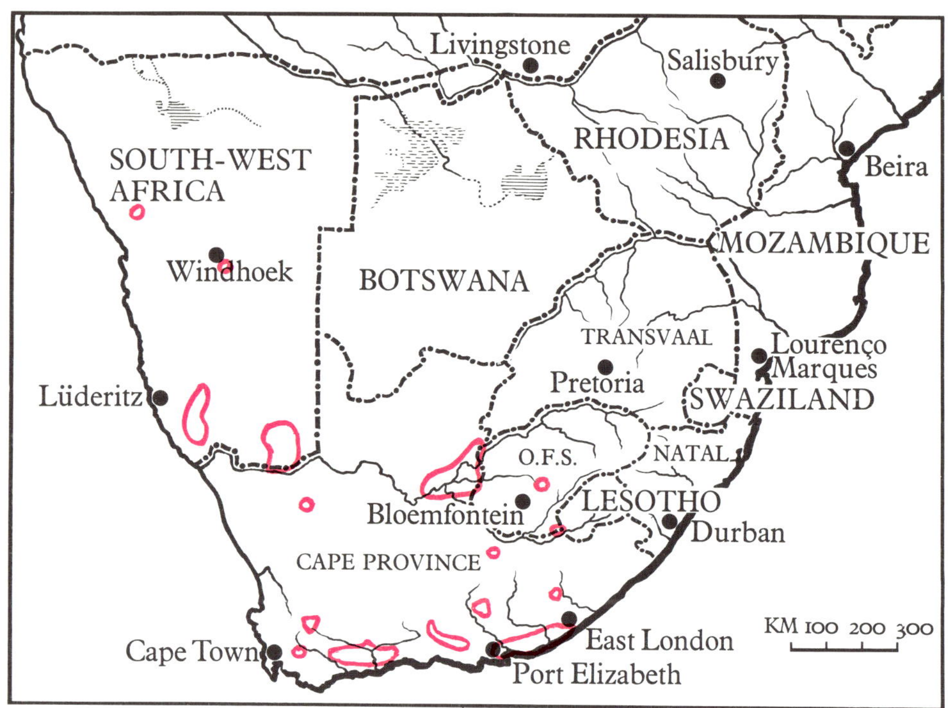

Hereroa

Hereroa Dinter et Schwantes
Named after the Hereros, natives of South West Africa, where the first plants were found.

Dinter et Schwantes, Z. Sukk. 3: 14, 15 (in key), *23*, 106. *1927*. - Bolus, Notes, 88: pl. 3 & 16B. 1928. - v. Pöllnitz, Aufteilung, 44. 1933. - Jacobsen, Sukk. 138; fig. 133, 134. 1933. - Pax, Natürl. Pfl. 215. 1934. - Jacobsen, Succ. Pl. 186; fig. 164, 165. 1935; Verzeichnis, 90. 1938. - Jacobsen, Herre, Volk, Mesembr. 73 (in key), 99. 1950. - Phillips, Genera, 220. 1951. - Jacobsen, Handbuch, 1404; fig. 1164, 1165. 1955. - Schwantes, Fl. Stones, 87, 338; fig. 28 (capsule), pl. 206. 1957. - Jacobsen, Handbook, 952 (in system of Schwantes), 962 (in key of Bolus), 973 (in key of Herre & Volk), 1170; fig. 1399, 1400. 1960; Lexikon, 430, t.168/2. 1970. - Friedrich in Merxmüller, Prodromus, 55. 1970.

Tufted to shrubby, decumbent perennials with woody not fleshy roots. Leaves opposite, somewhat connate at the base, oblong-narrow to terete, keeled or not keeled, towards the apex often peculiarly hatchet-like or with broad, tubercle-like teeth, often laterally compressed, dark-green, often with large, dark spots, about 6 cm long. Flowers often of several or in dichotomous cymes, often long pedicelled, with bracts, according to Schwantes flowers more ventricose than with *Bergeranthus*, yellow, sometimes open at night and then scented, to 4 cm in diam. Sepals 5, unequal, acuminate. Petals 2-4-series, acute or acuminate, if broad then narrower towards the base, yellow, outside reddish. Stamens erect, rarely column-like collected, often with papillae at the base. Ovary somewhat concave above. Glands when present, often separated in different parts. Placentas parietal. Stigmas 5, filiform, sometimes longer than the stamens. Capsule 5-10 locular; expanding keels diverging, with wings or membranous points, often very narrow; loculi-roofs present; tubercle small, more or less swollen,

very different as regards its form, often also crenulated, cap-like, often also along the outside with 1 or several small tubercles. Seeds pear-shaped, in the middle concave, light brown, small.

Species: 34, Aus, S.W.A.; Cape Province: Bushmanland, Karroo, Little Karroo; Eastern Cape Province: Grahamstown, Queenstown. (Type: *H. puttkameriana* (Dtr. et Berger) Dint. et Schwant.)

M.M.Page. 2.1923.

Hereroa bergeriana 169

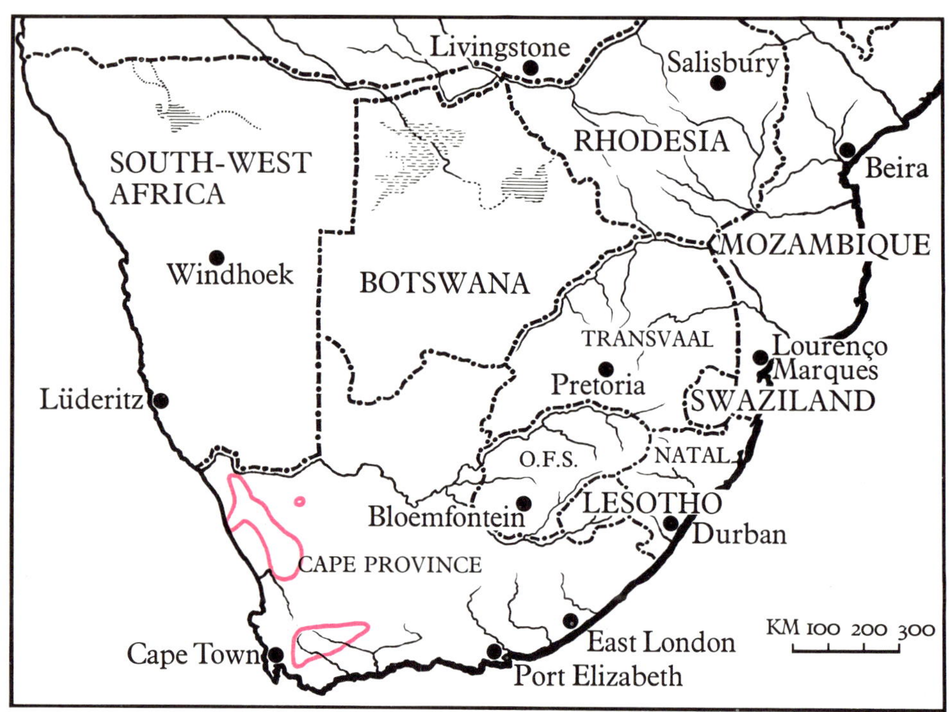

Herrea

Herrea Schwantes
Named in honour of H. Herre,
Stellenbosch.

Schwantes, Möllers Gärtnerztg. 42: 436. 1927; 43: 328. 1928. - N. E. Brown, Möllers Gärtnerztg. 43: 166, 400. 1928. - Schwantes, Mitt. Inst. Allg. Bot. Hamburg, 8: 165 (capsule & seed). 1929. - v. Pöllnitz, Aufteilung, 44. 1933. - Jacobsen, Verzeichnis, 92. 1938. - Jacobsen, Herre, Volk, Mesembr. 67 (in key), 99. 1950. - Phillips, Genera, 320. 1951. - Jacobsen, Handbuch, 1411; fig. 1166, 1167. 1955. - Straka, Anatomie & entwicklungsgeschichtl. Untersuchungen an Früchten paraspermer Mesembr., Nova Acta Leopoldina, 17. 1955. - Schwantes, Fl. Stones, 61, 341; fig. 16 (part of capsule). 1957. - *Bolus, Notes, 348 (with key); pl. 104-108. 1958.* - Jacobsen, Handbook, 954 (in system of Schwantes), 957 (in key of Bolus), 971 (in key of Herre & Volk), 1175 (key for species by Bolus); fig. 1401, 1402. 1960; Lexikon, 432, t.169/1 & 2. 1970.

Herbaceous, marcescent plants with perennial, large, rape-shaped rootstock and long prostrate branches, which may dry up and disappear when the fruit has ripened and been shed. Tubers cylindrical, fusiform, or irregular in shape, sometimes branched, up to 25 cm long and 6,7 cm in diam. Leaves opposite or subopposite or alternate (some-times all 3 positions occurring on the same branch), terete, subterete or semi-terete, viewed laterally obtuse or rarely acute, sometimes slightly concave above, rarely obtusely keeled. Peduncles straight or with age sigmoid-flexuose, the old flowers and fruits often pendent. Flowers expanding in the afternoon and in all those observed closing at sun-set, up to 12,5 cm in diam., yellow. Sepals 5, very rarely 4, flattened in the lower part, the marginal membranes of the inner ones colourless, fuscous, black or blackish, often with darker transverse veins, caudately elongated and there terete or subterete, obtuse or acute, shorter than, equalling, or longer than the petals. Petals very numerous, in several series (usually 4-5), spreading in slightly different planes, half-twisted in all buds seen, sometimes narrowed towards the base, the outer usually acuminate or acute or very rarely obtuse or truncate and obscurely emar-ginate, often pallid in the lower part, straw-, pale-yellow, sulphur or golden in the upper part, or very rarely white above and dull salmon on the lower sur-face, usually 0,5-2 mm broad, the innermost much narrower and passing into numerous loosely flexuose stamin-odes. Stamens at first inflexed and reaching the stigmas, then erect and finally spreading, filaments papillate near the base, 6-10 mm long. Nectary annular, inconspicuous. Ovary usually concave in the outer part and elevated towards the middle. Placentation parietal. Stigmas 12-24, slender, scarcely tapering up-wards, obscurely papillate. Fruit capsu-lar, but differing widely from all other capsular fruits in this family, except those of *Conicosia* and *Saphesia*, in that the ascending, elongated, tapering and wingless valves, converging towards the apex, are devoid of expanding keels and their separation depends not on moisture but entirely on desiccation. Their separation is slight, so that the free apical part scarcely becomes erect (as often in *Conicosia*) and the conical or domed shape of the upper part of the capsule is maintained throughout. For the rest of their length the separation is a mere split (sometimes a little wider when their margins are recurved) which may extend to the margin (the rim or "saucer" which is evolved from the mar-gin of the top of the ovary and which varies in breadth and depth). By manipu-lation, however, the valves may be separated to their base and the semini-ferous part be easily detached entirely from the receptacle, so that the whole

fruit may be divided into complete segments, each consisting of an entire
valve, portion of margin, and the two
entire adjacent membranes (dissepiments) attached to them, which in
typical *Herrea* reach just below the apex
of the valve, the actual apices being
compressed and closely contiguous before the separation occurs. The empty
receptacle remains whole, with the
central axis persisting. The loculi are
very narrow and without a roofing membrane. The capsules are very light and
easily blown about after disarticulation
from the peduncle, the seminiferous part
spongy or corky, pallid, among the
largest, and some the largest known, in
the family. Seeds globosely lenticular,
smooth and polished, various shades
of brown and usually with darker markings, usually 1,5 mm in diam., soon
detached and lying loose in the loculi, except those more firmly placed in the
seed-pouches, which were present in
nearly all the fruits examined.

Species: 24, Namaqualand, Vanrhynsdorp distr. Karroo. (Type: *H. nelii*
Schwant.)

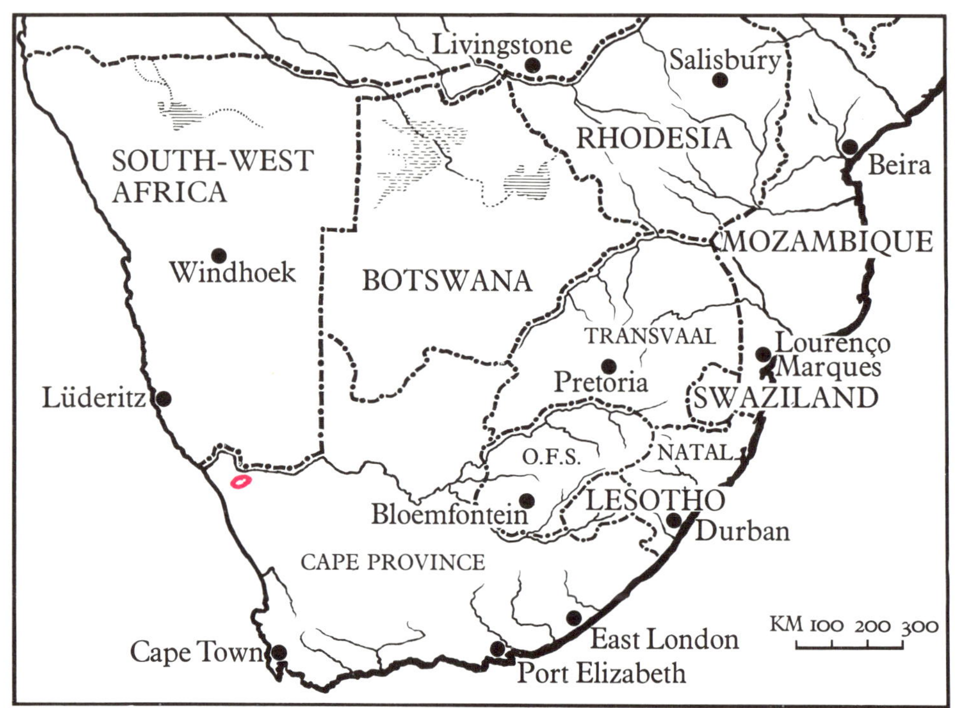

Herreanthus

Herreanthus Schwantes
Named in honour of H. Herre, Stellenbosch and the Greek *anthos* = flower.

Schwantes, Gartenwelt, 32: 514 with fig. 1928. - v. Pöllnitz, Aufteilung, 45. 1933. - Jacobsen, Sukk. 139; fig. 135. 1933. - Pax, Natürl. Pfl. 220. 1934. - Jacobsen, Succ. Pl. 186; fig. 106. 1935. - Jacobsen, Herre, Volk, Mesembr. 72 (in key). 99; fig. 6. 1950. - Phillips, Genera, 320. 1951. - Jacobsen, Handbuch, 1411; fig. 1168. 1955. - Schwantes, Kakt. & andere Sukk., 6: 107 with 2 figures. 1955. - Schwantes, Fl. Stones, 244, 340; pl. 66. 1957. - Jacobsen, Handbook, 953 (in system of Schwantes), 958 (in key of Bolus), 973 (in key of Herre & Volk), 1179; fig. 1403. 1960; Lexikon, 434, t.169/3. 1970.

Clumped, small plants, about 10 cm high, with fibrous roots and short branchlets. Leaves opposite, connate at the base for about 1 cm, semi-terete at the base, very thick, 3-angled, smooth, scarcely dotted, ending in a small mucro, upper surface flat, at lower side from base to middle, keel flattened, rather firm, light blue-green, often yellow-green, about 4 cm long and 1,5 cm thick. Flowers terminal, solitary, sessile or very short pedicelled, with 2 bracts, white, scented, after opening remaining open day and night for about 2 weeks, 3 cm in diam. Sepals 6, tubular-connate above the ovary. Petals in several series, tubularly connate, reflexed, white, posteriors changing to staminodes. Stamens in several series, conical, incurved. Ovary in the centre convex; glands not separated; placentas parietal. Stigmas 6, free, filiform. Capsule 6-locular with deep loculi; expanding keels parallel, with wings which are connate with each other; without loculi-roofs and tubercles. Seeds almost smooth, about 0,66 mm long, brown.

Species: 1, Namaqualand, near Steinkopf. (Type: *H. meyeri* Schwant.)

Hydrodea

Hydrodea N. E. Brown
From the Greek *hydrodes* = watery.

N. E. Brown, Gard. Chron., 78: 412 (in key), 1925; 84: 268. 1928. - v. Pöllnitz, Aufteilung, 45. 1933. - Pax, Natürl. Pfl. 215. 1934. - Jacobsen, Succ. Pl. 188, 189, fig. 188. 1935. - Dinter, Deutsche Kakt. Ges. 137, 200. 1937. - Jacobsen, Verzeichnis, 92. 1938. - Goossens, Blomplante, 148 (in key). 1940. - Jacobsen, Herre, Volk, Mesembr. 54 (in key), 99. 1950. - Phillips, Genera, 305. 1951. - Jacobsen, Handbuch, 1412, fig. 1169, 1170. 1955. - Schwantes, Fl. Stones, 20, 35, pl. 4. 1957. - *Jacobsen, Handbook*, 951 (in system of Schwantes), 957 (in key of Bolus), 966 (in key of Herre & Volk), 1179, fig. 1404, 1405. 1960. - Jacobsen, Lexikon, 434, t.169/4. 1970. - Friedrich in Merxmüller Prodromus, 84. (Mesembryanthemum). 1970.

Annual or biennial, small, branching, more or less procumbent, stems thick, cylindrical, papillose, forming mats or cushions a square of 0,75 m in area and up to 20 cm high; leaves opposite or alternate, mostly cylindrical, very fleshy, papillose, green, connate at the base, leaves on the floral branches mostly cylindrical, almost elliptical and clavate, obtuse, 3-4 cm long, 6-8 mm thick, very fleshy and juicy, mostly blood-red and therefore at Lüderitz called "bloody fingers" (i.e. blutige Finger); flowers terminal or axillary, short stalked or sessile, 20-22 mm in diameter, violet or lemon-yellow; calyx 5-lobed, with very long tips; petals forming a tube at base; stamens numerous, erect, filaments glabrous; corolla small, not or slightly exceeding the sepals; stigmas subulate; capsule 5-locular, valves of the expanded capsule erect, bifid, expanding keels parchment-like, stiff, parallel at the base, tips pointed, bent inside; seeds very small, ovoid, glabrous. Halophyte.

Species: 3, Swakopmund S.W.A., Lüderitz S.W.A., Isle of St. Helena. (Type: *H. sarcocalycantha* (Dint. et Berger) Dint.)

R.Luitharot
1971

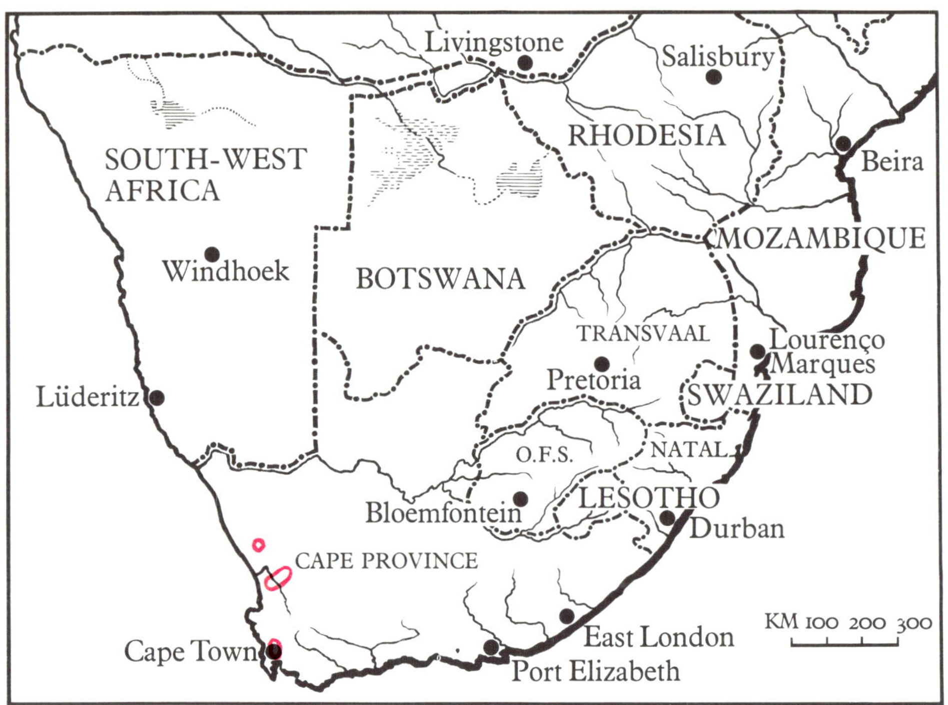

Hymenogyne

Hymenogyne (Haw.) N. E. Brown
From the Greek *hymen* = skin
and *gyne* = wife.

Haworth, Rev. pl. succ. 192. 1821. -
N. E. Brown, Gard. Chron. 78: 412 (in
key). 1925. - Phillips, Genera, 246. 1926.
- Bolus, Notes, 15; pl. 4A (capsule).
1928. - Schwantes, Mitt. Inst. Allg. Bot.
Hamburg, 8: 165 (capsule and seed).
1929. - N. E. Brown, Gard. Chron. 87:
71. 1930. - Bolus, Gard. Chron. 87: 267.
1930. - v. Pöllnitz, Aufteilung, 46. 1933. -
Jacobsen, Verzeichnis, 93. 1938. -
Goossens, Blomplante, 148 (in key).
1940. - Jacobsen, Herre, Volk, Mesembr.
58 (in key), 99. 1950. - Adamson,
Salter, Fl. Cape Pen. 376. 1950. - Bolus,
Notes, pl. 48, 49. 1950. - *Phillips,
Genera, 306. 1951.* - Jacobsen, Handbuch, 1416; fig. 1172. 1955. - Schwantes,
Fl. Stones, 35. 62. 342; fig. 17 (capsule
and seed). 1957. - Jacobsen, Handbook,
951 (in system of Schwantes), 956 (in
key of Bolus), 967 (in key of Herre &
Volk), 1181; fig. 1406, 1407. 1960; Lexikon, 434, t.170/1. 1970.

Annual herbs. Leaves opposite, radical
or clustered at the ends of the prostrate or
decumbent branches, flat, petiolate,
with the petiole dilated into a sheath at
the base, not connate. Flowers solitary,
pedicels long, erect, becoming prostrate
in fruit. Sepals 5. Petals numerous,
free or nearly so. Stamens numerous.
Style funnel-shaped with 12 papilla-like
stigmas on the surface of its rim. Capsule 9-12-locular, indehiscent when
wetted, separating in one-seeded carpels,
large, flat and broadly winged all round,
arranged in a single whorl around the
central axis; the tissue between the carpels drying up to wings when the fruit
ripens.

Species: 2, ranging from the Clanwilliam to the Cape district. (Type:
H. glabra (Ait.) Haw.)

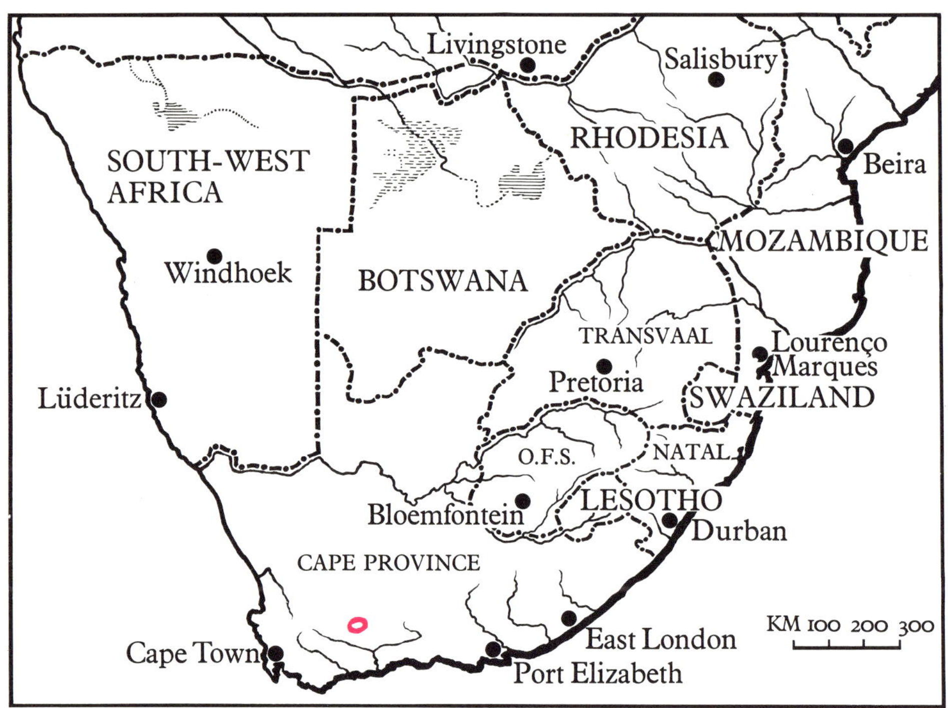

Imitaria

Imitaria N. E. Brown
From the Latin *imitari* = to imitate,
since the plant imitates the shape of
a *Conophytum*.

N. E. Brown, in Labarre (red.), Mesembr.
239; fig. 120. 1931. - v. Pöllnitz, Aufteilung, 46. 1933. - Jacobsen, Sukk. 141;
fig. 137. 1933; Succ. Pl. 190; fig. 169.
1935; Verzeichnis, 94. 1938. - Goossens,
Blomplante, 147 (in key). 1940. -
Jacobsen, Handbuch, 1382, 1416; fig.
1138. 1955. - Schwantes, Fl. Stones,
172, 340. 1957. - *Bolus, Notes, 388; pl.
103. 1958.* - Jacobsen, Handbook, 953
(in system of Schwantes), 961 (in key of
Bolus), 976 (in key of Herre & Volk),
1182; fig. 1408, 1960; Lexikon, 434,
t.170/3. 1970.

Gibbaeum nebrownii Tischer, J. Bot.
348. 1927; Kakt. & a. Sukk. 151. 1937. -
Jacobsen, Verzeichnis, 85 1938. - Jacobsen, Herre, Volk, Mesembr. 98. 1950.
- Nel, Gibbaeum Handbook, 22, 64.
1953.

Stemless clumps. Leaves united to
roundish bodies, the fissure dividing the
tip of the body into two convex lobes,
structure soft and fleshy, surface smooth,
velvety, set with hardly recognisable
fine hairs, intense grey-green or brownish, light green when cultivated, the tips
slightly translucent. Flower solitary
from the cleft, pink, about 2 cm in diam.
Otherwise like *Gibbaeum* but sepals
and petals form a tube.

Species: 1, Little Karroo (Eierpoort,
Dammetjies, in slate banks). (Type:
I. muirii N. E. Brown.)

R. Luithardt
1971

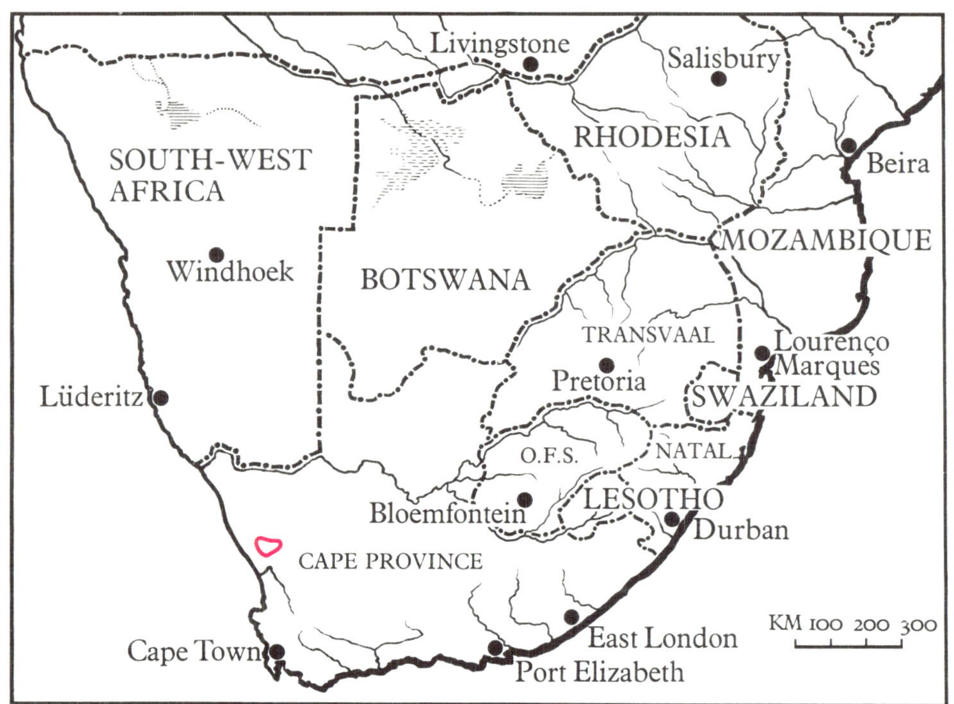

Jacobsenia

Jacobsenia L. Bolus
Named in honour of Dr H. Jacobsen,
Kiel, Germany.

Bolus, Notes, 255: pl. 90, 91. 1954. -
Jacobsen, Handbuch, 14; fig. 1173. 1955.
- Schwantes, Fl. Stones, 114, 341; pl.
37. 1957. - Jacobsen, Handbook, 952 (in
system of Schwantes), 960 (in key of
Bolus), 969 (in key of Herre & Volk),
1182; fig. 1409, 1410, 1411. 1960; Lexikon, 434; t.170/2. 1970.

Compact plants, 15-30 cm high, the
sterile branches clustered at the base,
the flowering ones erect and more or less
elongated with age (as in *Conophyllum,
Mitrophyllum* and *Mimetophytum*), herbaceous parts minutely papillate, glabrous or velvety; annual growths consisting of 2 pairs of leaves (as in *Conophyllum* etc.), but the pairs are similar
(not dimorphic) and are fully developed
at the same time, there being no definite
resting period; the pairs below the
flower similar to the other pairs. Leaves
cylindrical, one leaf of each pair a little
longer than the other. Flower stalked,
up to nearly 9 cm in diam., white or
lemon yellow, terminal, solitary. Sepals
5, 3 with membranous margins. Petals
3-4-series, acuminate. Stamens erect,
later on diffuse; staminodes o. Glands
5-angled, angles roundish, finely crenulated; stigmas 5, subulate. Capsule
5-locular, valve-wings broad, recurved;
expanding keels diverging, the margins
toothed, ending in awns; with loculi-
roofs but without tubercle. Seeds brown,
globose with navel.

Species: 2, Vanrhynsdorp distr.
(Nuwerus, Komkans). (Type: *J. kolbei*
(L. Bol.) L. Bol.)

Mes. Kolbii, L. Bolus.
2388/17 Dr. Kolbe, Namaqualan
Kirstenbosch Oct.,
1917.
495/16 Pillaus.
(ref. lost.)
Kirstenbosch

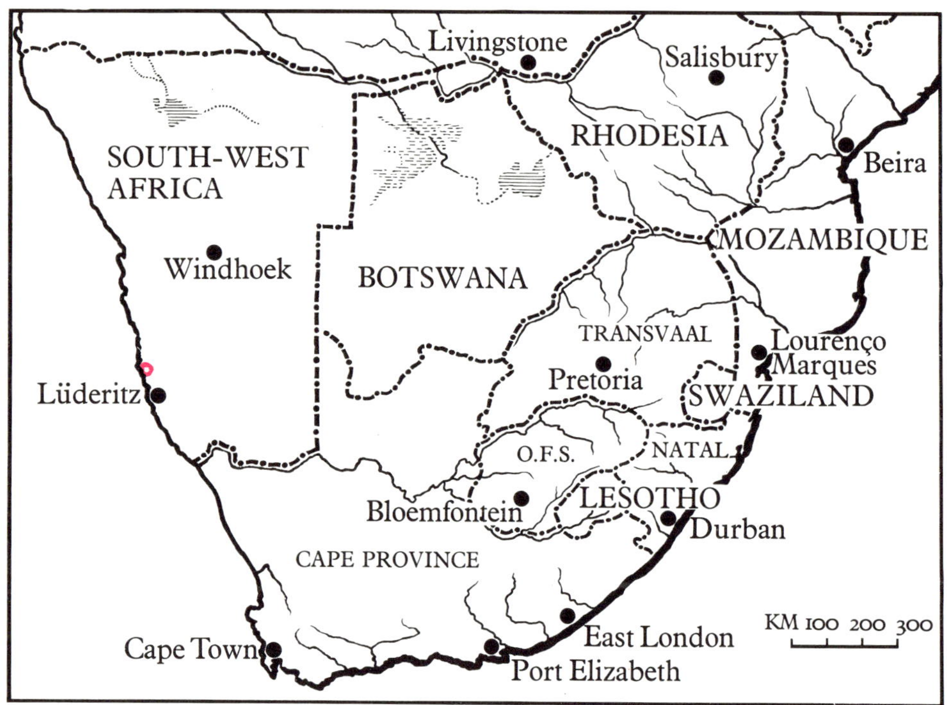

Jensenobotrya

Jensenobotrya Herre
Named in honour of the late
Mr Emil Jensen, and the Greek
botrya = bunch of grapes.

H. Herre, Sukkulentenkunde 4: 79 with 5 fig. 1951. - Jacobsen, Handbuch, 1417; fig. 1174, 1175. 1955. - Schwantes, Fl. Stones, 131, 339; pl. 41b, 42a & b. 1957. - Jacobsen, Handbook, 953 (in system of Schwantes), 958 (in key of Bolus), 971 (in key of Herre & Volk), 1184; fig. 1412, 1413, 1414. 1960. - Bolus, J. S. Afr. Bot. 32: 201, 234. 1966. - Jacobsen, Lexikon, 435, t.170/4. 1970. - Friedrich in Merxmüller, Prodromus, 58. 1970.

Procumbent shrub forming clumps, growing very old, with a thick stem and short woody branches, branchlets soft, later covered with old dry leaves, with 4-6 leaves, internodes 2-3 mm long. Leaves opposite, connate at the base, expanding towards the end, on the whole almost globose with 12-15 mm in diam., mostly reddish coloured, smooth not hairy, young leaves exceedingly juicy, older ones somewhat lax and thus very rugose. Flowers terminal, solitary, 2-2,5 cm in diam., pedicel 1 cm long. Sepals 5, unequal, 3 inner ones with membranous margins. Petals 2-series, free, acute, at the base white, towards the points magenta. Stamens numerous, diffuse; filaments at the base long and short haired; staminodes 0. Ovary flat or somewhat concave. Glands annular, darkgreen, crenulated. Stigmas 5, subulate, between the branches small styllodia. Placentas parietal. Capsule 5-locular, flat, to 1,2 cm in diam. with a more or less firm central column; expanding keels nearly without any colour, contiguous, later diverging with a long tape-like wing; without loculi-roofs and ubercles. Seeds oblong-ovate, about 1 mm long, glistening brown.

Species: 1, Lüderitz, S.W.A. (Spencer Bay, Delphinkopf). (Type: *J. lossowiana* Herre.)

Among all the members of the *Mesembryanthemaceae, Jensenobotrya* stands quite alone, owing to the fact that the clumps with their grape-like thick leaves are produced on a stem or branch which may be 100 to 200 years old. This can never happen with similar plants in the Karroo or any other place, as there is no fog which will keep the plants alive during prolonged droughts which always occur from time to time. This fact places this plant among the other two plant-wonders of South West Africa: the *Welwitschia* and the Naras (*Acanthosycios*). As far as we know *Jensenobotrya* occurs only on top of the Dolphin Hill, between Lüderitz and Walvis Bay which one can only reach from the sea or by helicopter from inland where a large lagoon is situated in front of it. Thus this unique plant is well protected by nature.

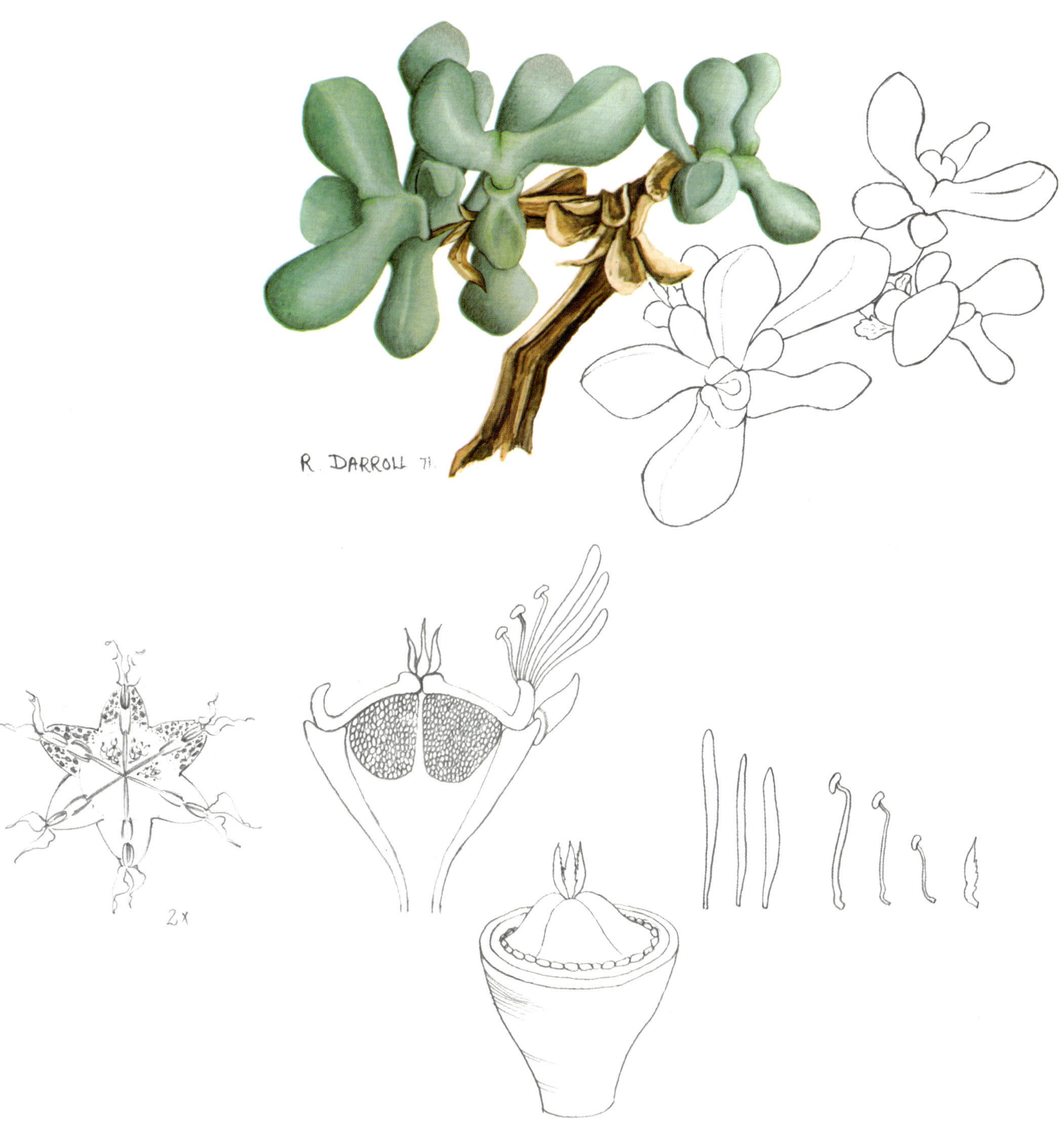

R. DARROLL 71
2x

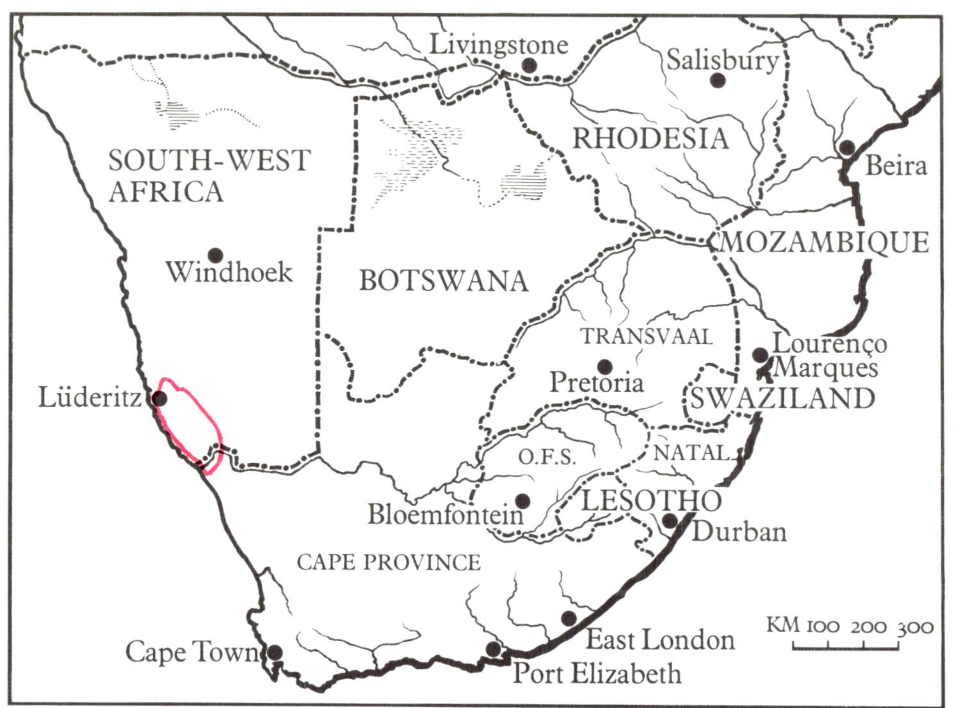

Juttadinteria

Juttadinteria Schwantes
Named in honour of the late
Mrs Jutta Dinter.

Schwantes, Z. Sukk. 3: 105. 1927; 2: 182. 1926. - v. Pöllnitz, Aufteilung, 46. 1933. - Jacobsen, Sukk. 141; fig. 138, 139. 1933. - Pax, Natürl. Pfl. 215. 1934. - Jacobsen, Succ. Pl. 190; fig. 170, 171. 1935; Verzeichnis, 94. 1938. - *Walgate in Bolus, Notes, 171 (with key of species); pl. 26-30. 1939.* - Jacobsen, Herre, Volk, Mesembr. 66 & 77 (in key), 99. 1950. - Phillips, Genera, 320. 1951. - Jacobsen, Handbuch, 1418; fig. 1176, 1177. 1955. - Schwantes, Fl. Stones, 136, 339; pl. 43a. 1967. - Jacobsen, Handbook, 953 (in system of Schwantes), 958 (in key of Bolus), 974 (in key of Herre & Volk), 1184; fig. 1415, 1416. 1960; Lexikon, 435, t.171/1, 3, 5. 1970. - Friedrich in Merxmüller, Prodromus, 58. 1970.

Subshrubs or caespitose plants or forming clumps, extremely succulent, with woody roots. Leaves decussate, very thick and half egg-shaped or broadly boat-shaped, very short or broadly linear or triangular or rhomboid-spatulate, connate and often vesicularly inflated, more or less semicylindrical, with a triangular transverse section towards the apex, with rather sharp or rounded edges and keel, upper leaf edges or often the keel also and the upper part of the lower side of the leaf covered with fairly blunt and broad protuberances or teeth; lower surface often prolonged forward like a chin; surface rather firm, white-grey, light yellow-green, bluish or whitish-green. Flowers solitary, more or less shortly pedicelled, without bracts, white, light to dark-violet red, often scented, opening in the afternoon, about 5 cm in diam. Sepals 4, 2 often toothed, lanceolate, carinate, 2 shorter ones with membranous margins. Petals 1-3-series, free or somewhat connate at the base, linear, very much reflexed, glittering on both sides, mostly longer than the sepals. Stamens numerous, filaments white, hairy at the base, column-like or conical-collected, often somewhat together at the base, anthers yellow; without staminodes. Ovary more or less flat or somewhat raised above, especially in the centre. Glands annular, crenulated. Placentas parietal. Stigmas 5-11, often 8, subulate, long, often somewhat longer than the stamens, free or with a short style (about 1-2 mm long), yellowish-green. Capsule 5-11-locular with a more or less firm central column; valve-wings well developed, following more or less broadly the keels as a limb, capsule inside flat; expanding keels only diverging near the apex; loculi-roofs rarely well-developed, mostly reduced to a limb; tubercle 0. Seeds more or less ball or pear-shaped, brownish, rough with roundish or oblong, often nearly wrinkle-like, glittering, smooth, prominent warts, to about 1 mm long.

Species: 12, South West Africa (Lüderitz, Aus), Cape Province (Richtersveld). (Type: *Juttadinteria kovisimontana* (Dint.) Schwant.)

Juttadinteria elizae 185

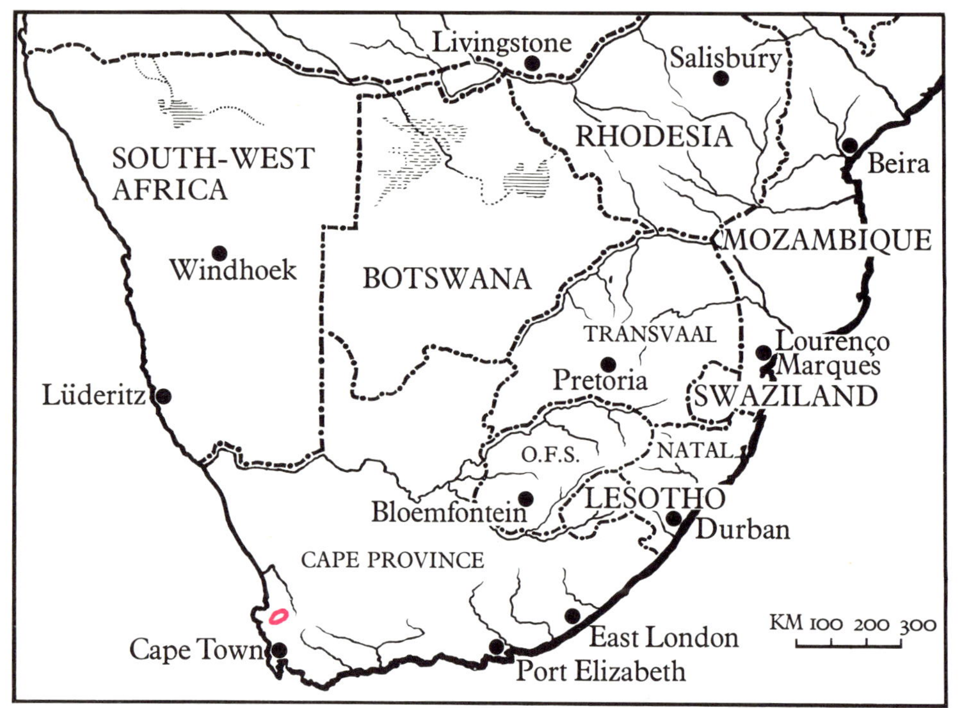

Kensitia

Kensitia Fedde
Named in honour of Dr L. Bolus
(neé Kensit).

Fedde, Fedde's Repert. 48: 11. 1940. -
Jacobsen, Handbuch, 1421. 1955. -
Schwantes, Fl. Stones, 339. 1957.
Bolus, Notes, *pl.* 115. 1958 - Jacobsen,
Handbook, 953 (in system of Schwantes),
960 (in key of Bolus), 970 (in key of
Herre & Volk), 1187; fig. 1417. 1960.
Piquetia N. E. Brown, Gard. Chron.
78: 433 (in key). 1925. - Phillips, Genera,
248. 1926. - Bolus, Notes, 8; pl. 1G
(flower). 1928 - v. Pöllnitz, Aufteilung,
60. 1933. - Pax, Natürl. Pfl. 218. 1934.
- Jacobsen, Verzeichnis, 159. 1938. -
Goossens, Blomplante, 147 (in key).
1940. - Jacobsen, Herre, Volk, Mesembr.
65 (in key), 120. 1950. - *Phillips, Genera,
312. 1951.* - Jacobsen, Lexikon, 436,
t.171/2. 1970.

Erect shrub, 30-60 cm high, many
branched, glabrous, branches forked,
reddish. Leaves opposite, somewhat connate at the base, the pairs of leaves
nearly equal in size, laterally compressed,
3-angled, dotted, glaucous, slightly incurved, the end acute with a fine point,
about 3 cm long and 1 cm thick. Flowers
solitary, terminal, to 5 cm in diam.,
pedicels 8-10 mm long. Sepals cupular
connate above the ovary. Petals various; outer petals numerous, spreading in
different planes, spatulate, clawed,
with obovate or elliptical blade, together
with the staminodes and the staminas
connate at the base and forming a firm
ring inside the saucer-like calyx-tube,
the claw white, the blade purple-rose,
oval or ovate slightly shorter than the
claw, the inner petals similar but much
shorter, the innermost filiform, united at
the base with the stamens and bent
inwards over them so that they might be
regarded as staminodes. Ovary inferior,
concave at the top. Placentas on the
outer walls. Stigmas 8-10, broad,

flattened, obtuse and minutely branched,
radiating closely upon the top of the
ovary in a star-like manner. Capsule 8-10-
locular, valves with the expanding
keels contiguous below, diverging above,
ending in awn-like points, without
wings; loculi roofed with flexible membranous wings, but without tubercles
at the openings. Seeds compressed,
subglobose.
Species: 1, Piketberg Mountain.
(Type: *K. pillansii* (Kensit) N. E.
Brown.)

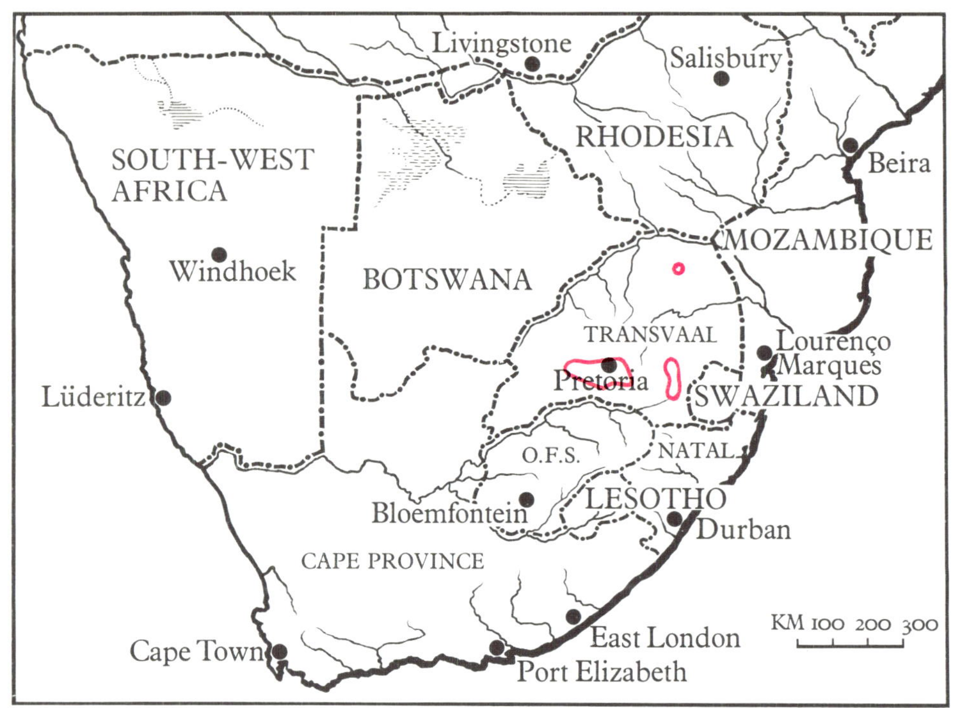

Khadia

Khadia N. E. Brown
Named after khadi, a drink which
the natives of the Transvaal prepare from
the roots of this plant.

N. E. Brown, Gard. Chron. 88: 279
(in key). 1930; 89: 279, 1931. - v. Pöllnitz, Aufteilung, 47. 1933. - Pax,
Natürl. Pfl. 215. 1934. - Bolus, Notes, 6;
pl. 1, 2, 3. 1936. - Goossens, Blomplante, 145 (in key). 1940. - Jacobsen,
Herre, Volk, Mesembr. 76 (in key), 100.
1950. - *Phillips, Genera, 306. 1951.* -
Jacobsen, Handbuch, 1421. 1955. -
Schwantes, Fl. Stones, 154, 339. 1957. -
Jacobsen, Handbook, 953 (in system of
Schwantes), 962 (in key of Bolus), 974
(in key of Herre & Volk), 1188 (with
key of species); fig. 1418. 1960; Lexikon, 436, t.171/4. 1970.

Very small succulent perennials, tufted,
with numerous very short branches
arising from a fleshy rootstock. Leaves
opposite, united at the base, crowded,
semi-terete below, trigonous at the
tips, smooth, pellucid dots. Flowers
solitary, pedicillate, pedicel often flattened, white-pink, about 4 cm in diam.
Sepals 5, forming a tube, 3 inner ones
with membranous margins. Petals 2-4-
series, diverging or reflexed. Stamens
numerous, anteriors recurved, posteriors
hairy, the outer ones often with barren
anthers; staminodes numerous, 3-4-
series, filiform, recurving. Ovary convex
above, later semi-globose. Glands
obscurely crenulated, dark-green.
Placentas on the outer walls. Stigmas
6-10, mostly 7-9, subulate or filiform,
linear acuminate, recurving. Capsule
6-10-locular, shortly and broadly obconic,
convex, with 6-10 raised and gaping
sutural ridges on the top; valves

with sharp keel; expanding keels about
$\frac{1}{2}$ as long as the valves, diverging from
the base, without wings or awns:
loculi somewhat acutely roofed with
stiffish loculi-wings; tubercle
present (small) or absent. Seeds slightly
compressed pear-shaped, minutely or
microscopically tuberculate.

Species: 6, only recorded from the
Transvaal. (Type: *K. acutipetala*
(N. E. Br.) N. E. Br.)

Khadia beswickii   189

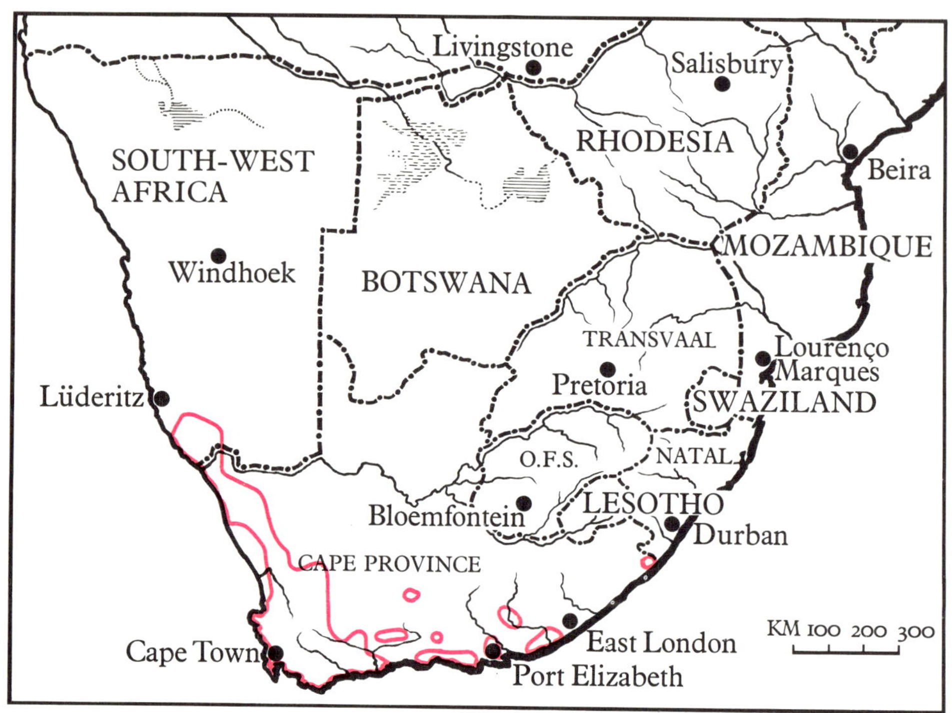

Lampranthus

Lampranthus N. E. Brown
From the Greek *lampros* = bright,
brilliant and *anthos* = flower.

N. E. Brown, Gard. Chron. 87: 72 (in key), 211. 1930. - v. Pöllnitz, Aufteilung, 47. 1933. - Pax, Natürl. Pfl. 216. 1934. - Jacobsen, Verzeichnis, 95. 1938. - Jacobsen, Herre, Volk, Mesembr. 63 (in key), 100; fig. 19. 1950. - Adamson, Salter, Fl. Cape Pen. 379. 1960. - Phillips, Genera, 320. 1951. - Jacobsen, Handbuch, 1422; fig. 1178-1181. 1955. - Schwantes, Fl. Stones, 64, 338; fig. 18. 1957. - Jacobsen, Handbook, 952 (in system of Schwantes), 959 & 961 (in key of Bolus), 969 (in key of Herre & Volk), 1190; fig. 1419-1423a. 1960; Lexikon, 437, t.172/1. 1970. - Friedrich in Merxmüller, Prodromus, 64. 1970.

Glabrous subshrubs with erect, spreading or prostrate, mostly compressed branches. Leaves opposite, usually shortly connate, always equal and similar, terete, semi-terete or 3-angled blunt or tapering, more or less curved, mostly not dotted, rarely with prominent scattered dots, often smooth, to about 5 cm long. Flowers terminal or axillary, in cymes or solitary, often pedicellated, rarely sessile, all colours except blue, about 6 cm in diam. Sepals 5. Petals free. Stamens erect, more or less diffuse, not column-like or conically collected; staminodes usually absent. Ovary flat or somewhat raised above. Glands not separated. Placentas parietal. Capsule 5- rarely 4-, 6- or 7-locular, with wings on the valves; expanding keels diverging; loculi-roofs present; tubercle none. Seeds nearly black, rough, with nipple, pear-shaped.

Species: 178. Cape Province: Namaqualand through the coastal area to the Eastern Province, and Natal. One species in Australia. (Type: *L. tenuifolius* (L) Schwant.)

Lampranthus vanzyliae 191

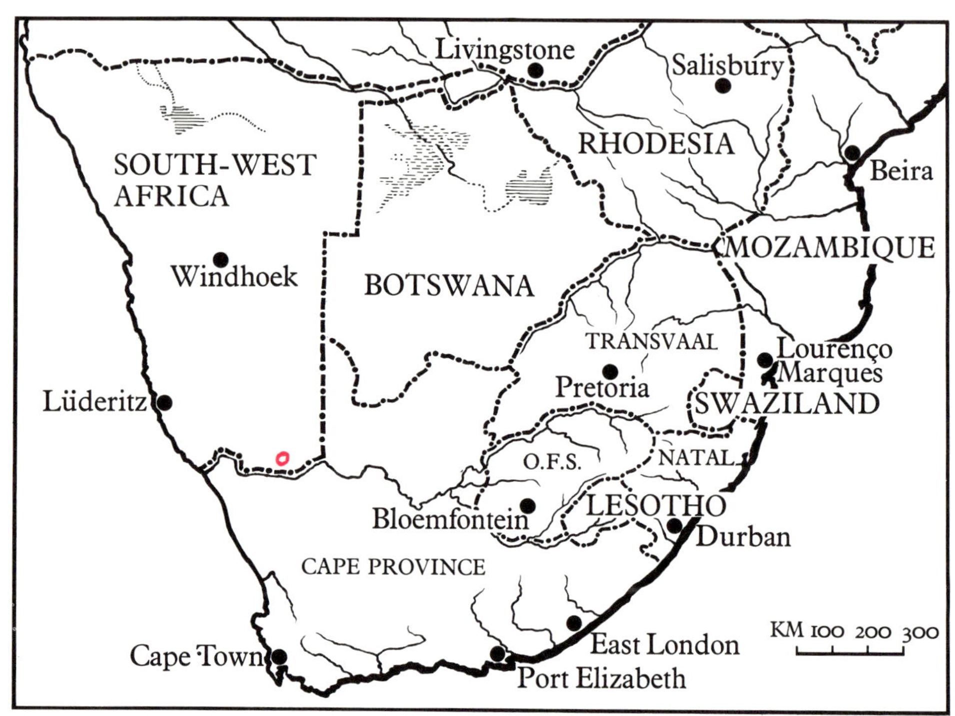

Lapidaria

Lapidaria Schwantes
From the Latin *lapis* = stone; lapidaria
= a group of stones.

Dinter et Schwantes, Möllers Gärtnerztg. 42: 223. 1927. - N. E. Brown,
Gard. Chron. 84: 472, 492. 1928. - Bolus,
Notes, 41; pl. 15 (coloured). 1928. -
Labarre (red.), Mesembr. 241; fig. 121,
122. 1931. - v. Pöllnitz, Aufteilung,
48. 1933. - Jacobsen, Sukk. 142; fig. 140.
1933. - Pax, Natürl. Pfl. 216. 1934. -
Jacobsen, Succ. Pl. 192; fig. 172. 1935;
Verzeichnis, 102. 1938. - Goossens,
Blomplante, 145 (in key). 1940. - Jacobsen, Herre, Volk, Mesembr. 71 (in key),
102; fig. 23. 1950. - *Phillips, Genera,
306, 1951.* - Jacobsen, Handbuch, 1453;
fig. 1182, 1183. 1955. - Schwantes, Fl.
Stones, 177. 340; fig. 29; pl. 53a. 1957. -
Jacobsen, Handbook, 953 (in system
of Schwantes, 959 (in key of Bolus), 915
(in key of Herre & Volk), 1214; fig.
1424, 1425. 1960; Lexikon, 447, t.172/2.
1970. - Friedrich in Merxmüller,
Prodromus, 66. 1970.

Stemless, small perennial with 6-8
leaves. Leaves shortly united at the base,
opposite, flat or slightly concave on the
face, lower side sharply keeled, towards
the apex obtuse 3-angled, not dotted,
with a minutely granulated surface,
white-green, edges reddish. Flowers
solitary, terminal, pedicellate, but with
the pedicel concealed in the bases of the
leaves, golden-yellow, to 5 cm in diam.
Sepals 7, rather equal, 3 broad marginated, dotted. Petals numerous,
3-series, somewhat connate at the base,
ascending-spreading, so as to form a
cup. Stamens numerous, erect, the
inner much shorter and more inflexed
than the outer, partly concealed in the
cup of the corolla; filaments bearded at
the base. Staminodes present. Ovary
inferior, flattish at the top. Placentas on
the outer walls. Glands obscure, annular,
crenulated. Stigmas 6-7 free, slender,
subulate or filiform, ascending, with
recurved tips. Capsule 6-7-locular, with
a winged pedicel, shortly and broadly
obconic, flat, with raised sutures at the

top; valves widely spreading or recurved
when expanded; expanding keels contiguous below, diverging at the apical
part, with broad membranous margins;
loculi-roofs present; tubercle none.
Seeds small, pear-shaped, light-brown.

Species: 1, South West Africa, near
Warmbad. (Type: *L. margaretae*
(N. E. Br.) Schw.)

MM.Page. 5.1924
Width 1.2cm
7. valves

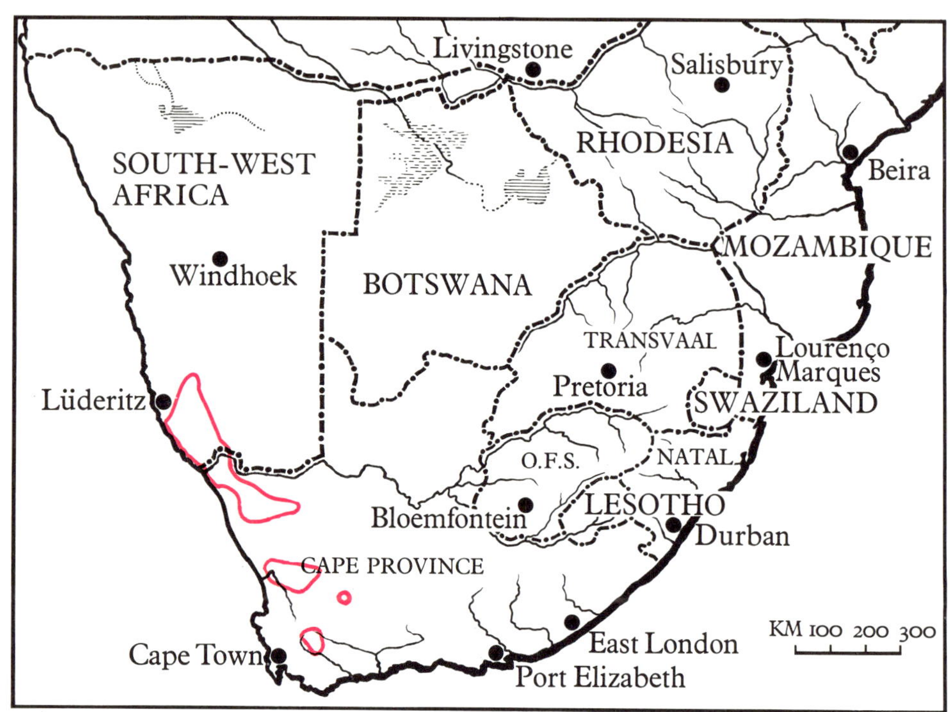

Leipoldtia

Leipoldtia L. Bolus
Named in honour of the late
Dr C. L. Leipoldt.

Bolus, Fl. Pl. S. Afr. 7: pl. 256. 1927. - N. E. Brown, Gard. Chron. 87: 72 (in key). 1930. - Pax, Natürl. Pfl. 216. 1931. - Jacobsen, Verzeichnis, 103. 1933. - Goossens, Blomplante, 147 (in key). 1940. - Jacobsen, Herre, Volk, Mesembr. 68 (in key), 102. 1950. - *Phillips, Genera, 307. 1951.* - Jacobsen, Handbuch, 1453. 1955. - Schwantes, Fl. Stones, 104, 338. 1957. - Jacobsen, Handbook, 952 (in system of Schwantes), 964 (in key of Bolus), 971 (in key of Herre & Volk), 1214. 1960; Lexikon, 447, 1970. - Friedrich in Merxmüller, Prodromus, 67. 1970.

Small or medium sized, erect or decumbent shrubs with dotted leaves. Leaves opposite, sheathed at the base, only slightly connate but easily separable, trigonous, turgid, dilated at the base, sometimes with raised scattered dots. Flowers solitary, very often 3-nate, in much-branched cymes, rarely sessile, peduncle terete, constricted under the calyx as if jointed, bracteate a little below the middle, persisting after flowering, hardened or sometimes spinescent, opening during daytime, whitish or light to dark violet-red, about 2,5 cm in diam. Sepals 5, very rarely 6, subequal, with membranous margins. Petals 1-2-series, linear-spathulate, obtuse. Stamens erect, column-like or conically collected, filaments of inner stamens bearded; staminodes none or only a few. Ovary inferior, slightly convex above. Placentas parietal. Glands crenulate. Stigmas 10-12, often crowded, narrowly subulate, acuminate. Capsule 10-12-locular; keels contiguous below, diverging upwards as the valves spread out flat, with wings, broadly semi-ovate; loculi-roof contiguous extending a little beyond the middle of the loculus; tubercle obovate, often saddle-like or incurved above. Seeds with nipple, rough.

Species: 20, South West Africa: Aus. Cape Province: Richtersveld (Holgat, Karrachab, Aughrabies, Sendlings-drift), Vanrhynsdorp, Pofadder, Worcester, Robertson, Oudtshoorn. (Type: *L. constricta* L. Bolus.)

M.M.Page. VII. 1924.

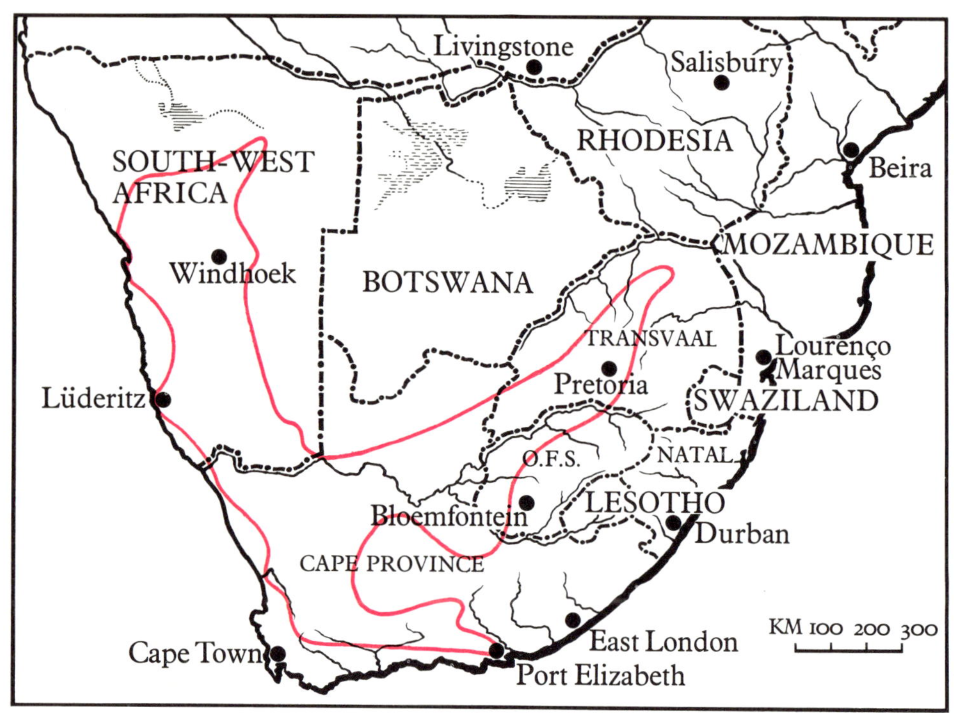

Lithops

Lithops N. E. Brown
From the Greek *lithos* = stone and
ops = appearance.

N. E. Brown, Gard. Chron. 71: 44; fig. 28, 33. 1922; 78: 413 (in key). 1925; 79: 80 (with fig.) 1926. - Phillips, Genera, 242, 1926. - Bolus, Notes, 12. 1928. - Tischer, Z. Sukk. 3: 220. 1928. - Labarre (red.), Mesembr. 31-33, 93, 94, 98, 123-138, 242; fig. 123-138. 1931. - v. Pöllnitz, Aufteilung, 48. 1933. - Jacobsen, Sukk. 143; fig. 141-160. 1933. - Pax, Natürl. Pfl. 216. 1934. - Jacobsen, Succ. Pl. 192; fig. 173-197. 1935; Verzeichnis, 104. 1938. - Goossens, Blomplante, 147 (in key), 102; fig. 2. 1950. - G. C. Nel. Lithops, 1947. - Jacobsen, Herre, Volk, Mesembr. 80 (in key), 102; fig. 2. 1950. - *Phillips, Genera, 307. 1951.* - Jacobsen, Handbuch, 1456, fig. 1184-1247. 1955. - Schwantes, Fl. Stones, 180; fig. 30, 31; pl. 54-65. 1957. - Jacobsen, Handbook, 953 (in system of Schwantes), 958 (in key of Bolus), 976 (in key of Herre & Volk), 1217; fig. 1426a-1500. 1960. - Dr H. W. de Boer & Dr B. K. Boom, Nat. Cact. & Succ. Journal, 19: 34 (analytical key with fig. all spec. et var.) 1964. - Jacobsen, Lexikon, 448, t.173-181. 1970. - Friedrich in Merxmüller, Prodromus, 68, 1970.

Clumps, nearly without stems, often with long tap-roots, branchlets each with 2 bodies, sometimes bodies also solitary or in small groups, more or less deep or flat in the ground, turbinate to terete. Leaves opposite, each body consists of 2 very connate leaves with truncate apex and with only a fissure open between the 2 leaves, free apex of the leaf truncate and with more or less translucent window or with foveolate windows or window-like spots, apex flat or somewhat raised, more or less rough, as regards the markings on the truncate apex as also the form of the bodies there are all changes possible, also as regards the colour which may be white-blue-grey to dove-grey to the darkest brown, often with red spots, which are very distinct without pocket-lens, often the marking is very obscure, too vague, too light or too dark, size of the bodies also very different from 2 cm to 5 cm in diam., margin sometimes absent or inner margin straight or with a triangular clear part at the junctions of two margins. Flowers solitary, rarely 2-3, from the centre of the fissure, pedicel with bractlets at base different in length, winged, opening in the afternoon, yellow to white or orange, size varied, about 1,5-4 cm in diam. Sepals 4-7, free to the ovary, slightly exserted above the leaves, not membranous. Petals many in 1-4-rows, linear, obtuse, rounded or with a notch at the apex, reflexed. Stamens many, column-like or conically collected, sometimes papillate on the lower part. Ovary inferior, style short or almost absent. Placentas parietal. Stigmas 4-7, filiform. Capsule 4-7-locular, small; valves with 1 stout central expanding keel, split at the apex, with broad marginate wings; loculi-roofs reduced to a limb or none, tubercle none. Seeds minute, with a nipple at one end, round or rotund, smooth or rugulose, coloured dark-brown or partly dark-brown, opaque or one part transparent.

Species: 79 and 10 varieties of species (Jacobsen). 40 and 6 varieties of species Dr. de Boer): South West Africa, the central districts of the Cape Province; Transvaal: near Pretoria.

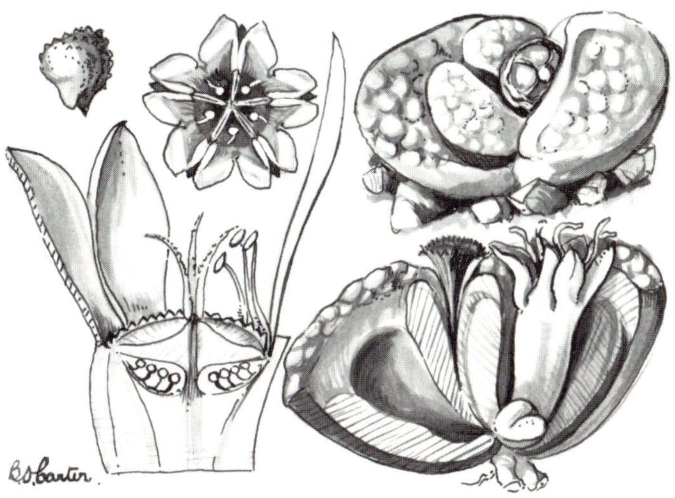

Lithops pseudotruncatella   197

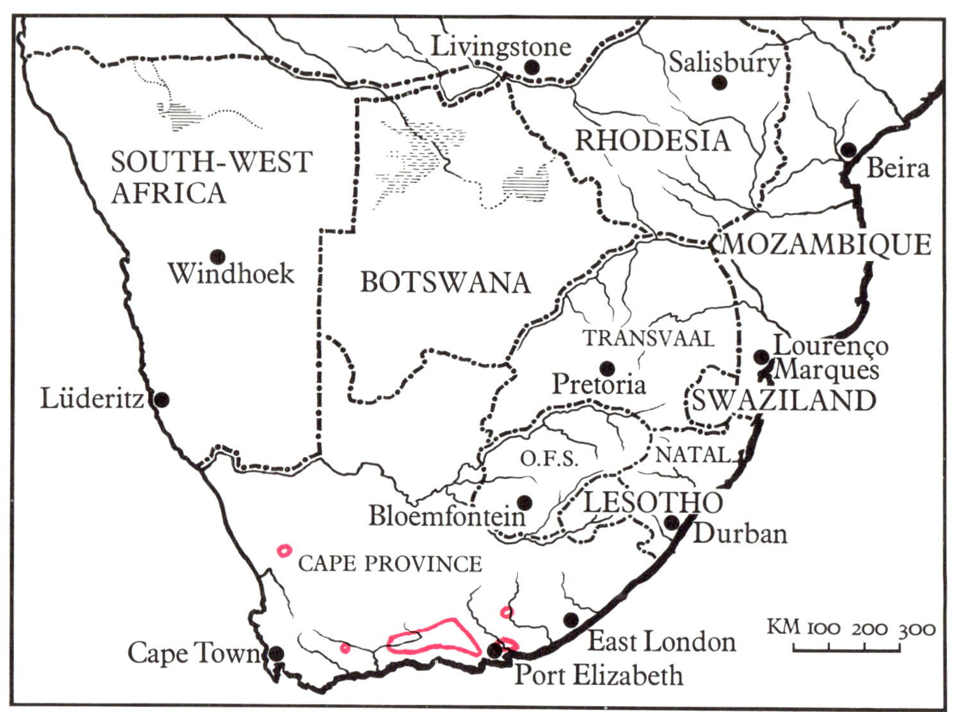

Machairophyllum

Machairophyllum Schwantes
From the Greek *machaira* = sabre
and *phyllon* = leaf.

Schwantes, Möllers Gärtnerztg. 42: 187.
1927. - v. Pöllnitz, Aufteilung, 48. 1933.
- Jacobsen, Sukk. 152. 1933. - Pax,
Natürl. Pfl. 220. 1934. - Jacobsen, Succ.
Pl. 206. 1935; Verzeichnis, 109. 1938. -
Jacobsen, Herre, Volk, Mesembr. 76
(in key) 103. 1950. - Phillips, Genera,
320. 1951. - Jacobsen, Handbuch, 1499;
fig. 1248. 1955. - Schwantes, Fl. Stones,
93, 338; pl. 24b. 1957. - Bolus, Notes,
pl. 136, 137. 1958. - Jacobsen, Handbook, 952 (in system of Schwantes), 962
(in key of Bolus), 974 (in key of Herre &
Volk), 1252; fig. 1501. 1960. - *Bolus,
Notes, J. S. Afr. Bot. 157 (with key).
1960.* - Jacobsen, Lexikon, 462, t.182/1.
1970.

Perissolobus N. E. Brown, Gard.
Chron. 88: 270 (in key). 1930; 89: 294.
1931, - v. Pöllnitz, Aufteilung, 59, 1933. -
Pax, Natürl. Pfl. 217. 1934. - Bolus,
Notes, 135. 1939. - Goossens, Blom-
plante, 146 (in key). 1940.

Plants densely compact, glabrous, with
age attaining 100 cm in diam. Vegetative
parts smooth and polished, pale glaucous
or canescent or more rarely green;
internodes enclosed in leaf-sheaths.
Leaves opposite, rather crowded, linear-
lanceolate ("dagger shaped") or rarely
oblong or subrhomboid, keeled in the
upper part, in profile either not narrowed
upwards and obliquely truncate, or one
leaf of a pair acute, or both leaves of
a pair abruptly acute, to 10 cm long and
2 cm broad. Flowers solitary or 1-2-
ternate, nocturnal or vespertine, to
6,5 cm in diam., pedunculate, peduncles
bracteate. Sepals 5-8, acute, acuminate,
or rarely obtuse, unequal, membranous
marginated. Petals densely 3-5-seriate,
yellow, golden or orange, outside copper-
red. Stamens remaining erect and
forming a cylinder throughout the life
of the flower, filaments epapillate;
staminodes absent. Ovary somewhat
raised, or rising abruptly from the margin
and approximate, or the outer part of
the ovary flat or concave with the lobes
in the middle approximate and obtusely
compressed. Glands rather small, distant or approximate or contiguous,
sometimes with a nectariferous depression
at the base. Stigmas 5-15, slender,
reaching the height, or a little less of
the longest stamens. Capsule 5-15-
locular, centre raised, sutures strongly
compressed, margins divergent, valves
shortly and very narrowly winged; expanding keels parallel below, widened
and diverging upwards, lacerate, the
awns long and nearly reaching the apex
of the valve (in *M. albidum* crossing
each other); loculi with a covering membrane; tubercle rather small and deeply
set. Seeds obovate, very minutely
granulate (1 mm long in *M. albidum*).

Species: 9, Cape Province: Barrydale,
Oudtshoorn, Prince Albert, George
and the Eastern Province. (Type:
M. albidum (L.) Schwant.)

Machairophyllum cookii 199

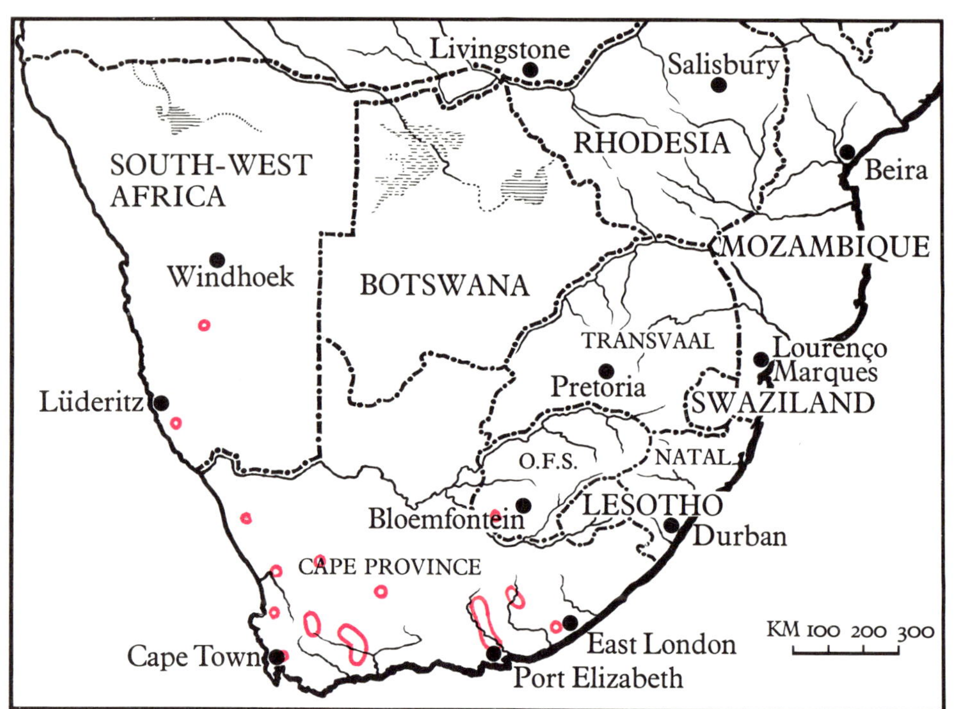

Malephora

Malephora N. E. Brown
From the Greek *male* = arm-hole and
pherein = to bear.

N. E. Brown, Gard. Chron. 81: 12 (in key). 1927. - Schwantes, Möllers Gärtnerztg. 43, 7. 1928. - v. Pöllnitz, Aufteilung, 49. 1933. : Pax, Natürl. Pfl. 216. 1934. - Jacobsen, Verzeichnis, 110. 1938. - Goossens, Blomplante, 146 (in key). 1940. - Jacobsen, Herre, Volk, Mesembr. 68 (in key), 103. 1950. - Phillips, Genera, 321. 1951. - Jacobsen, Handbuch, 1500. 1955. - Schwantes, Fl. Stones, 96, 340; pl. 92a (capsule) 1957. - Jacobsen, Handbook, 953 (in system of Schwantes), 964 (in key of Bolus), 971 (in key of Herre & Volk), 1253; fig. 1502, 1503; fig. 1186 (stamina), 1187 (calyx). 1960; Lexikon, 463, t./182/3 & 4. 1970 - Friedrich in Merxmüller, Prodromus, 56. (Hymenocyclus). 1970.

Crocanthus L. Bolus, Fl. Pl. S. Afr. 7: pl. 255. 1927.

Hymenocyclus Dinter et Schwantes, Möllers Gärtnerztg. 42: 27. 1927; Z. Sukk. 3: 14 (in key), 16. 1927. - Bolus, Notes, 94; fig. 21BC, 5 & 6. 1928. - v. Pöllnitz, Aufteilung, 46. 1933. - Jacobsen, Sukk. 140; fig. 136. 1933. - Pax, Natürl. Pfl. 220. 1934. - Jacobsen, Verzeichnis, 92. 1938. - Jacobsen, Herre, Volk, Mesembr. 64 (in key), 99. 1950. - Phillips, Genera, 320. 1951. - Jacobsen, Handbuch, 1413; fig. 1171. 1955.

Shrubs, erect, or decumbent, or prostrate, with distinct internodes. Leaves only slightly connate at the base, opposite, prismatic, oblong, semi-terete, smooth, not dotted, not papillate, often with a bluish coating of wax, up to 5 cm long. Flowers solitary or axillary, with short pedicels, without bracts, yellow, golden-yellow or red, to 5 cm in diam. Sepals 4-6. Petals free, thin, changing into staminodes. Stamens numerous, many series, erect or collected, at their base often hairy. Ovary flat. Glands not separated. Placentas parietal. Stigmas 8-11, very small, broad plumose. Capsule 8-11-locular, with valve-wings; expanding keels parallel, later diverging; loculi-roofs present; without tubercles or with reduced ones, but in some species one with 2 heads. Seeds flat, with tubercles in rows, very rough.

Species: 15, from Namaqualand through the Karroo to Fauresmith and in the Eastern Cape Province. (Type: *M. mollis* (Ait.) N. E. Brown.)

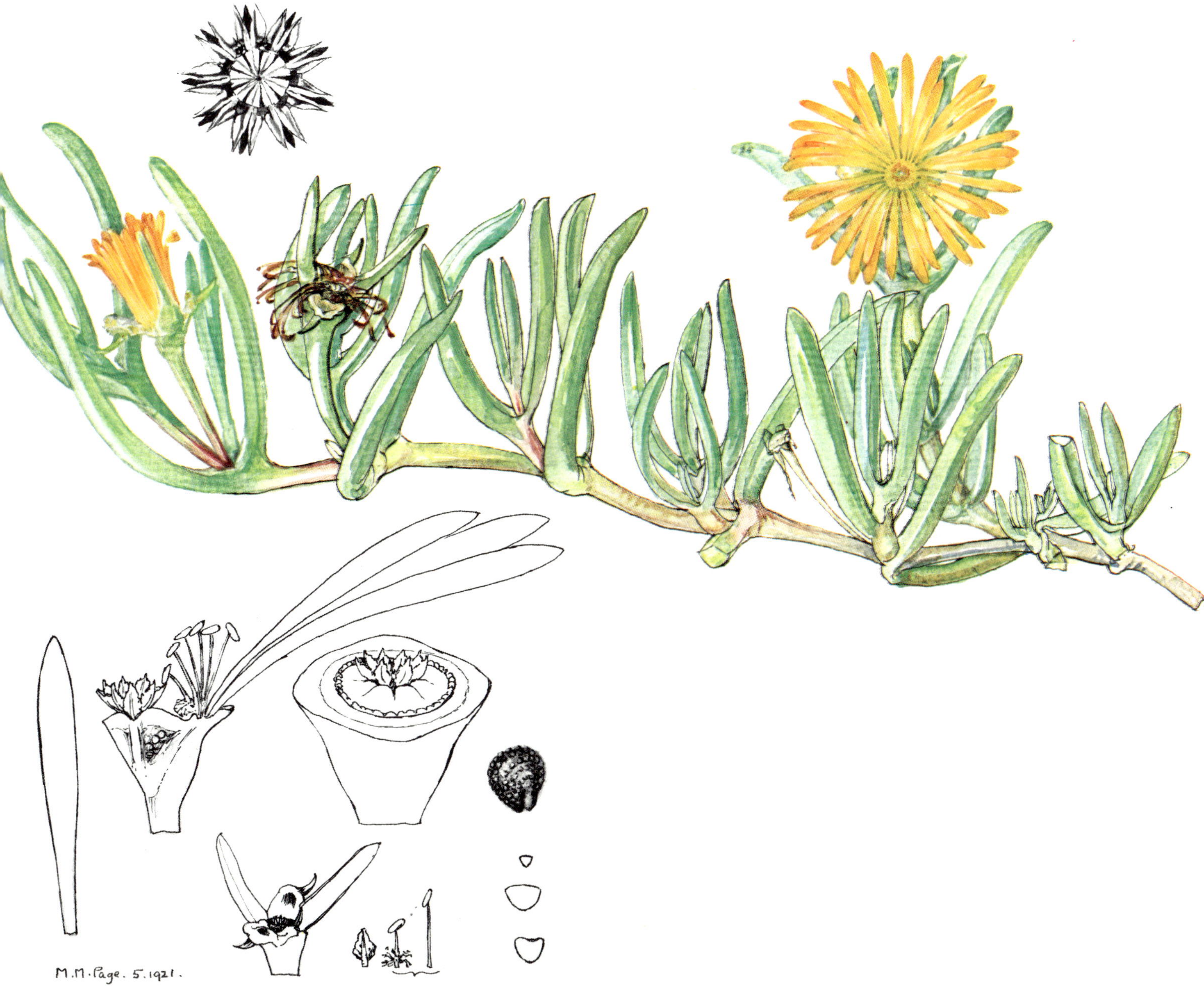

Malephora lutea 201

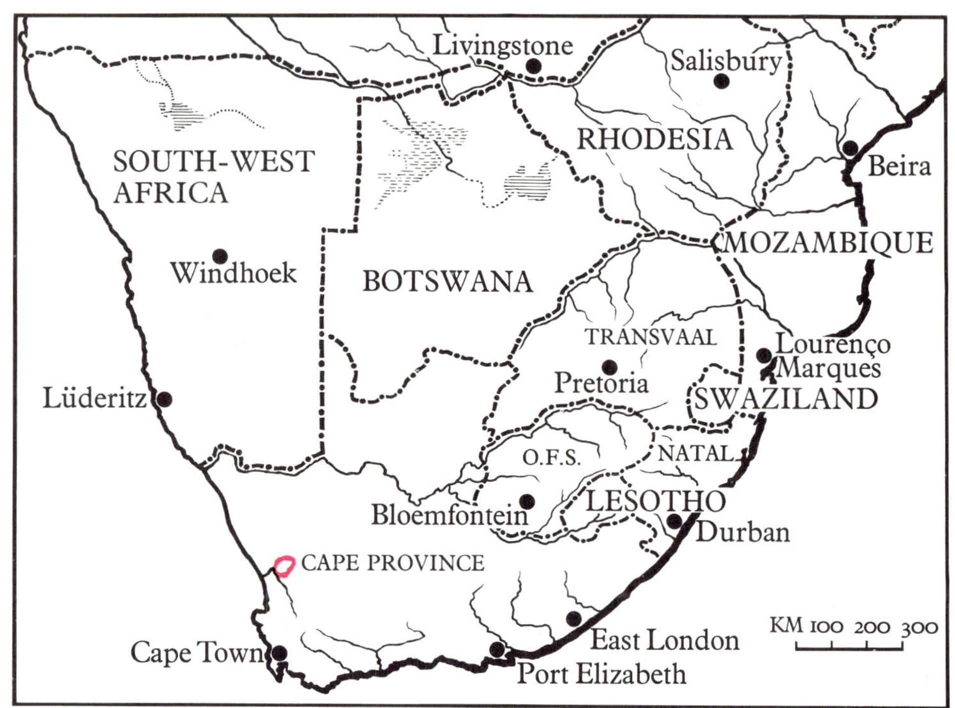

Maughaniella

Maughaniella L. Bolus
Named in honour of the late
Dr H. Maughan Brown.

Bolus, Notes, J. S. Afr. Bot. 27: 264.
1961.
Maughania N. E. Brown, *J. Cact. &
Succ. Soc. Am. 2: 390. 1930/31.* - Jacobsen, Verzeichnis, 110. 1938. - Jacobsen,
Herre, Volk, Mesembr. 104. 1950. -
Bolus, Notes, pl. 46, 47. 1950; 260.
1954. - Jacobsen, Handbuch, 1501; fig.
1249. 1955. - Schwantes, Fl. Stones,
126; pl. 40a & b. 1957. - Jacobsen, Handbook, 954 (in system of Schwantes),
960 (in key of Bolus), 973 (in key of
Herre & Volk), 1255; fig. 1504. 1960;
Lexikon, 464, t.182/2. 1970.
Monilaria luckhoffii L. Bolus, *Notes,
222, 1930*; 331. 1935; 260. 1954.

Perennial, with a swollen caudex, glabrous, with fibrous roots. Leaves opposite, forming a small half-spherical body,
composed of a series of truncate, firm
sheaths surrounding the body. Leaves
opposite, sub-erect, linear oblong, obtuse,
flat above, very rounded, soft and
pulpy in texture, finely papulose, glabrous, green, mostly reddish. Flowers between the leaves, solitary, terminal,
with a small pedicel, white to pinkish,
nearly 2 cm in diam. Sepals 5, free.
Petals free, narrow, acuminate. Stamens
not numerous. Placentas parietal. Stigmas
5-6, style about 1 mm long in the
young bud. Capsule 5-6-locular, very
thin like that of *Diplosoma*, shallowly
hemispherical below, acutely conical
above with prominent sutures; expanding keels dark brown, diverging from the
base, flattish and triangular in outline
at the basal part, with broad membranous marginal wings that are rounded
at the apex and overlap one another at
the base between the valves; loculus
roofs flat, sub-transparent; tubercle
none. Seeds numerous, minute, pearshaped, smooth.

Species: 1, Cape Province: Vanrhynsdorp, Knersvlakte and near Koekenaap. (Type: *Maughaniella insignis*
(N. E. Br.) L. Bol.)

B.O.Carter

11-10-1932

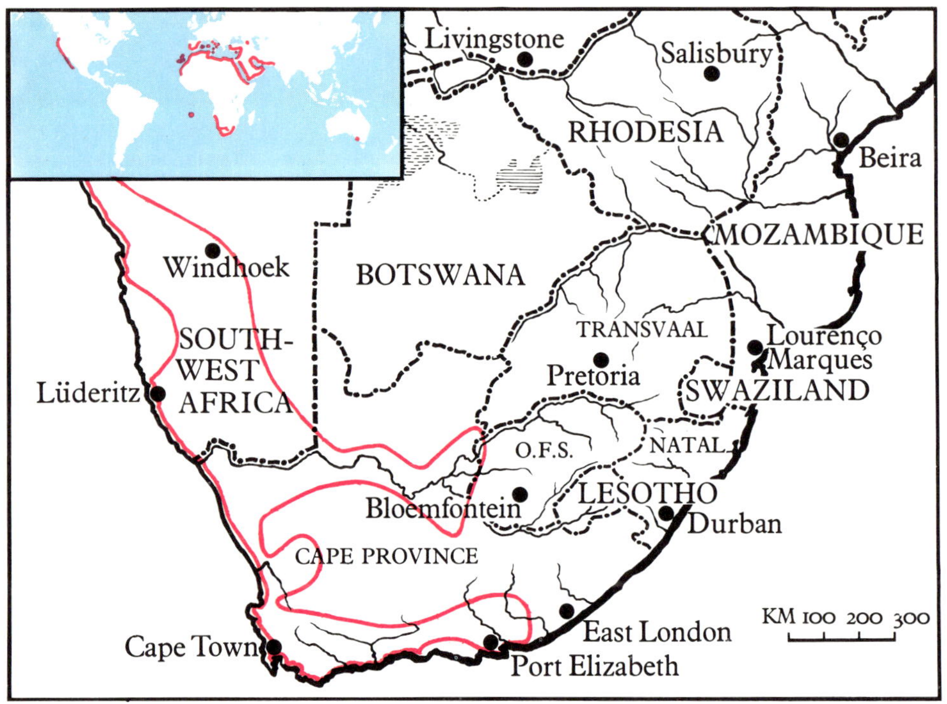

Mesembryanthemum

Mesembryanthemum (L.) emend.
L. Bolus
From the Greek *mesembria* = midday
and *anthemon* = flower.

Linnaeus, Spec. pl. 688. 1753. - emend.
L. Bolus, Notes, 164; fig. 25. 1939. -
Jacobsen, Herre, Volk, Mesembr., 54 (in
key), 104. - Adamson, Salter, Fl. Cape
Pen. 1950. - Jacobsen, Handbuch, 1503;
fig. 1250-1254. 1955. - Schwantes,
Fl. Stones, 13, fig. 2 (flowers), 3 (capsule); pl. 1a, 2a & b; 35. 1957. - Jacobsen, Handbook, 951 (in system of
Schwantes), 957 (in key of Bolus), 966
(in key of Herre & Volk), 1257, fig.
1505-1509. 1960; Lexikon, 464, t.183/1-3.
1970. - Friedrich in Merxmüller, Prodromus, 81. 1970.
Cryophytum N. E. Brown Gard.
Chron. 78: 412. 1925. - *Derenbergiella*
Schwant. Zeitchr. Sukk. 2: 138. 1925. -
Callistigma Dint. et Schwant., Gartenwelt, 32: 644. 1928.

Annual or biennial fleshy herbs, often
prostrate or with erect main stem and
branches, all parts covered with more or
less large papillae. Leaves at first opposite but later on alternate, sessile or
shortly petiolate, flat or terete, often
very large with undulated margin. Flower
solitary, sessile or shortly petiolate,
opposite the leaves or in branched cymes,
with or without bracts, all colours, except blue. Sepals 4-5, 2 often foliaceous,
forming a short tube. Petals numerous,
slender, linear or linear-filiform, united
at the base into a short tube, non-
petaloid staminodes absent or present.
Stamens numerous, erect, arising within
the corolla-tube, filaments several times
as long as the anthers. Ovary partly or
more than $\frac{1}{2}$ superior; placentas axile.
Glands 5, stigmas 5, rarely 4, erect, filiform. Capsule 5- (rarely 4) locular,
valves with the expanding keels contiguous, forming a central and often
acute keel, with broad inflexed or erect
submembranous wings on each side;
loculi open, without loculus-wings or
tubercles. Seeds numerous, small,
globose or compressed, sometimes D-
shaped, very minutely tuberculate, light
or dark-brown.

Species: About 74, South West
Africa, Namaqualand, Vanrhynsdorp
and through the coastal area to the Uitenhage and Albany districts, also from
Cradock, Graaff-Reinet, Prince Albert
and Oudtshoorn districts. Some from
the Mediterranean region, Arabia, Per-
sia, Baluchistan, Kurdistan and the
Atlantic Islands. (Type: *M. crystallinum*
(L.) N. E. Br.)

M.M.Page · XII.1922.

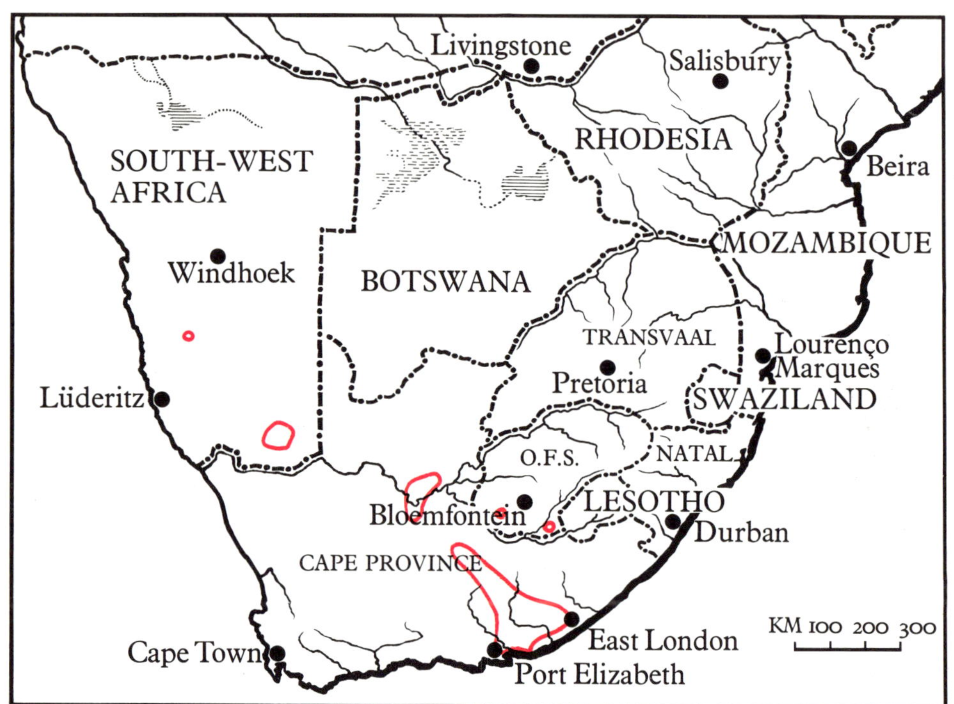

Mestoklema

Mestoklema N. E. Brown
From the Greek *mestos* = full and
klema = a small branch.

N. E. Brown, Gard. Chron. 100: 164.
1936. - Jacobsen, Herre, Volk, Mesembr.
118. 1950. - *Phillips, Genera, 308. 1951.*
- Jacobsen, Handbuch, 1549. 1955. -
Schwantes, Fl. Stones, 86. 1957. -
Jacobsen, Handbook, 952 (in system of
Schwantes), 961 (in key of Bolus), 969
(in key of Herre & Volk), 1293; fig.
1510. 1960; Lexikon, 467, t.184/1. 1970.
Friedrich in Merxmüller, Prodromus,
87. 1970.

Shrubs or shrublets, in one species with
a tree-like trunk, bushily and often
intricately branched, sometimes with a
tuberous rootstock; young branches
minutely papulose, slightly rough to the
touch when dried, and pallid, with the
flower cymes persisting and hardening
and becoming subspinose, but are not
pungent. Leaves opposite, not or slightly
united at the base, often leaving a tooth-
like projection on the stem when they
fall away, trigonous or subterete,
minutely or microscopically papulose
and glittering, bearing leaf-tufts in their
axils. Flowers small, pedicillate, in
terminal bracteate cymes that are 2 to 5
times dichotomously divided; bracts
small, leaf-like. Sepals 5, with some of
the lobes narrow and with membranous
margins. Petals in 1-series, linear.
Stamens numerous, collected into a cone,
the inner bearded; staminodes o. Ovary
small ½-superior, conical or convex and
ridged on the top; stigmas 5, erect,
shorter than the stamens, subulate.
Glands separated, 5 parts, crenulated,
dark-green. Placentas parietal. Capsule
small, 5 locular, ½-superior, valves recurved when expanded; expanding
keels contiguous below, diverging above,
with rather narrow membranous
acutely pointed wings; loculi-roofs present, sometimes reduced so that it
covers only half the loculi; tubercle none.
Seeds ovoid, smooth, brown.

Species: 7, South West Africa (Karas
Mts., Narudas-South); Cape Province:
Cradock, Griqualand-West, Hay,
Prieska, and Eastern Prov. Free State:
Fauresmith. Island: Réunion. (*M.
macrorrhizum*).(Type: *M. tuberosum*
(L.) N. E. Brown.)

Mestoklema tuberosum   207

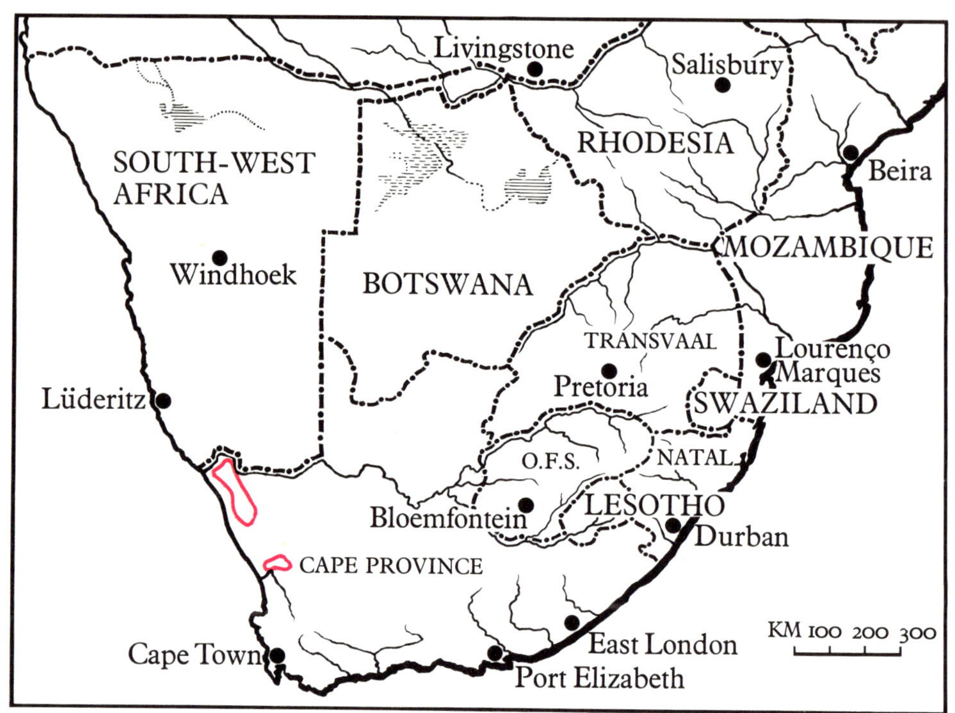

Meyerophytum

Meyerophytum Schwantes
Named in honour of the Rev. G. Meyer,
formerly of Kommaggas and Steinkopf.

*Schwantes, Möllers Gärtnerztg. 36: 436.
1927; Z. Sukk. 4: 320, 321 (in key), 322.
1928.* - v. Pöllnitz, Aufteilung, 51. 1933. -
Jacobsen, Sukk. 160; fig. 162. 1933. -
Pax, Natürl. Pfl. 221. 1934. - Jacobsen,
Sukk. Pl. 217; fig. 199. 1935; Ver-
zeichnis, 149. 1938. - Jacobsen, Herre,
Volk, Mesembr. 64 (in key), 118. 1950. -
Phillips, Genera, 321. 1951. - Jacobsen,
Handbuch, 1551; fig. 1255. 1955. -
Schwantes, Fl. Stones, 119, 34-1, pl.
39a. 1957. - *Bolus, Notes, 344, 345.
1958.* - Jacobsen, Handbook, 953 (in
system of Schwantes), 960 (in key of
Bolus), 970 (in key of Herre & Volk), 1294
(with key of species); fig. 1511. 1960. -
Jacobsen, Lexikon, 467, t.184/2. 1970.

Depacarpus N. E. Brown, Gard. Chron.
87: 71. 1930. - v. Pöllnitz, Aufteilung,
37. 1933. - Pax, Natürl. Pfl. 214. 1934. -
Phillips, Genera, 320. 1951.

Small shrublets, much branched. Leaves
opposite, papillate, two kinds of leaves:
(1) roundish or ovate, nearly completely
connate, fissure very short, mostly
surrounded by white, membranous
sheaths, the rests of the former pair of
leaves; (2) oblong-terete, only somewhat
connate at the base, recurved, at first
dark-green, later yellowish-green, often
reddish. Flowers terminal, solitary,
short pedicelled, bright violet-red,
to 4 cm in diam. Sepals 5, subequal, free,
3 with membranous margin; petals
3-4-seriate, linear. Stamens numerous,
column-like collected, erect, papillate
at the base. Ovary flat. Glands annular,
placentas parietal. Stigmas 5-7, short,
leafy. Capsule 5-7-chambered, very
short and broad obconical, with 5-7 sutures; expanding keels very broad and
flat, diverging from the base, with membranous wings; loculi-roofs stiff; tubercle of moderate size. Seeds ovate or
pear-shaped.

Species: 4, Cape Province, Namaqualand, Vanrhynsdorp, Vredendal. (Type:
M. meyeri Schw.)

Returned from S.A 9? Mris April 1929

Meyerophytum tinctum 209

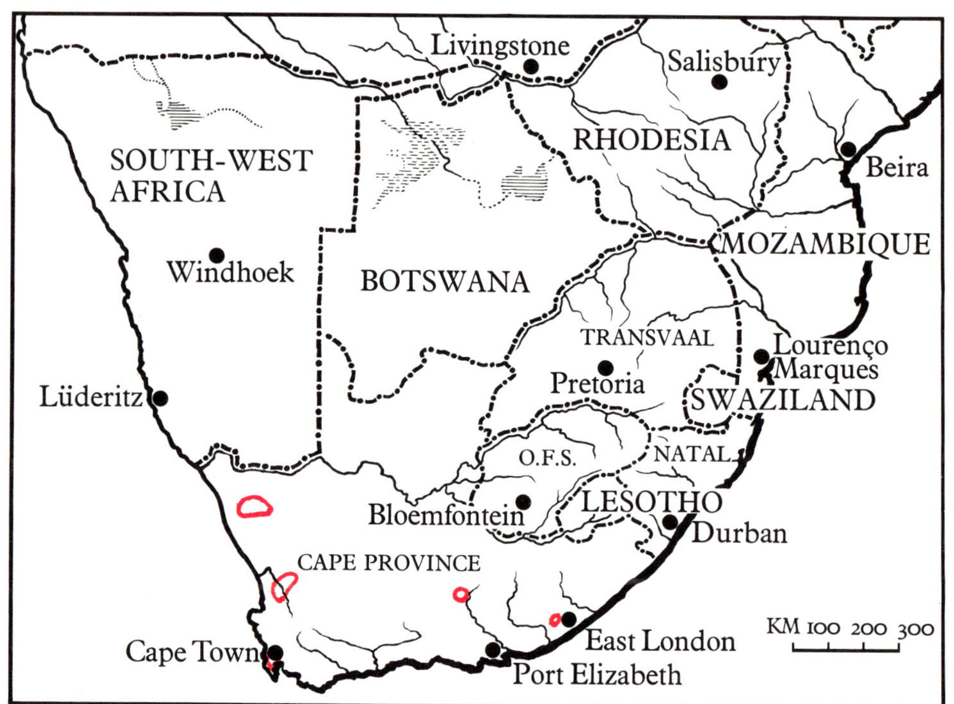

Micropterum

Micropterum Schwantes
From the Greek *micros* = small and
pteron = wing, on account of the very
undeveloped cell lid.

Schwantes, Möllers Gärtn. Ztg. 43: 17, 238. 1928. - N. E. Brown, Möllers Gärtnerztg. 166, 400. 1928. - Bolus, Notes, 78, fig. 9F (flower and capsule). 1928. - v. Pöllnitz, Aufteilung, 51. 1933. - Jacobsen, Verzeichnis, 149. 1933. - Jacobsen, Herre, Volk, Mesembr. 59 (in key), 118. 1950. - *Schwantes, Kakt. & Sukk. 213 (No. 2 & 3)* 1950. - Phillips, Genera, 321. 1951. - Jacobsen, Handbuch, 1551, fig. 1256. 1955. - Schwantes, Fl. Stones, 36, 340; fig. 5 (capsule), pl. 9a. 1957. - Jacobsen, Handbook, 953 (in system of Schwantes), 957 (in key of Bolus), 968 (in key of Herre & Volk), 1295; 1512. 1960; Lexikon, 468, t.184/4. 1970.

Annual, strongly papillose herbs with fibrous roots. Leaves opposite, lanceolate-spathulate and linear-spathulate, entire or rarely also pinnatifid leaves. Flowers terminal, solitary, pedicelled, small, yellow; without bracts, open during day-time or cleistogamous. Sepals 5, papillate. Petals 1-2- series, short, not numerous. Stamens, sometimes only a few, at first inflexed over the stigmas, later on erect, stigmas narrow subulate, acuminate or filiform. Capsule 5-locular, expanding keels parallel with broad wings; with very small loculi-roofs or without it; tubercle o. Seeds compressed triangled, angles much rounded, with very small tubercles.

Species: 7, Cape Province: Namaqualand, Vanrhynsdorp, South Western Cape, Eastern Province, Graaff-Reinet. (Type: *M. schlechteri* Schwant.)

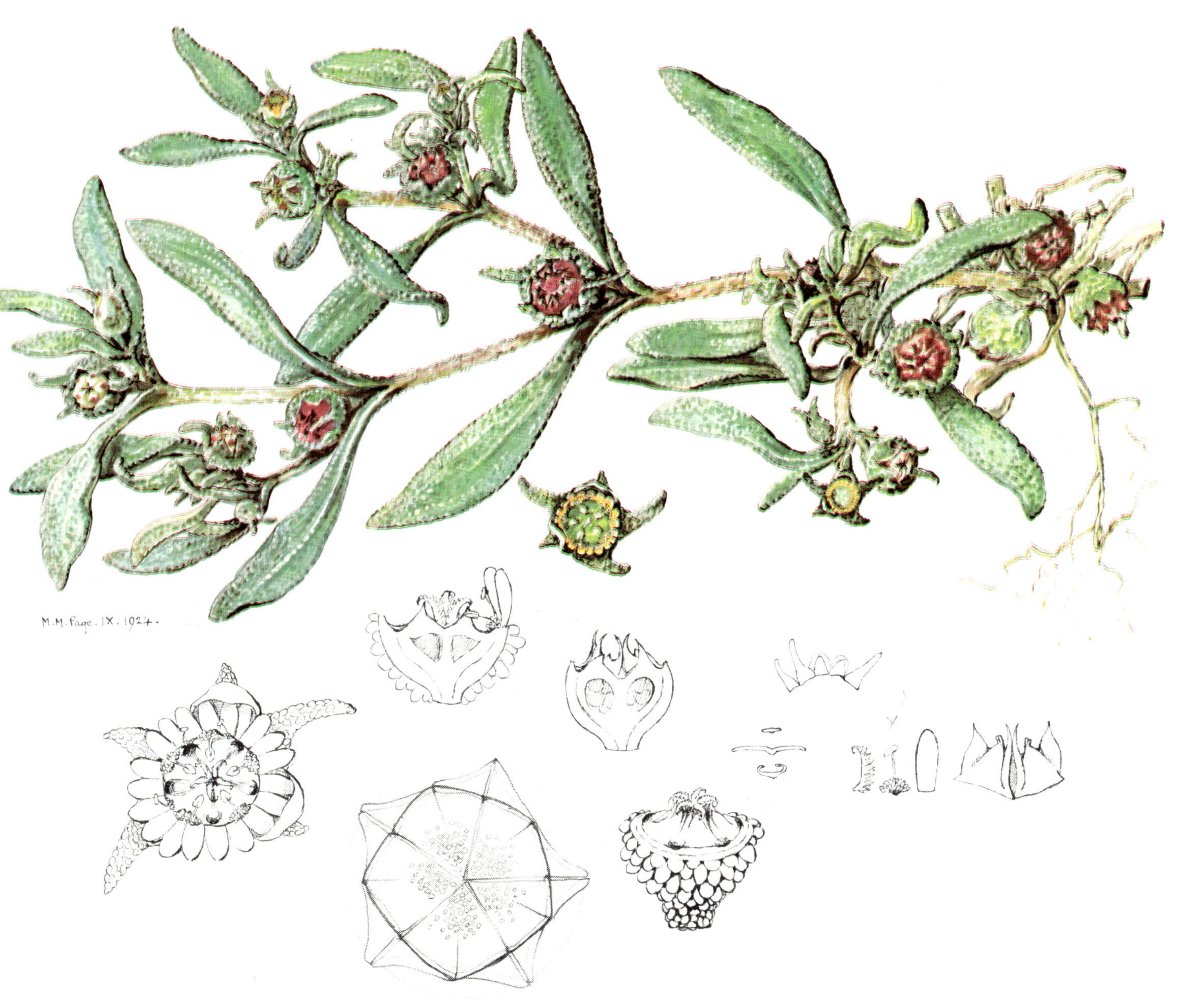

M·M·Page·IX·1924·

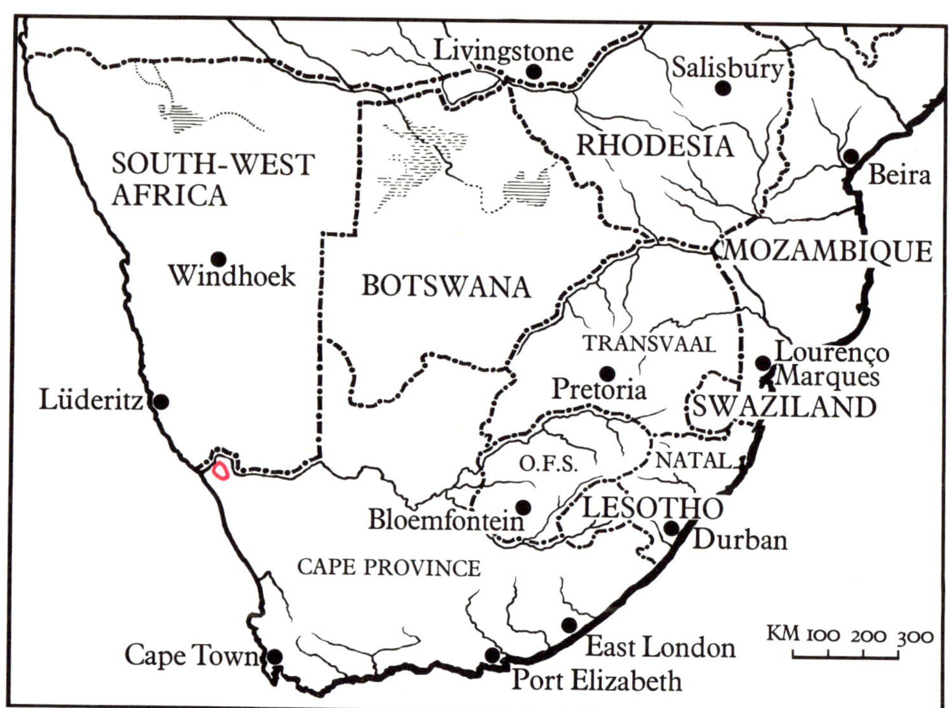

Mimetophytum

Mimetophytum L. Bolus
From the Greek *mimeomai* = to
imitate and *phyton* = plant, denoting
a plant which mimics others.

Bolus, Notes, 252; pl. 86, 87, 88, 1954. -
Jacobsen, Handbuch, 1553; fig. 1257.
1955. - Schwantes, Fl. Stones, 119, 341.
1957. - Jacobsen, Handbook, 953 (in
system of Schwantes), 962 (in key of
Bolus), 969 (in key of Herre & Volk),
1296; fig. 1513. 1960; Lexikon, 468,
t.185/1. 1970.

As regards the vegetative parts and the
flowers, this genus shows no differences
with *Conophyllum* and *Mitrophyllum*
except in the capsule which has a
tubercle.

Species: 2, Cape Province: Richtersveld, between Vlakmyn and Brakfontein.
(Type: *M. parvifolium* L. Bolus.)

Mimetophytum crassifolium 213

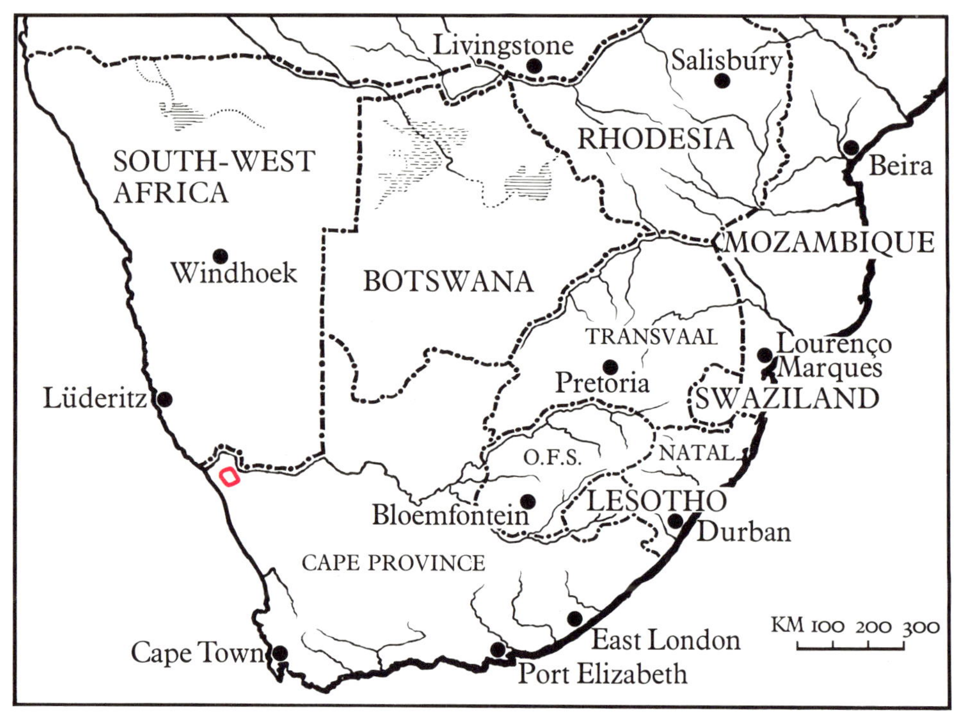

Mitrophyllum

Mitrophyllum Schwantes
From the Greek *mitra* = bishop's mitre and *phyllon* = leaf.

Schwantes, Z. Sukk. 3: 321. 1928; 2: 182. 1926. - N. E. Brown, Gard. Chron. 86: 208, 227, 248, 264; fig. 212, 222-224. 1929. - Labarre (red.), Mesembr. 262; fig. 142, 145. 1931. - v. Pöllnitz, Aufteilung, 51. 1933. - Jacobsen, Sukk. 161; fig. 163. 1933. - Pax, Natürl. Pfl. 217. 1934. - Jacobsen, Succ. Pl. 217; fig. 200. 1935; Verzeichnis, 150. 1938. - Goossens, Blomplante, 143. 1940. - Jacobsen, Herre, Volk, Mesembr. 64 (in key), 118; fig. 25. 1950. - *Phillips, Genera, 308. 1951.* - Bolus, Notes, 237; pl. 58, 59, 60. 1954. - Jacobsen, Handbuch, 1554; fig. 1258-1260. 1955. - Schwantes, Fl. Stones, 118, 341; pl. 38. 1957. - Jacobsen, Handbook, 953 (in system of Schwantes), 959 (in key of Bolus), 969 (in key of Herre & Volk), 1926; fig. 1514-1516. 1960; Lexikon, 468, t.185/2, 3, 4. 1970.

Shrublets or rarely stemless perennials, branches often decumbent, each branch-let develops yearly only 2 pairs of leaves, internodes often articulated. Leaves, each pair of leaves different: the first pair free except at the very base, flat above, rounded or keeled at the back, soft; the second pair united from a third part to nearly all of their length into a conical or rarely cylindric body, minutely crystalline-papulose, especially at the tips, resembling a bishop's mitre, fairly soft. Flowers solitary or developing into lax cymes when in fruit, terminal, pedicellate, with or without bracts, to about 3 cm in diam. Sepals 5, subequal, free. Petals numerous, in 3-4 series, linear. Stamens numerous, erect, in a column, later on spreading. Placentas on the outer walls. Glands narrow, annular. Ovary inferior, with 5-7 sutures on the flat top. Stigmas 5-7, subulate. Capsule very shortly and broadly obconic, with 5-7 prominent ridges on the flat top, 5-7-locular; expanding keels usually diverging from the base, with incurved tips, sometimes subparallel, broad and flat, with the inner edge raised, with membranous wings; loculi-roofs and tubercles none. Seeds ovoid or pear-shaped.

Species: 2, Cape Province: Namaqualand, Richtersveld. (Aughrabies, Brakfontein). (Type: *M. mitratum* Schwant.)

Mitrophyllum pillansii   215

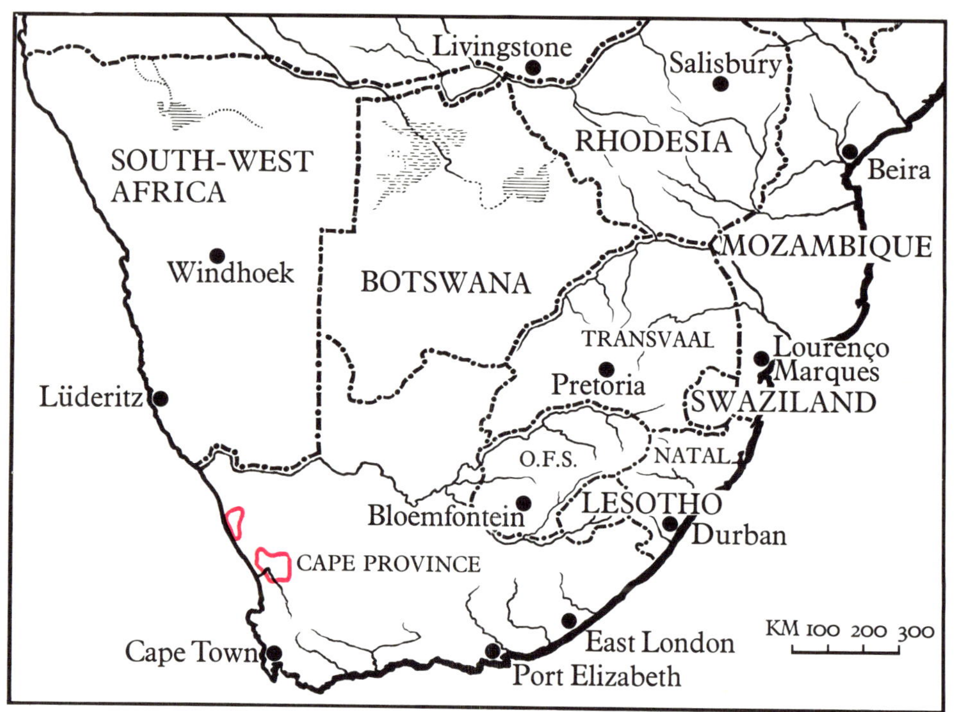

Monilaria

Monilaria Schwantes
From the Latin *monile* = string of pearls, *monilaria* = number of strings of pearls.

Schwantes, Gartenwelt, 33: 69. 1929; Z. Sukk. 2: 182. 1926; Z. Sukk. 3: 322. 1928. - N. E. Brown, Gard. Chron. 86: 32, fig. 19-21. 1929. - Labarre (red.), Mesembr. 270; fig. 147, 148. 1931. - v. Pöllnitz, Aufteilung, 54. 1933. - Jacobsen, Sukk. 162; fig. 164, 165. 1933. - Pax, Natürl. Pfl. 217. 1934. - Jacobsen, Succ. Pl. 219; fig. 201, 202, 1935; Verzeichnis, 151. 1938. - Goossens, Blomplante, 143 (in key), 1940. - Jacobsen, Herre, Volk, Mesembr. 64 (in key), 118. 1950. - *Phillips, Genera, 309. 1951.* - Jacobsen, Handbuch, 1558; fig. 1261, 1262. 1955. - Schwantes, Fl. Stones, 120, 341; pl. 39b. 1957. - Jacobsen, Handbook, 953 (in system of Schwantes), 960 (in key of Bolus), 970 (in key of Herre & Volk), 1299; fig. 1517, 1518. 1960; Lexikon, 470, t.186/1. 1970.

Dwarf, bushy succulent plants with crowded erect stout branches which are constricted at the nodes into short and often bead- or button-like joints, similar to a string of pearls. Leaves opposite, with glittering papillae, soft, of two kinds and only one pair of each kind produced annually; the first pair to appear when growth commences are rudimentary and sometimes scarcely more than a short fleshy sheath, with its semi-circular lips closed together into a bag-like or globular body but, by the growth of the second pair of leaves, the lips are pushed apart and remain surrounding the base of the joint formed by the second pair which are long and nearly cylindric but more or less flattened on the upper side and very shortly united at the base. Flowers solitary, terminal, pedicillate; pedicel long, papillate, with a pair of rudimentary leaves, sheathing the base, white, yellow or violet-red, about 4 cm in diam. Sepals 5, unequal, papulose or papillate, with the ovary shallow and convex beneath. Petals numerous, free, in 3-4-series, linear. Stamens numerous, erect in a column or later becoming more or less loose; hairy at the base or not. Ovary inferior, shallow, with 5-7 sutural ridges on the flat top. Glands annular. Placentas on the floor or outer wall of the chambers. Stigmas 5-7, stout, ovate to subulate. Capsule 5-7-locular, very short, acute, and broadly obconic, shallow, with 5-7 very prominent and thin sutural ridges on the flat top; valves reflexed or widely spreading; expanding keels more or less widely diverging, broad and flattish at the basal part, with the inner ridges raised, with narrow marginal wings; loculi shallow, somewhat acutely roofed with membranous loculi-roofs; tubercle absent. Seeds numerous in each loculus, ovoid, pointed at one end, smooth.

Species: 11, Cape Province, Namaqualand and Vanrhynsdorp district. (Type: *M. moniliformis* (Thunb.) Schwant.)

Monilaria salmonea   217

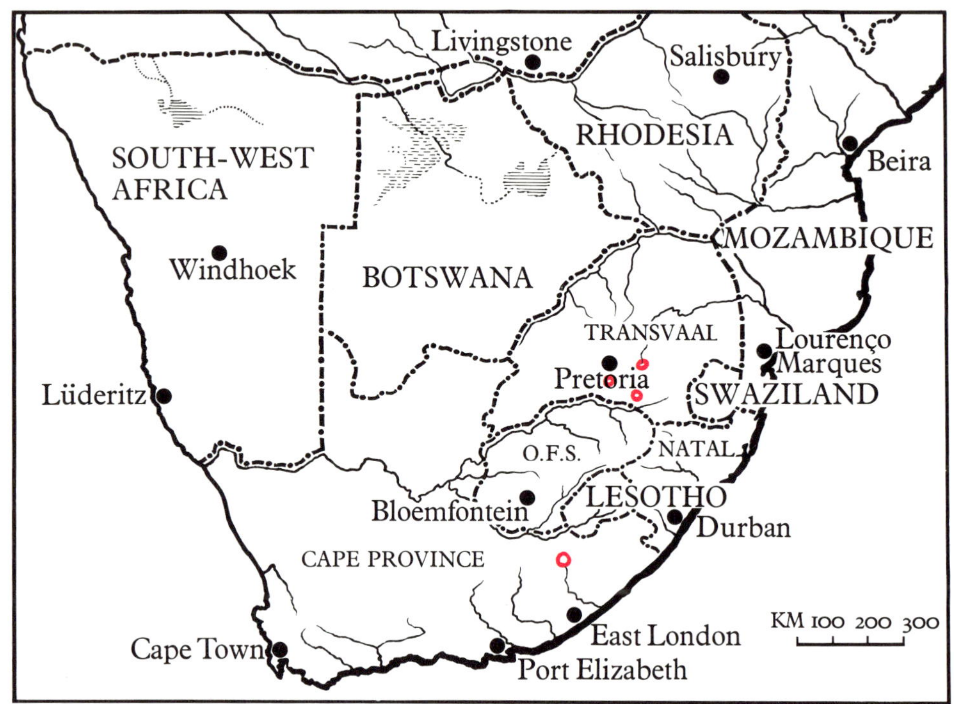

Mossia

Mossia N. E. Brown
Named in honour of Dr Moss.

N. E. Brown, Gard. Chron. 87: 71 (in key), 151. 1930. - v. Pöllnitz, Aufteilung, 55, 1933. - Pax, Natürl. Pfl. 217. 1934 - Jacobsen, Verzeichnis, 152. 1938. - Goossens, Blomplante, 145 (in key). 1940. - Jacobsen, Herre, Volk, Mesembr. 62 (in key), 119. 1950. - *Phillips, Genera, 309. 1951.* - Jacobsen, Handbuch, 1561; fig. 1263. 1955. - Schwantes, Fl. Stones, 342. 1957. - Jacobsen, Handbook, 958 (in key of Bolus), 969 (in key of Herre & Volk), 1301. 1960; Lexikon, 471. 1970.

Dwarf prostrate succulent perennial herb, with distinct internodes on the main stem. Leaves opposite, always more or less equal, only connate at the base, 3-angled to ovate, without pocket-lens not papillate, but sometimes with somewhat prominent, scattered dots, usually smooth. Flowers solitary, nearly sessile among the leaves or shortly pedicillate, white, open at night. Sepals 5, subequal, forming a very short tube above the ovary. Petals 1-2 series, inserted at the base of the tube, shorter than the sepals. Stamens numerous, erect, in a ring around the stigma, often without staminodes. Ovary inferior, concave at the top. Placentas on the floor of the chambers. Glands forming a shallow cup or bowl on top of the ovary, with 5 broad crenations at the margin. Stigmas 5, erect, filiform. Capsule 5-, rarely 4-6- or or 7-locular, very shortly and broadly obconic, small; expanding keels closely contiguous throughout and reaching to the apex of the valves, without marginal wings; loculi open or with very rudimentary loculi-roofs; tubercle none. Seeds ovoid, smooth.

Species: 1, Transvaal: near Johannesburg etc. Free State: Zastron. (Type: *M. intervallaris* (L. Bol.) N. E. Br.)

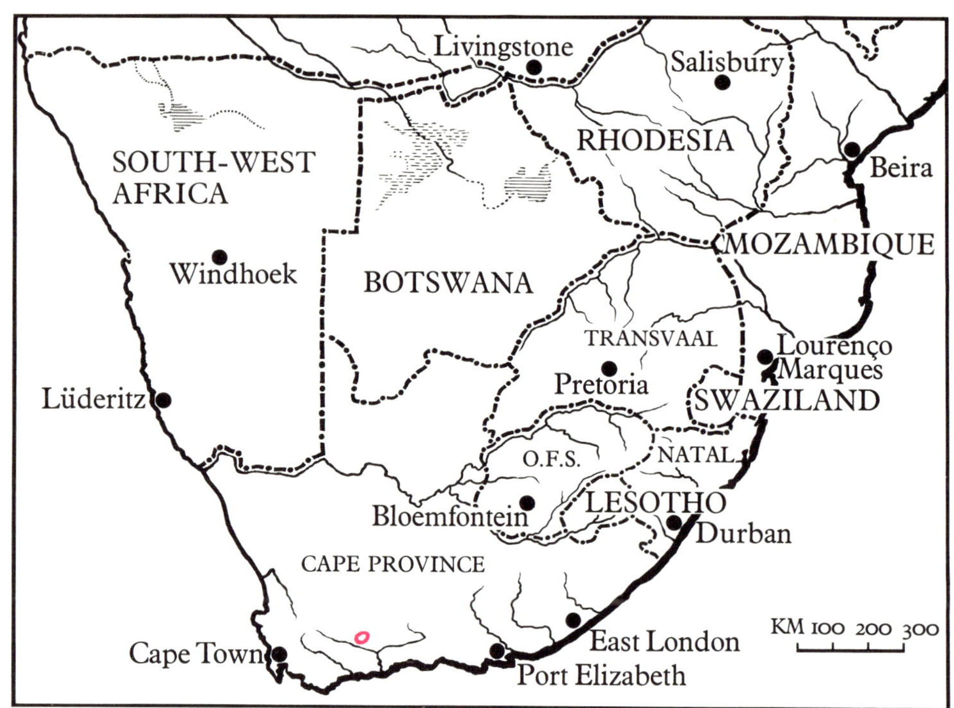

Muiria

Muiria N. E. Brown
Named in honour of the late
Dr John Muir, Riversdale.

N. E. Brown, Gard. Chron. 81: 116; fig. 59. 1927. - Tischer, Monatsschrift D. Kakt. Ges. 1: 233. 1929. - Labarre (red.), Mesembr. 98, 272; fig. 149, 150. 1931. - v. Pöllnitz, Aufteilung, 55. 1933. - Jacobsen, Sukk. 163; fig. 166, 167. 1933. - Pax, Natürl. Pfl. 217. 1934. - Jacobsen, Succ. Pl. 220; fig. 203. 1935; Verzeichnis, 153. 1938. - Goossens, Blomplante, 147 (in key). 1940. - Jacobsen, Herre, Volk, Mesembr. 80 (in key), 119. 1950. - *Phillips, Genera, 309.* 1951. - Jacobsen, Handbuch, 1561; fig. 1261. 1955. - Schwantes, Fl. Stones, 172, 340; pl. 52a. 1957. - Jacobsen, Handbook, 953 (in system of Schwantes), 957 (in key of Bolus), 976 (in key of Herre & Volk), 1302; fig. 1519. 1960; Lexikon, 471, t.186/3. 1970.

Stemless with short fibrous roots. Leaves connate to an ovate-conical body, with a very small fissure below the apex, very fleshy and somewhat soft, covered with velvet-like fine hairs, light-green to whitish-green, during the resting period enclosed in whitish, membranous sheaths, the rest of the former pair of leaves. Flower solitary, just exserted from the top of the growth, with the clavate pedicel included, bracts none, whitish to light violet-red, about 2 cm in diam. Sepals 6, free, oblong, with membranous tips. Petals, numerous, free, in several series, linear, outer ones nearly filiform. Stamens numerous, connivent-erect; filaments not hairy at the base. Ovary partly superior and very shallow, being partly immersed in the very stout clavate top of the pedicel. Placentas parietal. Glands 6-7 broad and nearly contiguous. Stigmas 6-7, short and stout, erect, reflexed. Capsule 6-7-locular, small, shallow, convex on the top, with the sutures between the valves raised into ridges; valves deltoid, when expanded spreading horizontally; expanding keels closely contiguous so as to form a stout central keel, with broad membranous marginal wings as long as the valves; loculi open without loculi-roofs or a tubercle. Seeds, many in a loculus, ovoid, slightly compressed, with a nipple at one end, smooth, brown.

Species: 1, Cape Province, Riversdale distr., Springfontein. (Type: *M. hortenseae* N. E. Brown.)

Muiria hortenseae 221

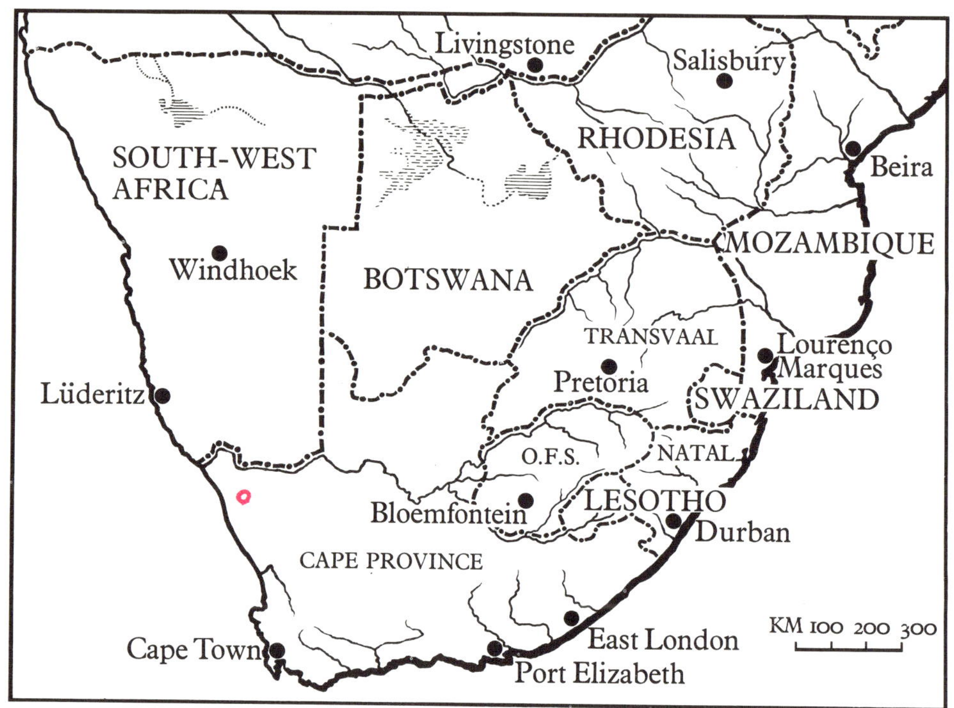

Namaquanthus

Namaquanthus L. Bolus
From Namaqualand and the Greek
anthos = flower.

Bolus, Notes, 257. 1954. - Jacobsen, Handbuch, 1563. 1955. - Schwantes, Fl. Stones, 332, 342. 1957. - Jacobsen, Handbook, 963 (in key of Bolus), 970 (in key of Herre & Volk), 1303; fig. 1521. 1960; Lexikon, 471, t.186/4. 1970.

Erect shrub, about 30 cm high, branches covered with leaves, oldest branches 1-1,5 cm in diam. Leaves fairly erect to spreading, terete, subcylindrical, keel and edges obscure, long united into a sheath, bluish green, up to 6 cm long and 1 cm in diam. Flowers during day-time, solitary, stalked, pedicel with 2 bracts, pink to purplish, 5-6 cm in diam. Sepals 4, unequal, outer ones about 1 cm long and 1,5 cm broad at the base. Petals 4-series, crowded, outer ones subspathulate, sometimes striate. Stamens collected in the beginning, later on spreading filaments purple, bearded. Ovary flat or convex above. Glands pink, finely crenulated. Stigmas 8-16, mostly 10-12, narrow subulate, purple. Capsule 8-16-locular, mostly 8-12, globose or obconical, sutures somewhat compressed with valve-wings reaching the tips of the valves; expanding keels dark brown, parallel at the base, later on diverging, lacerated, with loculi-roofs, tubercle none. Seeds obovate, echinate, brown, hairy, hairs up to 0,75 mm long, 2 mm long.

Species: 2, Cape Province: Namaqualand. (Type: *N. vanheerdei* L. Bolus.)

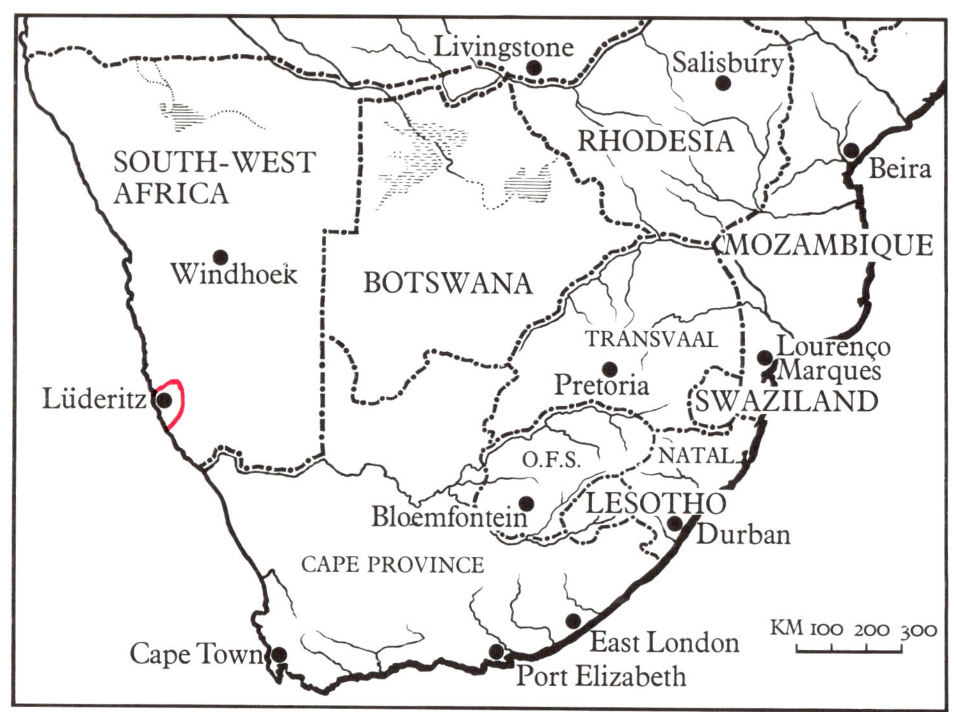

Namibia

Namibia Dinter et Schwantes
Named after the Namib, the desert along
the coast of South West Africa.

Dint. et Schwant., Z. Sukk. 2: 184.
1926; 3: 106. 1927. - v. Pöllnitz, Aufteilung, 56. 1933. - Pax, Natürl. Pfl. 217.
1934. - Jacobsen. Succ. Pl. 221. 1935;
Verzeichnis, 153. 1938. - Walgate in
Bolus Notes, 186; pl. 31. 1939. - Jacobsen, Herre, Volk, Mesembr. 77 (in key),
119. 1950. - Phillips, Genera, 321. 1951.
- Jacobsen, Handbuch, 1563. 1955. -
Schwantes, Fl. Stones, 137, 339; pl.
44a. 1957. - Jacobsen, Handbook, 953
(in system of Schwantes), 958 (in key
of Bolus), 975 (in key of Herre & Volk
1303 (with key of species); fig. 1522.
1960; Lexikon, 471, t.186/2, 1970. -
Friedrich in Merxmüller, Prodromus,
62, 67 (Juttadinteria) 1970.

Forms clumps, with short branchlets,
each with 4-6 leaves. Leaves opposite,
connate at the base, very thick, semi-
ovate, upper surface 3-angled with
recurved, obtuse apex, lower side boat-
shaped with rounded keel, softer than
Juttadinteria and *Dracophilus*, whitish-
grey or light grey-green with fine points
on the dots (difference from *Juttadinteria*). Flowers solitary, terminal,
sessile or shortly pedicellated, open
during day-time, white to light violet-
red, about 3 cm in diam. Sepals 5, unequal, outside fine papillate. Petals 1-
series, larger than the sepals, linear,
apices more or less reflexed; staminodes 0.
Stamens conically collected, filaments
bearded at the base. Ovary convex.
Glands annular, crenulated. Placentas
parietal. Stigmas 9-25, yellowish-green,
narrow-subulate, as long as the stamens.
Capsule 9-25-locular, top convex;
expanding keels parallel at the base,
later diverging, with oblong wings, without awns; loculi-roofs none or reduced
to a limb, tubercle none. Seeds brown,
smooth, ovate with nipple.

Species: 3, Lüderitz, S.W.A.
(Halenberg, Pomona, Rote Kuppe,
Prince of Wales Bay). (Type: *N. cinerea*
(Marl.) Dint. et Schwant.)

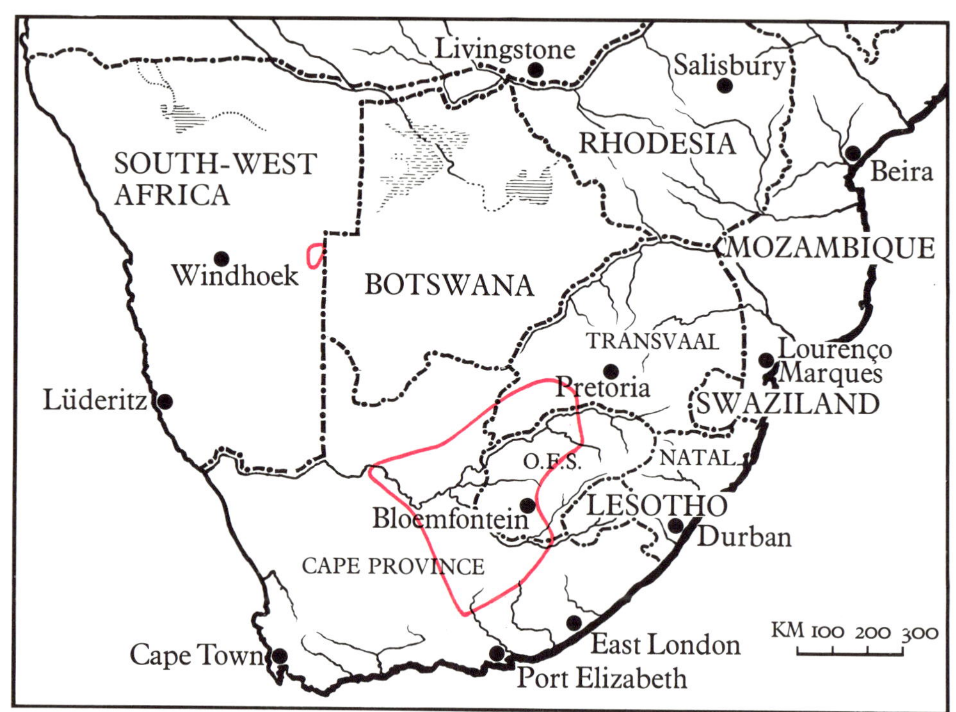

Nananthus

Nananthus N. E. Brown
From the Greek *nanos* = dwarf and
anthos = flower.

N. E. Brown, Gard. Chron. 78: 433 (in key). 1925; J. Bot. 66: 78. 1928. - Labarre (red.), Mesembr. 274; fig. 151-153. 1931. - N. E. Brown, Gard. Chron. 88: 278 (in key). 1930; 89: 92, 133; fig. 48, 49, 64-70. 1931. - v. Pöllnitz, Aufteilung, 56. 1933. - Jacobsen, Sukk. 164. 1933. - Pax, Natürl. Pfl. 217. 1934. - Jacobsen, Succ. Pl. 222. 1935; Verzeichnis, 153. 1938. - Goossens, Blomplante, 144 (in key). 1940. - Jacobsen, Herre, Volk, Mesembr. 75 (in key), 119. 1950. - Jacobsen, Handbuch, 153. 1955. - Schwantes, Fl. Stones, 156, 339; pl. 47a & b. 1957. - *Bolus, Notes, 382; pl. 130-134. 1958.* - Jacobsen, Handbook, 953 (in system of Schwantes), 958 (in key of Bolus), 974 (in key of Herre & Volk), 1305 (with key of species); fig. 1523a & b. 1960. Lexikon, 472, t.187/1. 1970. - Friedrich in Merxmüller, Prodromus, 88. 1970.

Aloinopsis Schwantes (partly); *Prepodesma* N. E. Brown, Gard. Chron. 88: 279 (in key). 1930; 89: 389; fig. 195, 196, 1931. - v. Pöllnitz, Aufteilung, 61. 1933. - Pax, Natürl. Pfl. 218. 1934. - Jacobsen, Succ. Pl. 124. 1935 - Bolus, Notes, 134. 1938.

Dwarf tufted glabrous plants with a tuberous rootstock, tuber up to 13 cm long and 4,5 cm in diam. Leaves opposite, ascending or spreading, viewed from above the lower part square or oblong with the sides parallel, the upper part widened, the one side usually bulging more than the other, widest near the middle or well above it, linear-lanceolate, ovate or broadly ovate, sub-obtuse, acute or acuminate, sometimes conspicuously aristate, obtusely keeled upwards, viewed laterally widest near the middle or well above it, truncate obtuse or subacute, punctate, the dots often white or whitish, to about 5 cm long. Flowers solitary, usually expanding soon after noon, 2-3 cm in diam., peduncles very short, bracteate at the base. Sepals 5, free, some with membranous margins. Petals 2-3-seriate, spreading in one plane, yellow, with or without a central red stripe. Stamens numerous, collected into a broad cone which is completely exposed to view to its base, the inner ones bearded at the base; with or without staminodes. Ovary inferior, flattish to conical, sometimes with prominent sutures. Placentas on the floor or outer walls of the chambers. Glands annular. Stigmas 6-10, slender, papillate to the apex. Capsule sub-globose below, nearly flat or convex or semiglobose above, sutures not or scarcely compressed, valves strongly recurved in the fully expanded capsule, amply winged, adjacent wings cohering or free; expanding keels erect, contiguous below, diverging upwards and nearly reaching apex of valve; loculi-roofs very narrow, spreading or erect and back to back; tubercle none. Seeds broadly oval or broadly obovate, smooth, the embryo reddish brown and darker than the rest of the seeds.

Species: 9, Cape Province: Karroo, Prieska, Kimberley; Free State, Transvaal and Bechuanaland. (Type: *N. aloides* (Haw.) Schwant.)

Nanathus aloides Haw. var. *latus*   227

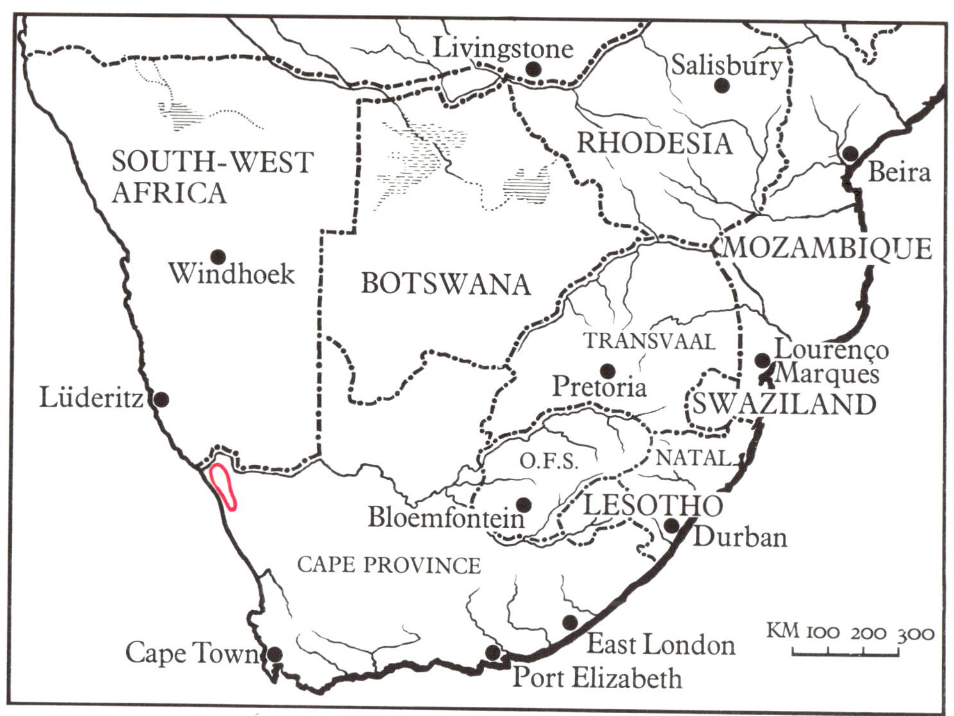

Nelia

Nelia Schwantes
Named in honour of the late Prof.
G. C. Nel, Stellenbosch.

Schwantes, Gartenflora, 77: 129 with 2 fig. 1928. - Möllers Gärtnerztg. 43: 92. 1928. - v. Pöllnitz, Aufteilung, 56. 1933. - Jacobsen, Sukk. 165; fig. 170. 1933. - Pax, Natürl. Pfl. 220. 1934. - Jacobsen, Succ. Pl. 223; fig. 205. 1935; Verzeichnis, 154. 1938. - Jacobsen, Herre, Volk, Mesembr. 72 (in key), 119. 1950. - *Phillips, Genera, 310. 1951.* - Jacobsen, Handbuch, 1568; fig. 1269. 1955. - Schwantes, Fl. Stones, 75, 339; fig. 20 (flower); pl. 18a. 1957. - Jacobsen, Handbook, 953 (in system of Schwantes), 958 (in key of Bolus), 973 (in key of Herre & Volk), 1308; fig. 1524. 1960; Lexikon, 473, t.187/4. 1970.

Sterropetalum N. E. Brown, Gard. Chron. 83: 266. 1928. - Labarre (red.), Mesembr. 292; fig. 163. 1931. - v. Pöllnitz, Aufteilung, 68, 73, 1933. - Pax, Natürl. Pfl. 219. 1934.

Tufted stemless perennials. Leaves opposite, shortly united at the base, subterete or triquetrous, acute or obtuse, flat or slightly convex above, lower side to about the middle sharply keeled, smooth, shining, not dotted, blue-green with a reddish lustre, about 3 cm long and 1 cm broad and thick. Flowers 1-3 with the lateral flowers also on short bracteate peduncles, pedicels short, peduncle solitary, terminal at the side of a new pair of leaves; flowers white, towards the middle yellowish, after opening remain open day and night for about 3 weeks, about 2 cm in diam. Sepals 5, shortly dish-like connate above the ovary, unequal, 3 smaller with broad membranous margins and 2 larger ones without it. Petals numerous, in several series, shorter than the sepals, linear-spathulate, emarginated, with stout short hairs at the base, erect, stiff, whitish, posteriors gradually changing into many staminodes, which together with the petals are tubularly connate at the base. Stamens connivent to $\frac{1}{2}$ their length with the upper part, abruptly deflexed and tortuous, very short, bearded at the basal part. Ovary inferior, flattish or very slightly convex at the top. Placentas at the floor of the chambers. Glands none. Stigmas 5, very small, ascending-spreading, broadly 3-angled, green. Capsule shortly obconic, flattish at the top with the sutures slightly raised at the central part only, 5-locular; valves recurved when expanded; expanding keels yellow, contiguous into a very stout convex central keel with broad marginal wings that are not united in pairs between the valves; loculi-roofs reduced to a limb; tubercle none. Seeds somewhat flat, pear-shaped to cordate, light-yellow with minute warts, about 1 mm long.

Species: 4, Namaqualand (Kommaggas, Richtersveld). (Type: *N. meyeri* Schwant.)

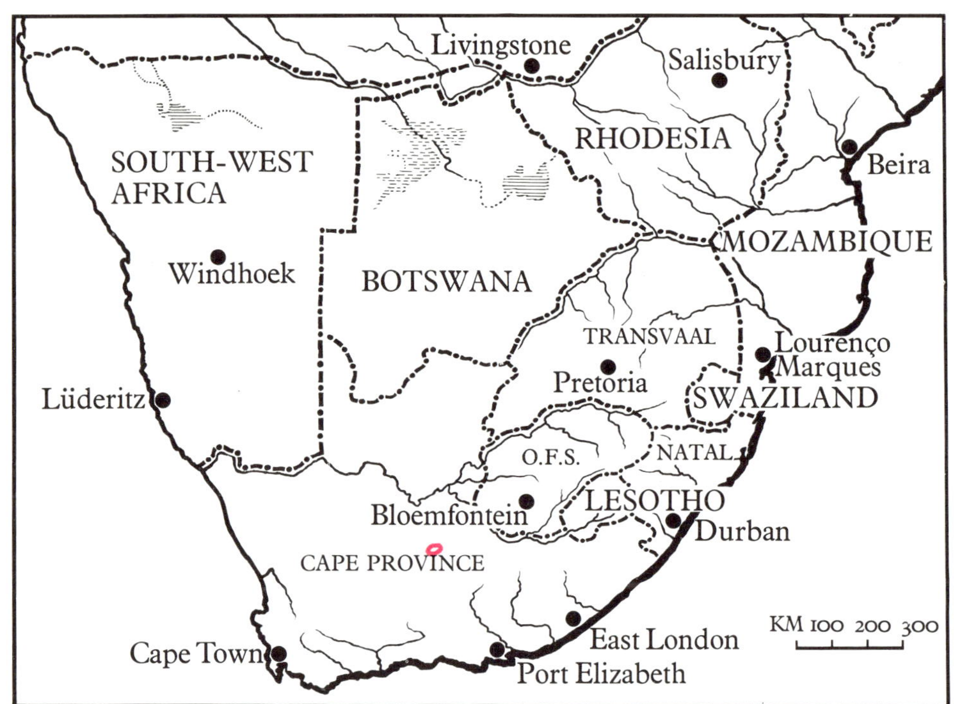

Neohenricia

Neohenricia (L. Bolus), L. Bolus
Named in honour of Dr M. Henrici,
Fauresmith, O.F.S.

Bolus, J. S. Afr. Bot. 4: 51. 1938. -
Jacobsen, Herre, Volk, Mesembr. 70 (in
key), 119. 1950. - Phillips, Genera,
321. 1951. - Jacobsen, Handbuch, 1569.
1955. : Schwantes, Fl. Stones, 339.
1957. - Bolus, Notes, 391; pl. 135. 1958 -
Jacobsen, Handbook, 953 (in system of
Schwantes), 959 (in key of Bolus), 972
(in key of Herre & Volk), 1309; fig. 1525.
- 1960; Lexikon, 473, t.187/2. 1970.
Henricia L. Bolus, Notes, 39. 1936. -
Kakteen & Andere Sukk. 152, 153. 1937.
- Jacobsen, Verzeichnis, 90. 1938.

Freely branching, dwarf plantlet forming
mats. Leaves 4 together on the branches,
opposite, somewhat connate at the
base, ascending or rather erect, upper
surface flat, lower side convex, not
keeled, from the sides broadened towards the apex, obtuse, truncate or
rounded and densely covered with lighter
coloured warts and larger groups of
warts, sometimes arranged in rows, the
other part of the leaves is smooth,
rather firm, more or less light brown,
thus mimicking the Karroo soil and
gravel, among which it grows, so well
that it is difficult to discover it. About
1 cm long. Flowers solitary, pedicel
slender, terete, about 2 cm long, without
bracts, white, opening at night, about
1,2 cm in diam. Sepals 5, subequal,
obtuse, red-brown. Petals 2-seriate, free,
somewhat lax, obtuse. Stamens few,
filaments white, without papillae. Ovary
somewhat convex. Glands annular,
obscurely crenulated. Placentas parietal.
Stigmas 5, slender, golden-yellow.
Capsule 5-locular, with deep loculi similar to *Stomatium*; expanding keels
contiguous at the base, later somewhat
diverging, with wings; loculi-roofs reduced to a limb, sometimes well developed; tubercle none. Seeds rough,
with nipple.

Species: 1, Cape Province: Victoria
West, Hanover. Free State: Fauresmith.
(Type: *N. sibbettii* (L. Bol.) L. Bolus.)

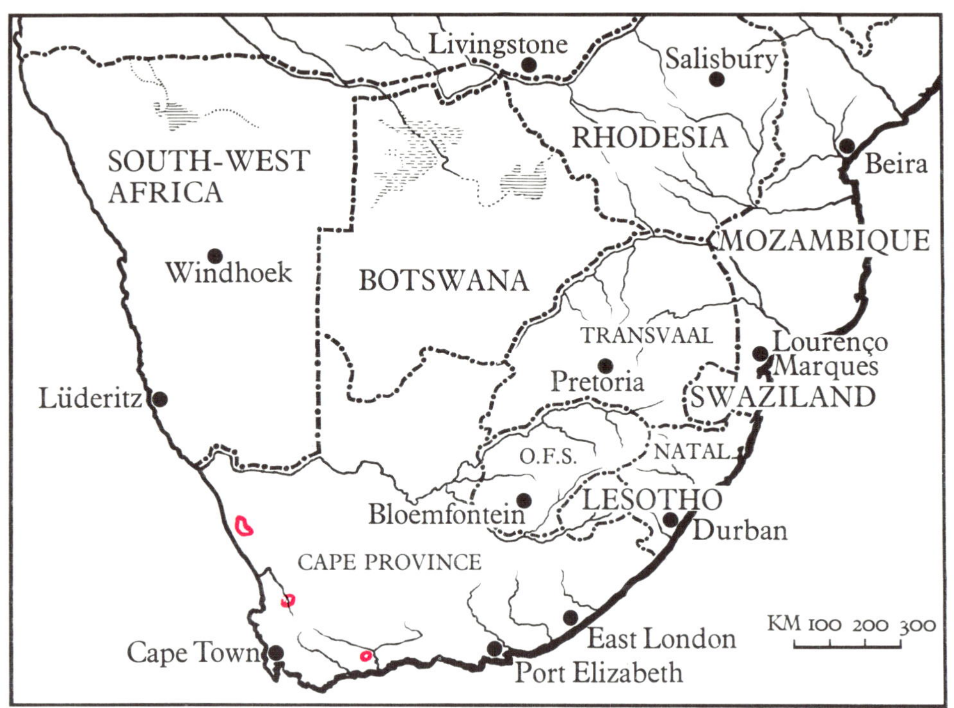

Octopoma

Octopoma N. E. Brown
From the Latin *octa* = eight and *poma*
= cover.

N. E. Brown, Gard. Chron. 87: 72 (in key), 126. 1930. - v. Pöllnitz, Aufteilung, 57. 1933. - Pax, Natürl. Pfl. 217. 1934. - Jacobsen, Verzeichnis, 155. 1938. - Goossens, Blomplante, 146 (in key). 1940. - Jacobsen, Herre, Volk, Mesembr. 67 (in key), 119. 1950. - *Phillips, Genera, 311. 1951.* - Jacobsen, Handbuch, 1574. 1955. - Schwantes, Fl. Stones, 338. 1957. - Jacobsen, Handbook, 952 (in system of Schwantes), 963 (in key of Bolus), 971 (in key of Herre & Volk), 1326. 1960; Lexikon, 474, 1970.

Very small bushily branched perennial succulents, branches succulent when young becoming woody with age, with distinct short internodes. Leaves opposite, united at the base and often continuous with the stems, always short, not papillate, rarely with prominent dots, mostly smooth. Flowers terminal, 1-3 to a branch, subsessile or shortly pedicellate. Sepals 4-5, unequal, free. Petals numerous, in 2-3 series, free, cuneately linear. Stamens erect, collected into a compact column or cone, sometimes becoming lax, surrounded by a few staminodes, filaments sometimes bearded. Ovary inferior, convex at the top. Glands in a crenulate ring. Placentas on the outer walls of the chambers. Stigmas 7-10 subulate. Capsule 7-10-locular, shortly obconic, convex at the top, with slightly raised sutures; expanding keels contiguous at the basal half, diverging above, awned or awnless at the tips, without wings; loculi roofed with membranous wings; tubercle present. Seeds rough, with nipple.

Species: 3, Cape Province: Namaqualand, the Clanwilliam and Riversdale districts. (Type: *O. octojuga* (L. Bol.) N. E. Brown.)

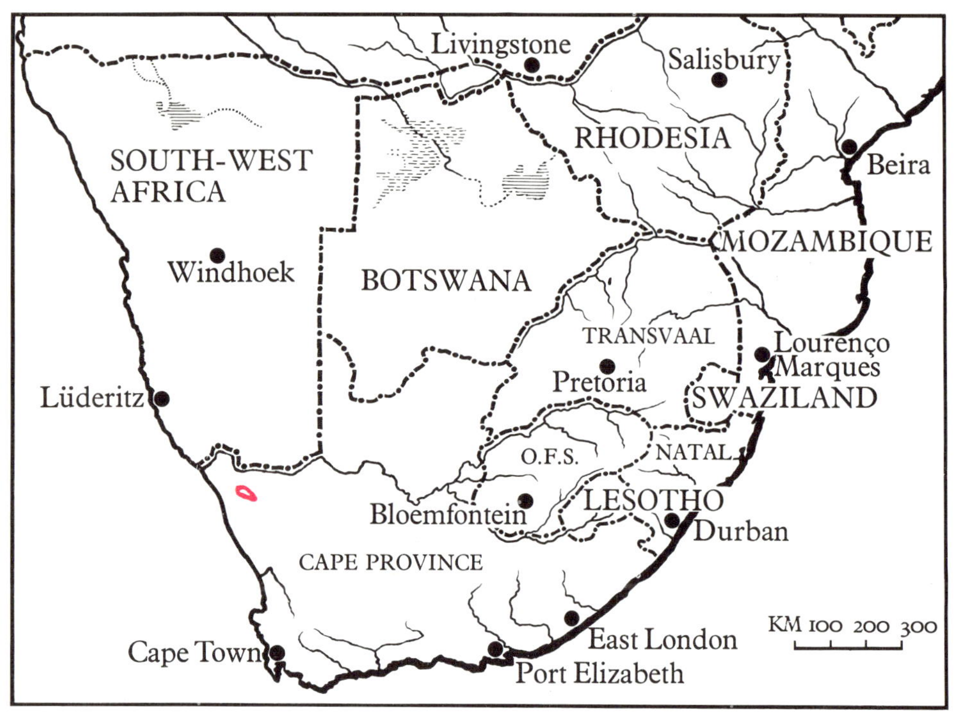

Odontophorus

Odontophorus N. E. Brown
From the Greek *odus* = tooth and *phorus* = bearer.

N. E. Brown, Gard. Chron. 81: 12 (in key). 1927. - Bolus, Notes, 101; pl. 4F (capsule), fig. 25. 1928. - v. Pöllnitz, Aufteilung, 58. 1933. - Jacobsen, Sukk. 166; fig. 171. 1933. - Pax, Natürl. Pfl. 217. 1934. - Jacobsen, Succ. Pl. 224; fig. 206. 1935; Verzeichnis, 155. 1938. - Goossens, Blomplante, 147 (in key). 1940. - Jacobsen, Herre, Volk, Mesembr. 66, 74 (in key), 119. 1950. - Phillips, Genera, 321. 1951. - Jacobsen, Handbuch, 1574; fig. 1271, 1272. 1955. - Schwantes, Fl. Stones, 113, 338; pl. 34a. 1957. - Jacobsen, Handbook, 952 (in system of Schwantes), 963 (in key of Bolus), 973 (in key of Herre & Volk), 1327; fig. 1530, 1531. 1960; Lexikon, 474, t.188/1. 1970.

Dwarf shrubs with fleshy roots, branches ascending or prostrate and forming clumps, only one species shrubby with long prostrate branches (*O. marlothii*); growths with 1-2 pairs of leaves. Leaves opposite, very thick and soft-fleshy, grey-green, tuberculate, soft-hairy, oblong, triquetrous, keeled, obtuse or obliquely acuminate, along the edges with tubercle-like, thin, flexible teeth, dotted, to about 4 cm long and 1,5 cm broad and thick. Flowers solitary, sessile or short pedicellate or pedicel up to 3,5 cm long with 2 bracts, whitish to orange-yellow, to about 3 cm in diam. Sepals 4-5, the inner ones with membranous margins. Petals 3-6-seriate, free, narrow. Stamens erect, in 6-series, the outer ones with papillae at the base, the inner ones papillate up to the middle and incurved. Staminodes none. Glands crenulate or dentate. Placentas parietal. Stigmas 8-11, subulate. Capsule 8-11-locular, semi-globose, soft-hairy; expanding keels parallel, toothed, ending in free membranous points or narrow wings; loculi-roofs stiff, punctate; tubercle large, nearly closing the opening. Seeds flatly compressed, ovate, light-brown.

Species: 6, Cape Province, Namaqualand, Steinkopf district. (Type: *O. marlothii* N. E. Br.)

Fully open
12.50.
18-7-'30
B.O.Carter.

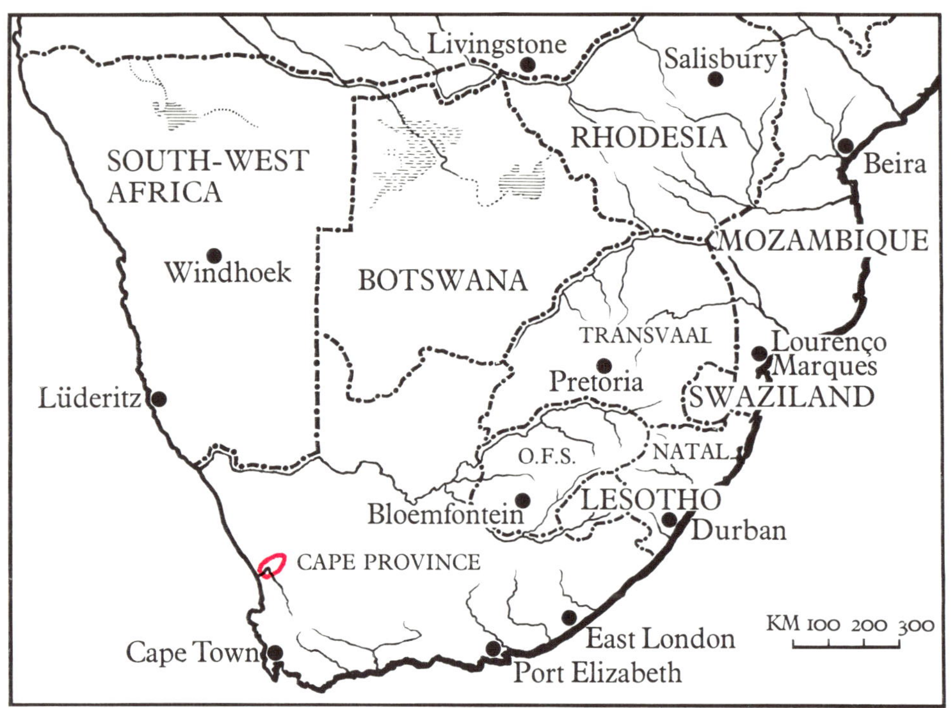

Oophytum

Oophytum N. E. Brown
From the Greek *oon* = egg and
phyton = plant.

N. E. Brown, Gard. Chron. 78: 413
(in key). 1925; 79: 48. 1926. - Phillips,
Genera, 241. 1926. - Bolus, Notes, 39;
pl. 44 (coloured). 1928. - Tischer,
Z. Sukk. 3: 225. 1928. - v. Pöllnitz, Aufteilung, 58. 1933. - Jacobsen, Sukk.
167; fig. 172. 1933. - Pax, Natürl. Pfl.
217. 1934. - Jacobsen, Succ. Pl. 225; fig.
207. 1935; Verzeichnis, 155. 1938. -
Goossens, Blomplante, 147 (in key).
1940. - Jacobsen, Herre, Volk, Mesembr.
79 (in key), 119. 1950. - *Phillips, Genera,*
311. 1951. - Jacobsen, Handbuch,
1576; fig. 1273. 1955. - Schwantes, Fl.
Stones, 247, 340; pl. 67b. 1957. - Bolus,
Notes, pl. 138-140. 1958. - Jacobsen,
Handbook, 953 (in system of Schwantes),
959 (in key of Bolus), 975 (in key of
Herre & Volk), 1329; fig. 1532. 1960;
Lexikon, 474, t.188/2. 1970.

Perennials, less than 2,5 cm high, tufted,
succulent; growths of small ovoid soft
and pulpy bodies, with a small fissure,
with its lips closed or gaping, often
whitish because enclosed in leaf-
sheaths of the older bodies. Leaves opposite, completely or very long fused to
an ovoid or oblong-ovoid body, somewhat dotted, often fiery-red, rarely hairy
(pocket-lens), about 2 cm in diam.
Flowers solitary, with bracts exserting
from the orifice as long as the flowers,
inside white with a red margin, about
2,5 cm in diam. Sepals 6-7 produced
above its union with the ovary into a short
green (not membranous) tube. Petals
numerous, 2-3-series, probably free at
the base, anteriors broad and obtuse or
emarginate. posteriors narrow and
acuminate. Stamens numerous, nearly
erect or somewhat incurved, papillate at
the base of the filaments. Ovary somewhat convex with 6 ridges on the top.
Glands distinct annular. Placentas on

the floor of the chambers. Style o.
Stigmas 6 or 5, filiform, acuminate.
Capsule 5-6-locular, with broad, membranous valve-wings, often connate
between the valves; expanding keels
contiguous; without loculi-roofs and
tubercle. Seeds numerous in one chamber, small, compressed, with a nipple
at the narrow end.

Species: 3, Cape Province: Namaqualand, Vanrhynsdorp district. (Type:
O. oviforme N. E. Br.)

Oophytum nanum   237

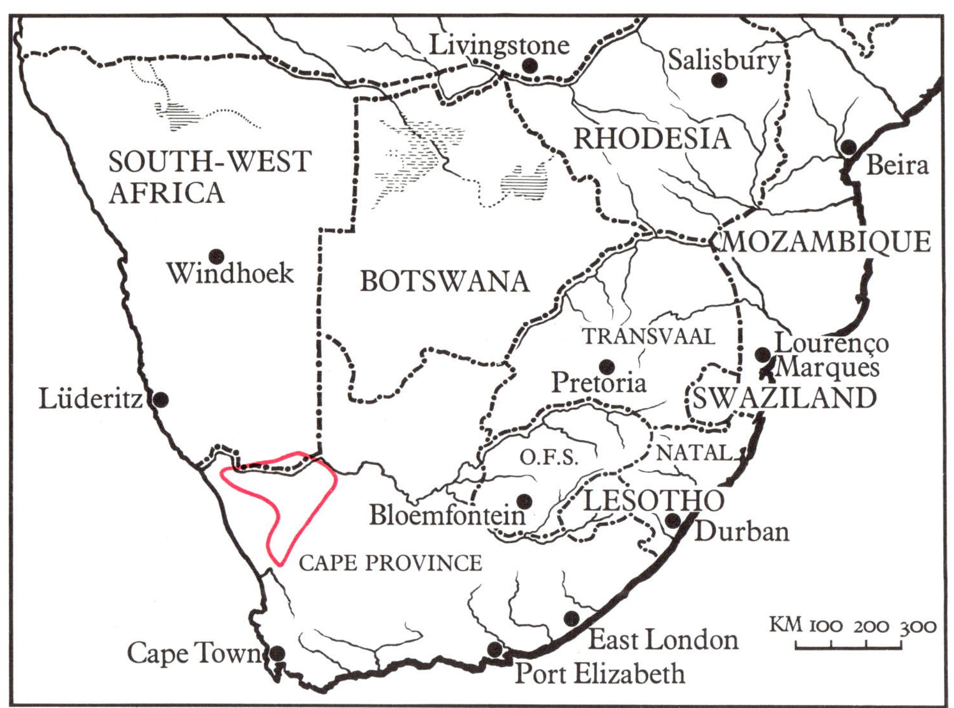

Ophthalmophyllum

Ophthalmophyllum Dinter et Schwantes
From the Greek *ophthalmos* = eye and *phyllon* = leaf.

Dinter et Schwantes, Z. Sukk. 3: 21. 1927. - Schwantes, Möllers Gärtnerztg. 42: 64. 1927. - Tischer, Succulenta, 10: 200. 1928. - *Bolus, Notes, 36; pl. 13 (coloured). 1928.* - v. Pöllnitz, Aufteilung, 58. 1933 - Jacobsen, Sukk. 167; fig. 173. 1933; Succ. Pl. 225; fig. 208-215. 1935; Verzeichnis, 156. 1938. - Jacobsen, Herre, Volk, Mesembr. 79 (in key), 119. 1950. - Phillips, Genera, 321. 1951. - Jacobsen, Handbuch, 1576; fig. 1274-1286. 1955. - Schwantes, Fl. Stones, 324, 340; fig. 34; pl. 87c. 1957. - Jacobsen, Handbook, 953 (in system of Schwantes), 959 (in key of Bolus), 975 (in key of Herre & Volk), 1329 (with key of species by Dr A. Tischer); fig. 1533-1546. 1960; Lexikon, 475, t.188/3 & 4. 1970. - Friedrich in Merxmüller, Prodromus, 89. 1970.

Clumps or tufts with very short, mostly invisible branchlets, each with one pair of leaves, nearly fused, hence forming a body, often enclosed in old whitish membranous sheaths, the rests of former pairs of leaves; there may be up to 7 bodies on one plant. Leaves or better bodies, formed from 2 opposite, fused leaves, rarely spherical, often oblong, rarely smooth, mostly dotted or hairy, often papillate, very soft, at the apex with a gaping fissure or orifice, showing 2 or more free points of the leaves, rarely only with a small orifice, apices with one more or less large, smooth and translucent window, often reddish or also blood-red, about 3 cm long and 2 cm in diam., the flanks sometimes covered with transparent dots. Flowers more or less shortly pedicellated, with fleshy bracts, exserted, sheath-like connate, solitary, white, light to dark-violet, about 3 cm in diam. The free apices of the bracts are club-shaped, thickened, 1-1,2 cm long, more or less flat above, 3-angled in diam. with rounded edges, without clorophyll and not hairy, e.g. also with a small window, as with the ordinary leaves and with the free points far above the ovary; pedicels hairy terete, included into the body as also the ovary which is not laterally compressed and is 0,6 cm thick. Sepals 4-7, equal, reddish brown, more or less membranous, marginated, forming a very narrow tube, about 0,4-0,6 cm long above the ovary. Petals numerous, several series, fused to a tube, rounded or obtuse, somewhat reflexed at the apex. Stamens numerous, translucent white, not hairy. Ovary flat to conical. Glands in 6 distinctly separate parts. Placentas parietal. Stigmas 4-7, often 6, filiform, yellowish green, erect, apices reflexed, connate at the base and forming a style. Capsule 4-7- often 6-locular, expanding keels contiguous with broad valve-wings; loculi-roofs reduced to a limb or none. Seeds very numerous, very small, brownish-yellow with very short, tubercle-like processes, not very rough.

Species: 19, South West Africa, southern part, Namaqualand, Bushmanland, Vanrhynsdorp district. (Type: *O. friedrichiae* (Dint.) Dint. et Schwant.)

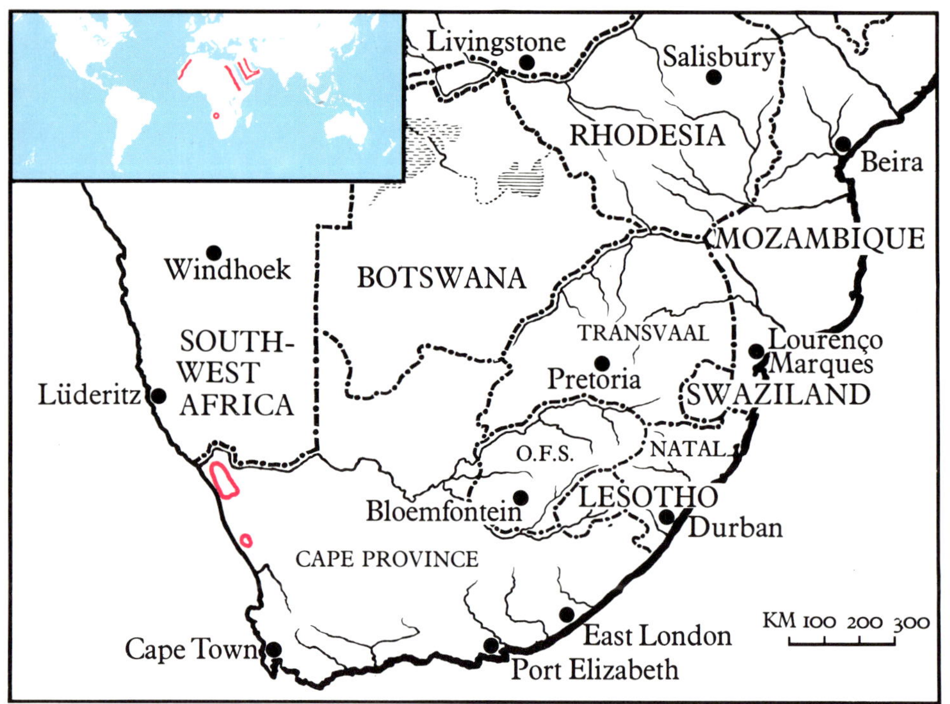

Opophytum

Opophytum N. E. Brown
From the Greek: *opo* = sap, juice and
phyton = plant.

N. E. Brown, Gard. Chron. 78: 412
(in key). 1925. : v. Pöllnitz, Aufteilung,
58. 1933. - N. E. Brown, Gard. Chron.
84: 253. 1928. - Pax, Natürl. Pfl. 217.
1934. - Jacobsen, Verzeichnis, 157.
1938. - Goossens, Blomplante, 148 (in
key). 1940. - Jacobsen, Herre, Volk,
Mesembr., 56 (in key), 120. 1950. - Phillips, Genera, 321. 1951. - Jacobsen,
Handbuch, 1584. 1955. - Schwantes,
Fl. Stones, 35. 1957. - *Jacobsen, Handbook*, 951 (in system of Schwantes),
957 (in key of Bolus), 966 (in key of
Herre & Volk), 1336. 1960. - Jacobsen,
Lexikon, 477. 1970. - Friedrich in
Merxmüller, Prodromus, 84. (Mesembryanthemum). 1970.

Annual, low, sometimes prostrate or
erect, fleshy herbs with distinct internodes between the pairs of leaves, about
4-10 cm long, 1 cm thick, papillose;
leaves containing a considerable amount
of water (name!), united into a sheath
at the base, opposite or alternative,
cylindrical or only convex on the lower
surface, 2-7 cm long, about 1 cm thick,
light-green, often reddish; pedicels up to
4 cm long or subsessile; flowers solitary
or 1-3 together, terminal or lateral,
up to 2 cm in diameter, whitish or
yellowish; calyx 5-lobed, more or less
unequal, up to 5 cm long, tube about
1 cm long, minute papillate; corolla
well exceeding the sepals, tube about
5-7 mm long; staminodes and stamens
numerous; filaments in several series,
light green, segments ca. 3- seriate; ovary
5-chambered, lobes sometimes slightly
compressed; stigmas 5, slender, obtuse,

whitish; capsule 5-loculi, pallid, up to
1,6 cm long, expanded 2,2 cm in diameter, valve-wings erect, bifid at the
apex, recurved, expanding keels finely
serrated, at first parallel, later on diverging; seeds numerous, more or less
globose, pallid, micropyle sometimes
brown, 1 mm long.

Species: 7, Richtersveld, Steinkopf,
Southern Cape, Mossamedes (Angola),
Arabia, Egypt, Western Sahara. (Type:
O. aquosum (L. Bol.) N. E. Brown.)

section of stem
open fruit
B. O. Carter
Sept. 1927.

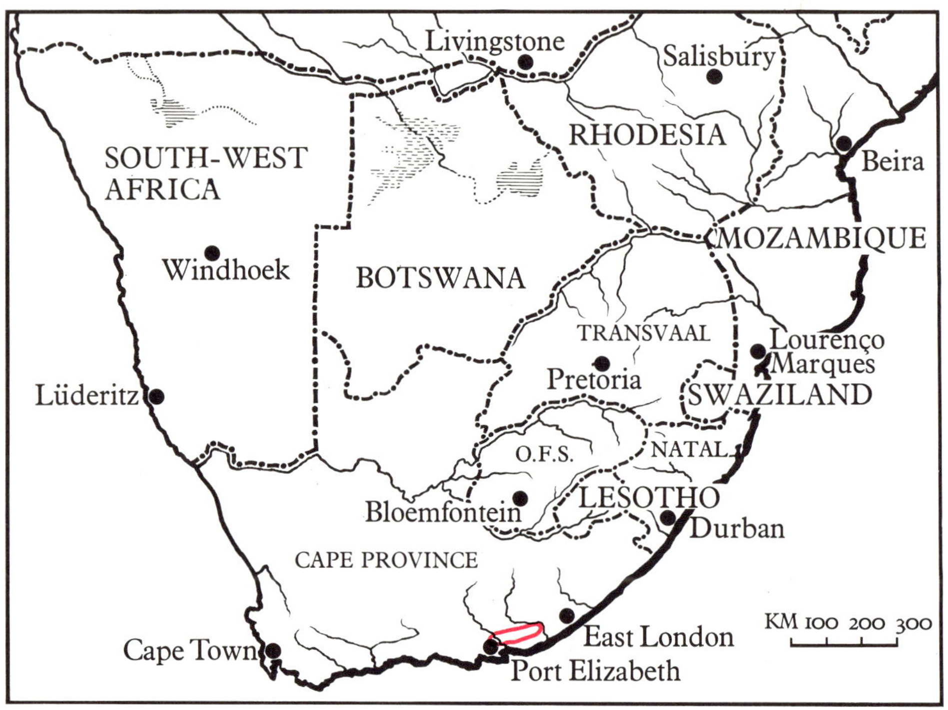

Orthopterum

Orthopterum L. Bolus
From the Greek *orthos* = erect and
pteron = wing, on account of the erect
wings on the cell lid of the capsule.

Bolus, S. Afr. Gard. 17: 281. 1927.
Notes, 77, 130; fig. 8, 9 D(capsule). 1928.
- v. Pöllnitz, Aufteilung, 58. 1933. -
Pax, Natürl. Pfl. 221. 1934. - Jacobsen,
Succ. Pl. 230; fig. 216. 1935. - Bolus,
Notes, 108; pl. 17 K. 1937. - Jacobsen,
Verzeichnis, 157. 1938. - Jacobsen, Herre,
Volk, Mesembr. 74 (in key), 120. 1950.
- Phillips, Genera, 321. 1951. - Jacob-
sen, Handbuch, 1585; fig. 1287. 1955. -
Schwantes, Fl. Stones, 129. 340. 1957. -
Jacobsen, Handbook, 953 (in system
of Schwantes), 962 (in key of Bolus),
973 (in key of Herre & Volk), 1337; fig.
1547. 1960; Lexikon, 477, t.188/5
1970.

Perennial, short stemmed, forming
clumps, branchlets with 6-8 leaves.
Leaves opposite, somewhat unequal,
nearly erect, linear lanceolate, with a
small spine at the apex, flat above or somewhat convex, lower side rounded, towards the apex keeled, along the keel and
the margins 1-2 tubercles with translucent, recurved teeth, smooth, light-
green, covered with dark green dots,
about 3 cm long and 9,8 cm broad and
thick. Flowers, sessile, golden-yellow,
outside reddish, opening afternoon,
about 5 cm in diam. Sepals 5, subequal.
Petals 3-4-series, linear-spathulate.
Stamens erect. Ovary obtuse-coni-
cal above. Glands 5 distant parts,
greenish-brown. Placentas parietal. Stigmas 5-6, long and thin. Capsule 5-6-
locular, valves recurved when open; expanding keels along the margins of the
valves, only half as long as these, parallel,

ending in a broad wing which ends
with an awn-like point, loculi-roofs
erect to about 0,8 cm high, touching each
other and thus half-spherically vaulted;
tubercle present but somewhat obscure;
the valves along the middle-line of the
inside with a sail-like membrane to
about 3 cm high. Seeds pear-shaped with
very fine spines.

Species: 2, Eastern Cape Province.
(Type: *O. waltoniae* L. Bolus.) Resembling *Faucaria* and closely related to it.

n. sp. Orthopterum coega.
1078 Bolus
31
Leaf surface magnified.
open 3-30 p.m.
from the wild.
nat.
tip
mid
base
S.O.Baxter
21-7-1931
×2
×4
×4

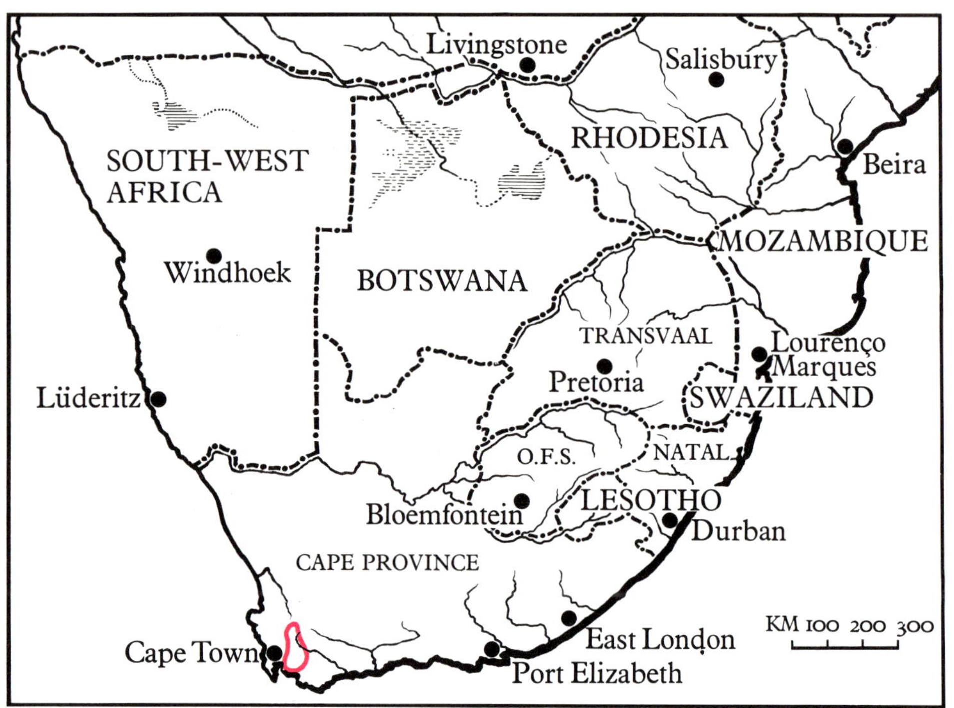

Oscularia

Oscularia Schwantes

From the Latin *osculum* = little mouth,
oscularia = a collection of little mouths.

*Schwantes, Möllers Gärtnerztg. 42: 187.
1927.* - N. E. Brown, Gard. Chron. 87:
71 (in key). 1930. - v. Pöllnitz, Aufteilung, 59. 1933. - Jacobsen, Sukk. 168;
fig. 174. 1933. - Pax, Natürl. Pfl. 217.
1934. - Jacobsen, Succ. Pl. 230; fig. 217.
1935; Verzeichnis, 157. 1938. - Goossens, Blomplante, 145 (in key). 1940. -
Jacobsen, Herre, Volk, Mesembr. 60 (in
key), 120. 1950. - Phillips, Genera,
321. 1951. - Jacobsen, Handbuch, 1585;
fig. 1288, 1289. 1955. - Schwantes, Fl.
Stones, 68, 338; pl. 13b. 1957. - Jacobsen, Handbook, 952 (in system of
Schwantes), 961 (in key of Bolus), 968
(in key of Herre & Volk), 1337; fig.
1548, 1549. 1950; Lexikon, 477,
t.189/2. 1970.

Small subshrubs with erect or spreading
branches. Leaves opposite, somewhat
united at the base, subequal, 3-angled,
short, narrowed towards the base, expanded towards the apex, with a short
point, the margins and the keel edge
dentate or the latter entire, rarely with
prominent scattered dots, mostly smooth,
blue- to grey-green with a waxy coat,
to about 2 cm long and 1,3 cm broad.
Flowers solitary or 3-nate, mostly
pedicelled and with bracts (one pair),
rarely sessile, white to pink, after
opening always open, almond-scented,
to about 1,5 cm in diam. Sepals 5, free.
Petals linear, short, free. Stamens in
column or conically collected, often
with 1-3-series of staminodes. Ovary
somewhat convex to conical above.
Glands with 5 distant parts, crenulated,
dark-green. Placentas parietal. Stigmas
usually 5, concealed by the stamens.
Capsule 5-locular, rarely with 4-, 6-, or
7-loculi; expanding keels diverging
from the base, more than half as long as
the valves, with narrow wings, loculi-
roofs present, without a tubercle.
Seeds nearly black, pear-shaped, with
nipple, rough.

Species: 3, South-Western Cape.
(Type: *O. deltoides* (L.) Schwant.)

M.M.Page. XII. 1922.

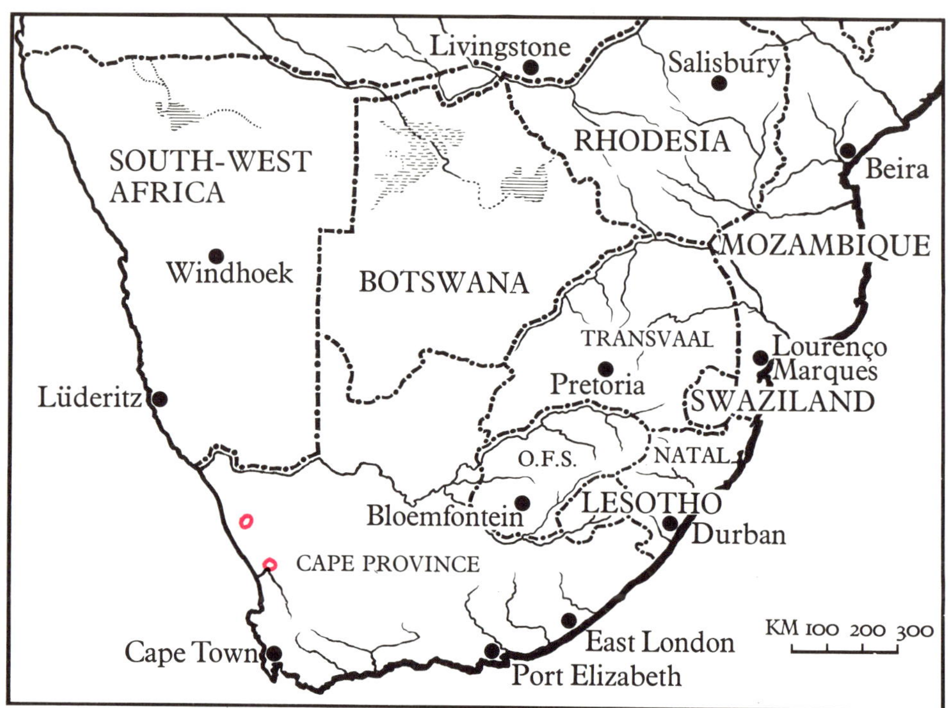

Ottosonderia

Ottosonderia L. Bolus
Named in honour of Dr Otto Sonder
from Hamburg, one of the editors of
the *Flora Capensis*.

Bolus, Notes, 292. 1958. - Jacobsen, Handbook, 952 (in system of Schwantes), 963 (in key of Bolus), 1339. 1960; Lexikon, 478. 1970.

Robust, glabrous shrubs, sterile branches short, floral branches elongated. Leaves opposite, as seen from the side tapering to obtuse, surface flat, back surface rounded with keel reduced to a limb, shining green, to about 6 cm long and 5 mm broad and thick. The inflorescence of this genus is strikingly different from that of any other genus in this family. The first year's growth of a flowering branch consists of a pair of foliage-leaves, followed by a pair of bracts and a solitary pedunculate flower. The axils of the bracts are gemmiferous, each bud developing into the similar pair of bracts and flower which constitute the second year's growth. This process continues for several years, so that under normal conditions a wide spreading cyme is produced, with the scars of the old bracts in its lower part and the hardened remains of more or less incomplete ones in the upper portion, the remains of the fruit persisting even longer than the bracts. The latter appear to be connate for more than half their length, but the actual sheath (measuring from the axil and its bud to the spical pustule) is much shorter, the false sheath being a continuation of the leaf-tissue clothing the internode, as is frequently seen in the section *Uncinata* of the genus *Ruschia*. Sepala 4, apiculate, the outer ones compressed, keeled, the inner ones acute to acuminate, margins with brown membranes. Petals 1-2 seriate, angustate, obtuse, purplish-pink. Stamens and staminodes ca. 5-seriate, collected. Ovary convex. Glands annular. Placentas parietal. Stigmas 6-8 narrow subulate. Capsule 6-8-locular, when open valves erect without wings; expanding keels erect, thick; loculi-roofs well developed; tubercle large. Seeds narrow obovate, brown, 1,25 mm long.

Species: 2, Cape Province, Namaqualand, Vanrhynsdorp district. (Type: *O. monticola* (Sond.) L. Bolus.)

246

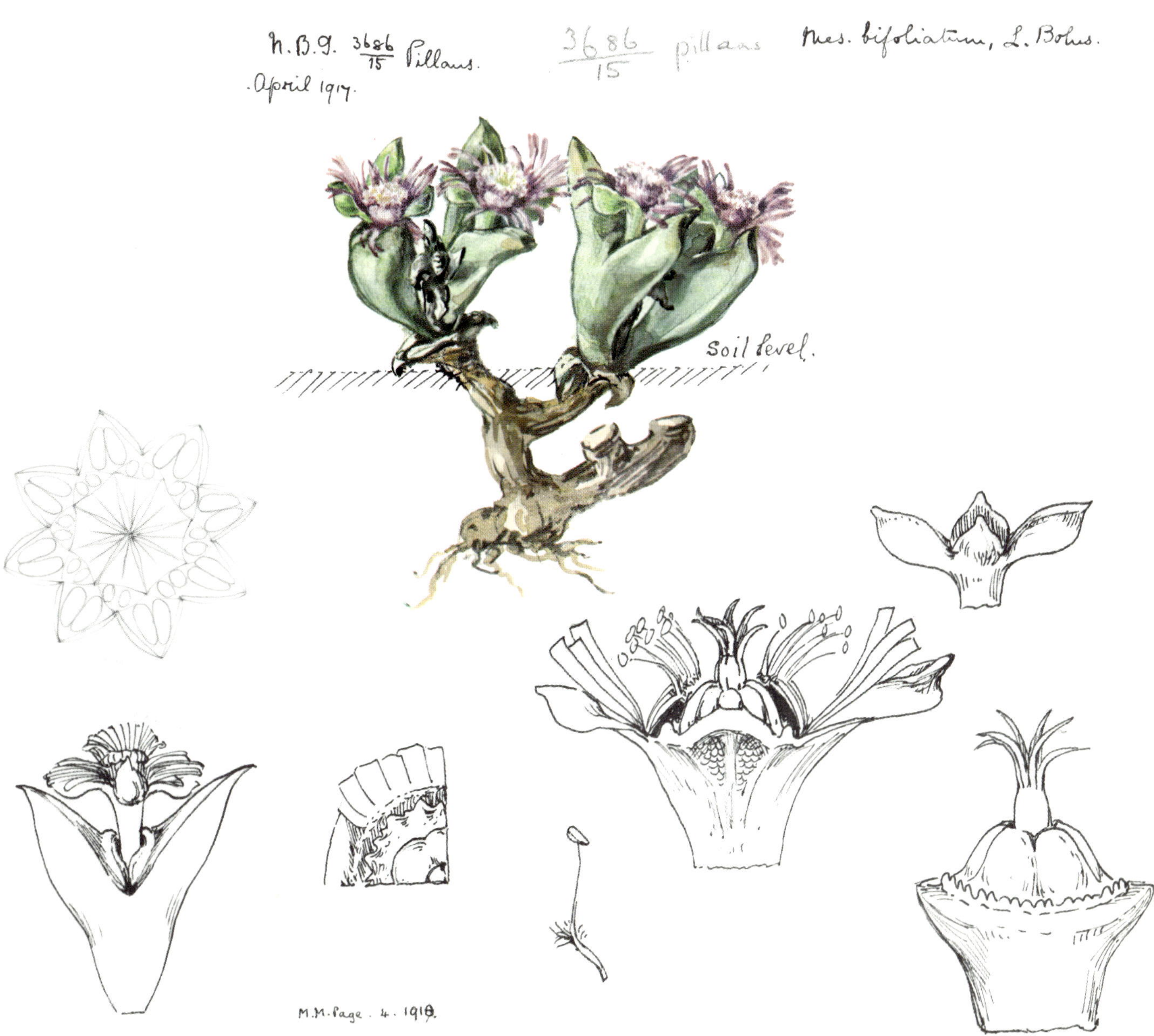

Ottosonderia monticola 247

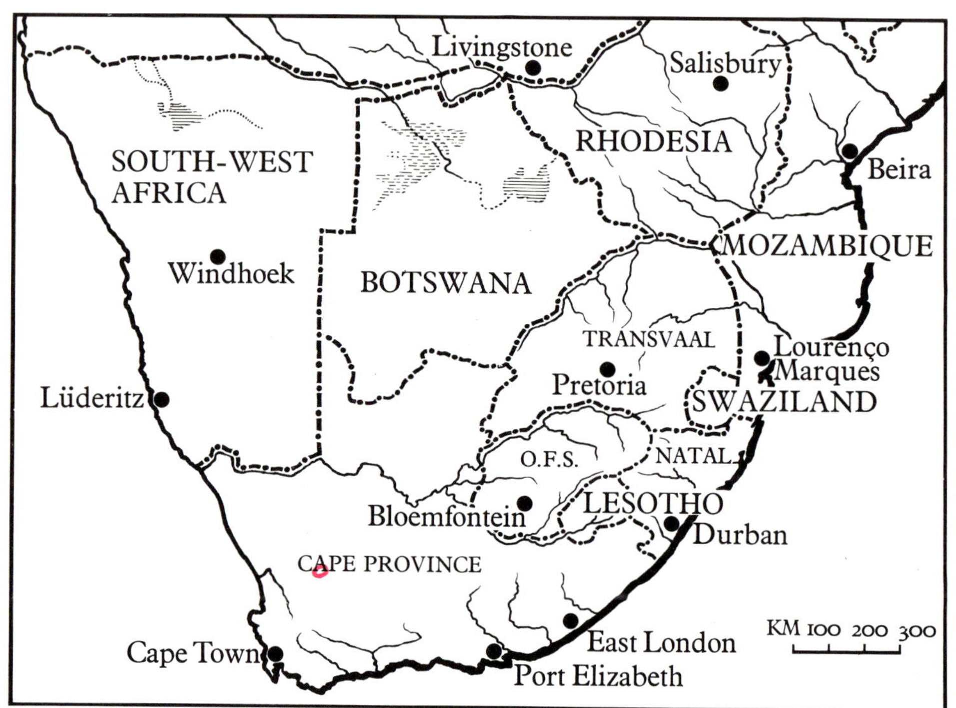

Pherolobus

Pherolobus N. E. Brown
From the Greek *pherein* = bring forth and *lobos* = hump.

N. E. Brown, Möllers Gärtnerztg. 43: 400. 1928. - v. Pöllnitz, Aufteilung, 59. 1933. - Pax, Natürl. Pfl. 217. 1934. - Jacobsen, Verzeichnis, 158. 1938. - Goossens, Blomplante, 149 (in key). 1940. - Jacobsen, Herre, Volk, Mesembr. 59 (in key), 120. 1950. - *Bolus, Notes, 231; pl. 50. 1950.* - Phillips, Genera, 321. 1951 - Jacobsen, Handbuch, 1587. 1955. - Schwantes, Fl. Stones, 50, 340. 1957. - Jacobsen, Handbook, 953 (in system of Schwantes), 959 (in key of Bolus), 967 (in key of Herre & Volk), 1339. 1960; Lexikon, 478. 1970.

Annual herb, 5-10 cm high, with decumbent branches and long internodes. Leaves opposite, spathulate, surface convex, entire, papillate. Flowers pale lilac, at the base purplish. Sepals 5, hairy, nearly equal. Petals flaccid, light purple, darker at the base. Stamens, filaments, anthers and pollen dark-red or purple. Ovary with 5 processes or fleshy lobes, lobes roundish or rounded at the apex, pink, shiny and glittering, to 3 mm long, in the beginning erect, later on adpressed, viewed from the side, concealing the short alternate stigmas; when old nearly distant and showing the stigmas completely; ovary in the old flower nearly semiglobose, lobes obtuse, only a bit compressed. Placentas parietal. Stigmas 5, persistent (on the fruit). Capsule 5-locular; expanding keels contiguous, ending in an awn; loculi-roofs present, without a tubercle. Seeds rough, without tubercles, only with prominent papillae, brown.

Species: 1, Cape Province: Calvinia, Hantam Mts. (Type: *P. maughani* N. E. Br.)

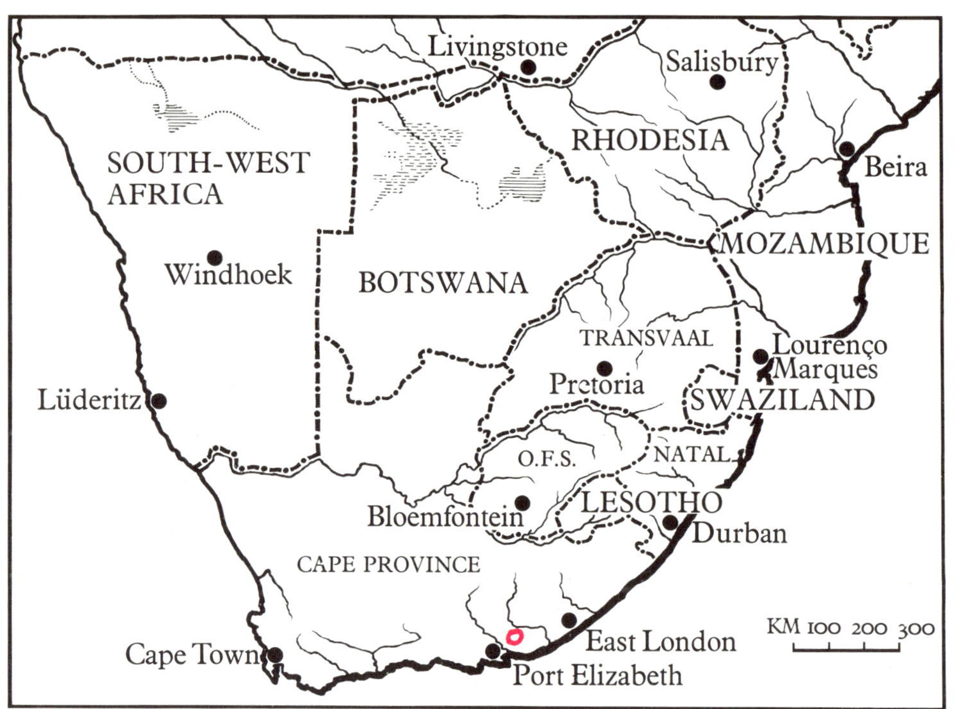

Platythyra

Platythyra N. E. Brown
From the Greek *platy* = broad, flat,
and *thyra* = door.

N. E. Brown, Gard. Chron. 78: 412 (in key). 1925; 84; 313. 1928. - v. Pöllnitz, Aufteilung, 60, 1933. - Pax, Natürl. Pfl. 218. 1934. - Jacobsen, Verzeichnis, 159. 1938. : Goossens, Blomplante, 148. 1940. - Jacobsen, Herre, Volk, Mesembr. 55 (in key), 120. 1950. - *Phillips, Genera, 313. 1951.* - Jacobsen, Handbuch, 1588. 1955. - Schwantes, Fl. Stones, 34. 1957. - Jacobsen, Handbook, 951 (in system of Schwantes), 956 (in key of Bolus), 966 (in key of Herre & Volk), 1340; fig. 1550. 1960; Lexikon, 478, t.189/1. 1970.

Perennial; rootstock a cluster of long fleshy roots; stem herbaceous, prostrate, triangular or quadrangular, with distinct internodes, glabrous. Leaves opposite, flat, petiolate, ovate, or lanceolate. Flowers solitary, axillary or in the forks of the branches, pedicillate, without bracts, yellow. Sepals 4, very unequal, with 2 large and leaf-like, outer ones much larger than the inner ones, forming a short tube above the ovary. Petals very much shorter than the sepals, linear, united into a short tube at the base, arising in the middle of the produced calyx-tube. Stamens numerous, in many series, arising from the corolla-tube, erect. Ovary partly superior. Placentas axile. Stigmas 4, small, stout, obtuse, sessile. Capsule 4-locular, valves recurving when wetted, very broad, each with 2 thin and deeply parallel expanding keels, toothed all along their tops, with an inflexed thin membranous flap or wing turned back from their tips and attached to the valve about midway between the keels and the margin of the valve and forming a sort of pocket on the outer side of each keel; loculi open, no loculi-roofs or tubercles. Seeds moderately large for the group, reniform D-shaped, minutely tuberculate.

Species: 1, Eastern Cape Province (Uitenhage) (Type: *P. haeckeliana* (Berger) N. E. Br.)

395/16. Pillans. {Oct.1917. Kirstenbosch.
M. Haeckelianum. Berger.
=1575 Pillans.
E. Pillans, ZoutKloof
W. Hankey
May 1909.
Growing in a pot;
usually n.± procumbent.
in garden beds.
M.M.Page.10.1917.
Platythyra haeckeliana   251

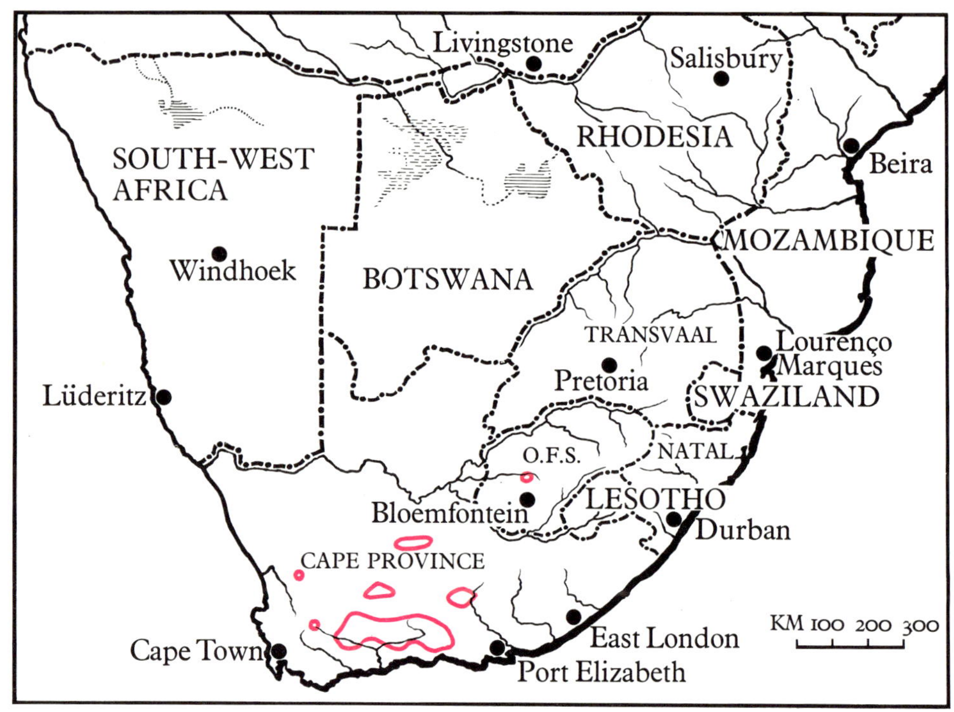

Pleiospilos

Pleiospilos N. E. Brown
From the Greek *pleios* = full
and *spilos* = dots.

N. E. Brown, Gard. Chron. 78: 433 (in key). 1925; 80: 89 with fig. 1926. - Schwantes, Z. Sukk. 3: 22. 1927. - N. E. Brown, Gard. Chron. 83: 216. 1928. - Labarre (red.), Mesembr. 33, 95, 277; fig. 154, 155. 1931. - v. Pöllnitz, Aufteilung, 60. 1933. - Jacobsen, Sukk. 169; fig. 175. 1933. - Pax, Natürl. Pfl. 218. 1934. - Jacobsen, Succ. Pl. 231, fig. 218, 219, 220. 1935; Verzeichnis, 159. 1938. - Goossens, Blomplante, 147 (in key). 1940. - Jacobsen, Herre, Volk, Mesembr. 75 (in key), 120; fig. p. 3 (capsule), fig. 20. 1950. - *Phillips, Genera, 313. 1951.* - Jacobsen, Handbuch, 1589; fig. 1290-1300. 1955. - Schwantes, Fl. Stones, 140, 339; fig. 127; pl. 45b, 46; pl. 5 (coloured). 1957. - Jacobsen, Handbook, 953 (in system of Schwantes), 963 (in key of Bolus), 974 (in key of Herre & Volk), 1340; fig. 1551-1562. 1960; Lexikon, 478, t.189/1, 2, 3. 190/1, 2, 3. 191/1, 2. 1970.

Punctillaria N. E. Brown, Gard. Chron. 78: 433 (in key) 1925; 80: 212 with fig. 1926. - Phillips, 243. 1926. - Schwantes, Z. Sukk. 3: 23. 1927. - Bolus, Notes, 47 with coloured plate, 18, 82; pl. 12, 13, 14. 1928. - N. E. Brown, Gard. Chron. 84; 492. 1928. - Labarre (red.), Mesembr. 280; fig. 156-158. 1931. - v. Pöllnitz, Aufteilung, 62. 1933.

Dwarf stemless succulent perennials with 1-4 pairs of leaves on each branchlet and with invisible internodes. Leaves opposite, more or less connate at the base, equal, ovate very stout and thick, as broad as long, or less than twice as broad as long, firm, grey-green or brownish, usually conspicuously dotted, edges etc. often reddish, varying much as regards the size, rock-like, flat above, lower side very much rounded and keeled, obtuse or acute. Flowers solitary or 2-4 to a growth, terminal between the leaves, sessile or shortly pedicellate, bracteate, yellow to orange, to about 8 cm in diam. Sepals 5-6, free nearly or quite down to their junction with the ovary, or produced into a short tube or cup. Petals numerous, free, linear, arising round the margins of the top of the ovary. Stamens numerous, inserted into the calyx tube or, as with *Pl. bolusii*, on the disc of the nectary glands, along the margin of the calyx. Staminodes 0. Ovary inferior, conical. Glands annular, dark-green, sometimes undulate. Placentas parietal. Stigmas 9-15, diverging, yellowish, filiform, sessile. Capsule 9-15-locular sub-hemispherical; valves reflexed when wetted; expanding keels parallel, towards the apex diverging, then free and the free part bearing a triangular membranous wing, with the wings uniting with the wings of the adjacent valves in pairs or becoming free and standing erect between the valves; loculi acutely roofed by the rather rigid loculus-wings, or separated elements of the chamber-partitions, turned back at the opening and somewhat resembling the mouth of a trumpet, so that the whole of roofs is raised into a sort of crown-like structure above the level of the expanding keels with or without tubercles at the mouth of the loculi. Seed compressed, ovoid, with a nipple at one end, microscopically tuberculated, dark brown.

Species: 35, Cape Province, Karroo, Free State. (Type: *P. bolusii* (Hook. f.) N. E. Br.)

Pleiospilos bolusii 253

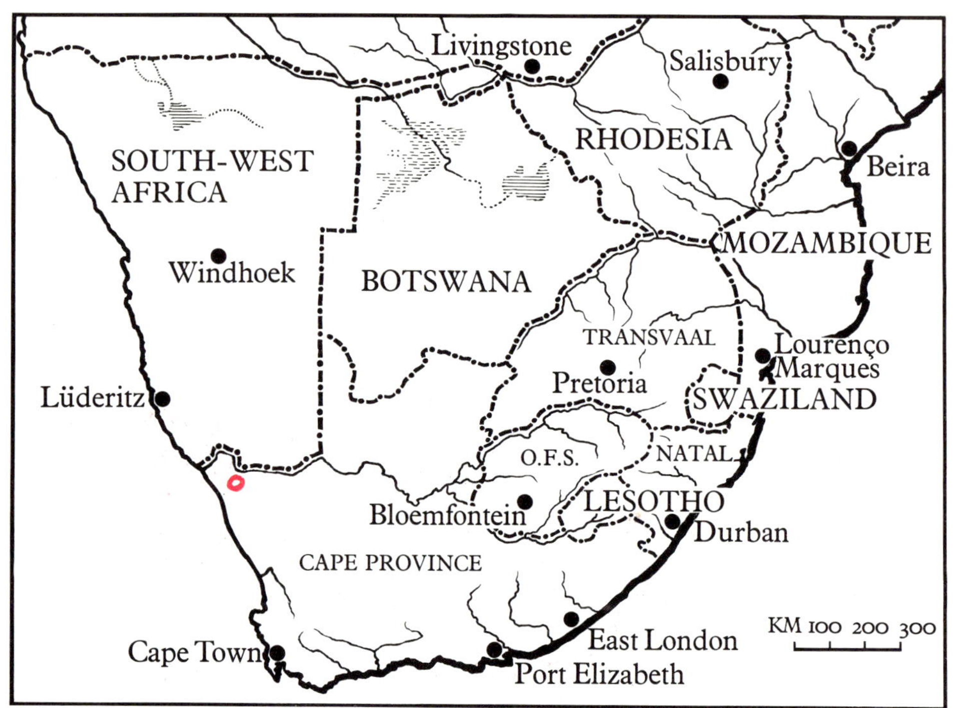

Polymita

Polymita N. E. Brown
From the Greek *poly* = many and
mita = very small petals.

N. E. Brown, Gard. Chron. 87: 72 (in key). 1930. - v. Pöllnitz, Aufteilung, 61. 1933. - Pax, Natürl. Pfl. 218. 1934. - Jacobsen, Verzeichnis, 162. 1938. - Jacobsen, Herre, Volk, Mesembr. 65 (in key). 1950. - Phillips, Genera, 321. 1951 - Jacobsen, Handbuch, 1598. 1955. - Schwantes, Fl. Stones, 338. 1957. - Jacobsen, Handbook, 952 (in system of Schwantes), 962 (in key of Bolus), 970 (in key of Herre & Volk), 1348. 1960; Lexikon, 480. 1970.

Robust, stiff, glabrous, branching shrub, branches ascending. Leaves long united into a sheath with impressed longitudinal lines, the tip spreading, at the end acute, mucronate, margins and keel somewhat horny and sharp about 10 mm long and 6 mm diam. Flowers solitary or in cymes, often pedicellate, white, about 2 cm in diam. Sepals 5. Petals filiform-linear, very thin. Stamens incurved, concealed by the longer and more numerous staminodes, which are also incurved and surround them in 3-4-series. Placentas parietal. Stigmas 8-12, filiform-subulate, stout. Capsule 8-12, rarely 7-locular; expanding keels towards the apex suddenly diverging, with broad terminal wings; loculi-roofs present; with tubercle. Seeds smooth, with nipple.

Species: 1, Cape Province: Namaqualand, Copper Mts. (Type: *P. albiflora* (L. Bol.) L. Bol.)

254

Polymita albiflora   255

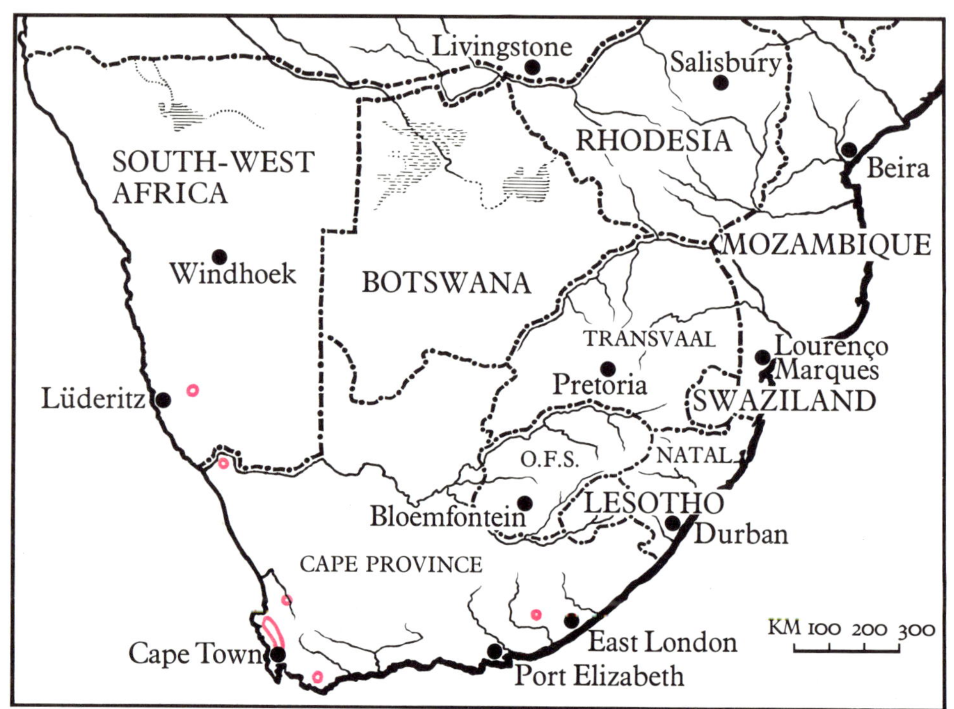

Prenia

Prenia N. E. Brown

From the Greek *prenes* = prostrate, referring to the habit.

N. E. Brown, Gard. Chron. 78: 412, 413 (in key). 1925; 84: 268. 1928. - Bolus, Notes, 82; fig. 11E, 1928. - v. Pöllnitz, Aufteilung, 61. 1933. - Pax, Natürl. Pfl. 218. 1934. - Jacobsen, Succ. Pl. 235, 1935; Verzeichnis, 162. 1938. - Goossens, Blomplante, 142, 148 (in key). 1940. - Jacobsen, Herre, Volk, Mesembr. 55/56 (in key), 121. 1950. - Adamson, Salter, Fl. Cape Pen. 374. 1950. - *Phillips, Genera, 314. 1951.* - Jacobsen, Handbuch, 1599. 1955. - Schwantes, Fl. Stones, 27, 35; pl. 6a. 1957. - Jacobsen, Handbook, 951 (in system of Schwantes), 956 (in key of Bolus), 966 (in key of Herre & Volk), 1348. 1960; Lexikon, 480. 1970. - Friedrich in Merxmüller, Prodromus, 93. 1970.

Papillate succulent perennials, either with a short branching main stem bearing tufts of crowded leaves at the end from which decumbent flowering branches arise, or with long prostrate stems having distinct internodes. Leaves opposite, sessile, several times longer than broad, linear-lanceolate, flat or channelled above and keeled or rounded on the back, sometimes blue-green, not connate, covered with a grey waxy coat. Flowers in terminal lax few-flowered cymes or occasionally solitary, without bracts, opening at noon, white, pink, purplish, pink or yellow, about 4 cm in diam. Sepals 4-5 produced above its union with the ovary into a short tube, 2 often leafy and larger than the others. Petals numerous, 2-3-series, united at the base into a short tube, arising just above the union of the calyx with the ovary very different, also filaceous. Stamens numerous, arising from the tube of the corolla, erect; staminodes o. Ovary partly superior. Placentas axile. Stigmas 5, filiform, as long as the stamens. Capsule 4-5-locular, obconic $\frac{1}{2}$ superior; expanding keels closely contiguous, forming an acute central keel, with large marginal wing-like submembranous flaps standing erect or folded in towards the keel; loculi open, without loculi-roofs or tubercles. Seed compressed, somewhat D-shaped in outline, minutely tuberculate.

Species: 5, Cape Province: South-Western districts, Namaqualand. (Type: *P. pallens* (Ait.) N. E. Brown.)

M.M.Page. XII. 1922.

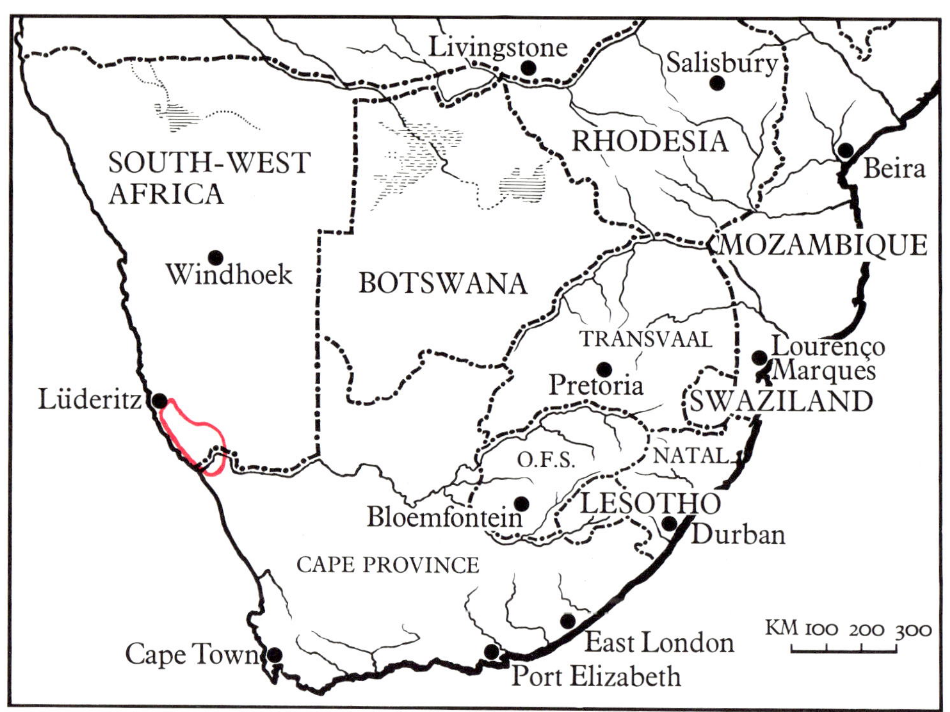

Psammophora

Psammophora Dinter et Schwantes
From the Greek *psammos* = sand and
phorein = to bear.

Dinter et Schwantes, Z. Sukk. 2: 188. 1925/26. - Bolus, Fl. Pl. S. Afr. 7: pl. 280 (coloured). 1927. - v. Pöllnitz, Aufteilung, 61. 1933. - Jacobsen, Sukk. 171; fig. 176. 1933. - Pax, Natürl. Pfl. 221. 1934. - Jacobsen, Succ. Pl. 235; fig. 221, 22. 1935; Verzeichnis, 162. 1938. - Jacobsen, Herre, Volk, Mesembr. 59 (in key), 121. 1950. - Phillips, Genera, 321. 1951. - Jacobsen, Handbuch, 1599; fig. 1301-1304. 1955. - Schwantes, Fl. Stones, 161, 338; pl. 49a & b. 1957. - Jacobsen, Handbook, 952 (in system of Schwantes), 958 (in key of Bolus), 968 (in key of Herre & Volk), 1349; fig. 1563-1566. 1960; Lexikon, 480, t.192/1. 1970. - Friedrich in Merxmüller, Prodromus, 94, 1970.

Low tufted, very succulent plants with almost woody, densely leafy shoots; branches growing above or buried in the soil so that only the leaves are visible. Leaves short, thick decussate, more or less triangular or nearly semicylindrical, in transverse section, with a rounded keel and margins, tapering to a point or obtuse, with the lower side drawn forward like a chin; surface dull glossy, bluish-grey-green, sticky; in its natural habitat the leaves protected from excessive transpiration by a coating of sand and dust, later on surface peeling off, to about 4 cm long and 1 cm broad and thick. Flowers terminal, solitary, stalked, the clumps with one pair of bracts on the base of the pedicel which is rather long, light-violet or white, to about 2,5 cm in diam. Sepals 4, subequal. Petals linear, obtuse. Stamens numerous, collected, whitish; staminodes present. Ovary turbinate. Glands crenulated with undivided glandular wall. Placentas parietal. Stigmas 5-6, filiform as long as the stamens. Capsule (5-)6-7-locular, flat on top, globose to 5-angled; expanding keels contiguous, with rather long wings, light yellow with crenulated margins and the points toothed; without loculi-roofs and tubercles. Seeds light-brown with a slight orange tint, pear-shaped with minute tubercles.

Species: 5, from Lüderitz, S.W.A., along the coast to Alexander Bay and in the Richtersveld. (Type: *P. nissenii* Dinter et Schwantes.)

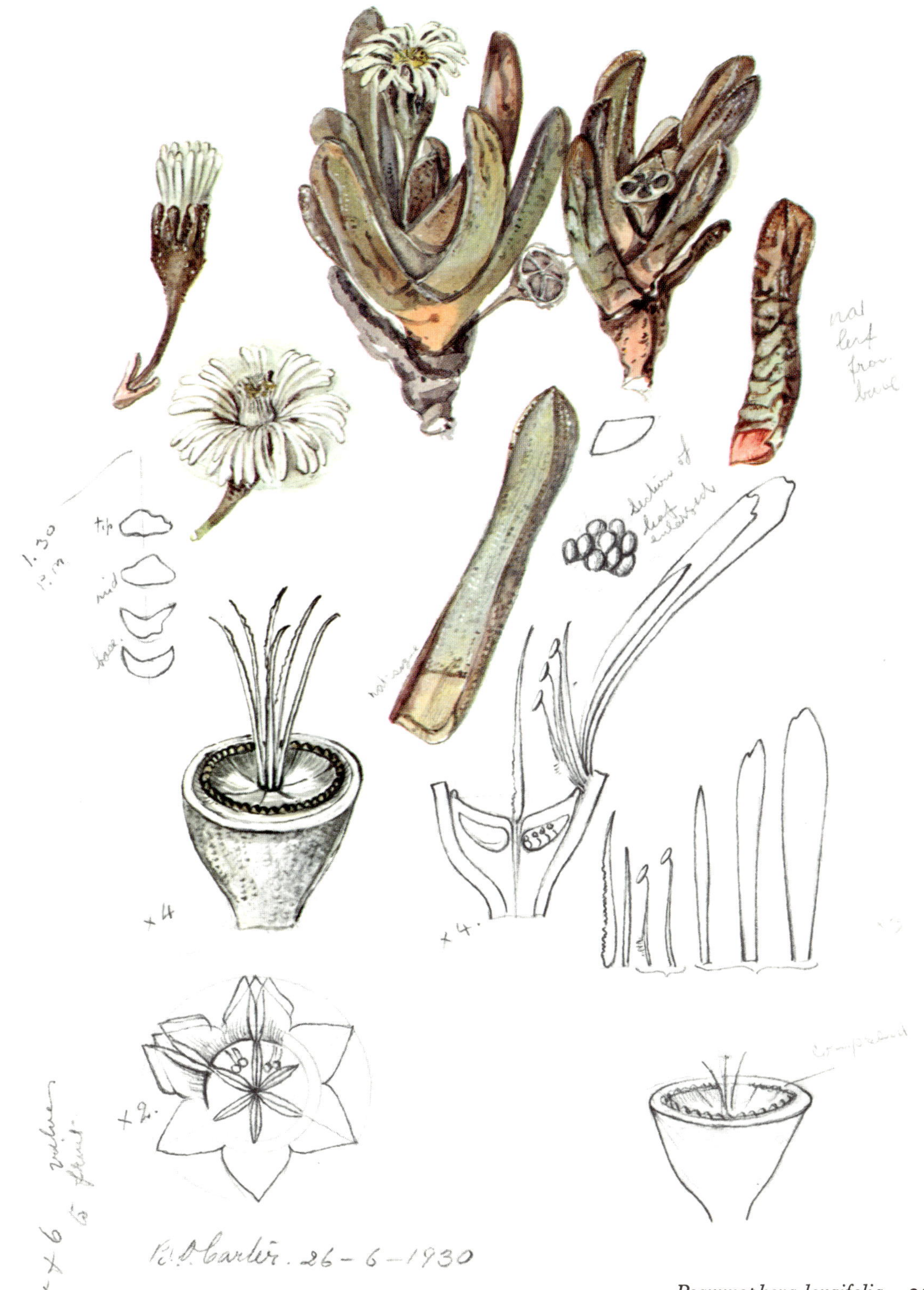

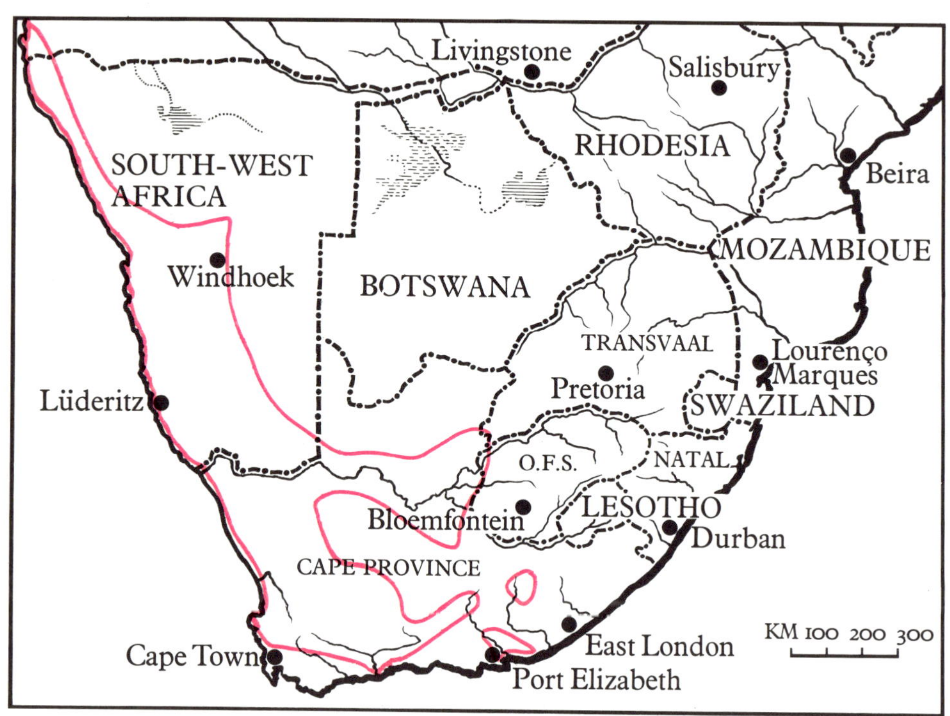

Psilocaulon

Psilocaulon N. E. Brown
From the Greek *psilos* = bare and
kaulon = stalk.

N. E. Brown, Gard. Chron. 78: 433 (in key). 1925. - Phillips, Genera, 247. 1926. - N. E. Brown in Burtt-Davy, Fl. Pl. & Ferns, Tvl. & Swaziland, 1: 157. 1926. - Bolus, Notes, 8; fig. 1D, 66, 87, fig. 15B. 1928. - v. Pöllnitz, Aufteilung, 62. 1933. - Pax, Natürl. Pfl. 218. 1934. - Jacobsen, Succ. Pl. 236. 1935; Verzeichnis, 163. 1938. - Goossens, Blomplante, 142 (in key). 1940. Jacobsen, Herre, Volk, Mesembr. 56 (in key), 121; fig. 1. 1950. - Adamson, Salter, Fl. Cape Pen. 375. 1950. - *Phillips, Genera, 314. 1951.* - Jacobsen, Handbuch, 1602; fig. 1305. 1955. - Schwantes, Fl. Stones, 29, 35; pl. 8a, 1957. : Jacobsen, Handbook, 951 (in system of Schwantes), 956 (in key of Bolus), 966 (in key of Herre & Volk), 1351; fig. 1567, 1568. 1960; Lexikon, 480, t.192/2, 3, 4, 1970. - Friedrich in Merxmüller, Prodromus, 95. 1970.

Brownanthus Schwant. Zeitschr. Sukk.-kunde, 3: 14, 20. 1927. - Jacobsen, Lexikon, 367, t.148/1. 1970. Sunk by Friedrich (l.c.) into *Psilocaulon*.

Much-branched succulent annuals or perennials, often erect bushes but sometimes decumbent; branches continuous or jointed, sometimes beaded, rarely papillate. Leaves opposite, sessile, free or slightly united at the base, small, slender, semi-terete or terete, withering and deciduous, often leaving naked branches, according to L. Bolus (l.c.)

the lowest part of the leaf is persistent and remains as a scale on the stem. Flowers small, terminal in cymes or solitary on lateral branches, short stalked, white or reddish or yellow, less than 2 cm in diam. Sepals 4-5 free almost down to their union with the ovary or with a very short tube above that union. Petals numerous, linear, united at the base into a tube which is sometimes very short. Stamens numerous, erect. Ovary conical. Placentas axile. Glands 5, distant. Stigmas 4-5, filiform, reflexed, without style. Capsule 4-5-locular, valves with the expanding keels closely continguous into 1 central keel and their margins with membranous wings inflexed upon the keel; loculi without wings or tubercles. Seed compressed, somewhat D-shaped or obtusely triangular in outline, nearly smooth or a little rough.

Species: 72, South West Africa, Namaqualand, Karroo, Western and Eastern Cape Province. (Type: *P. articulatum* (Thunb.) N. E. Brown.)

M.M.Page. X. 1921.

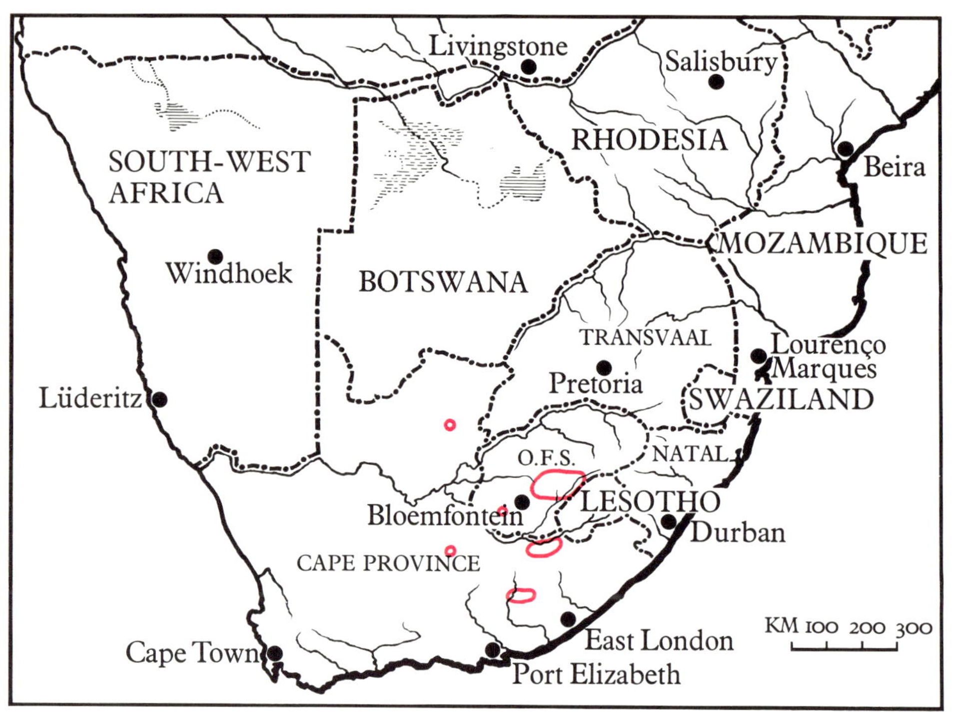

Rabiea

Rabiea N. E. Brown
Named in honour of the
Rev. W. A. Rabie, Orange Free State.

N. E. Brown, Gard. Chron. 88: 279 (in key). 1930; 89: 53. 1931. - Bolus, Notes, 63 (Mesembr. & Nananthus), pl. 7; fig. 22. 1928. - v. Pöllnitz, Aufteilung, 63. 1933. - Pax, Natürl. Pfl. 218. 1934. - Jacobsen, Succ. Pl. 237, fig. 223. 1935; Verzeichnis. 168. 1938. - Goossens, Blomplante, 145 (in key). 1940. - Jacobsen, Herre, Volk, Mesembr. 75 (in key), 122. 1950. - *Phillips Genera, 315. 1951.* - Jacobsen, Handbuch, 1611; fig. 1306. 1955. - Schwantes, Fl. Stones, 157, 339. 1957. - *Bolus, Notes, 360; pl. 109-114. 1958.* - Jacobsen, Handbook, 953 (in system of Schwantes), 959 (in key of Bolus), 974 (in key of Herre & Volk), 1358 (with key of species); fig. 1569-1572. 1960; Lexikon, 484, t.193/1. 1970.

Dwarf compact glabrous plants; rootstock fleshy or tuberous; branchlets crowded, 4-6 leaved, internodes enclosed in the leaf-sheaths. Leaves ascending or spreading, two of a pair more or less dissimilar, upper surface acute or acuminate, margins parallel, or one margin straight and the other outwardly curved, viewed laterally tapering upwards and keeled, rarely terete, or one of a pair at least compressed and dilated near the apex, often with raised and conspicuous white or reddish brown dots, to about 5 cm long. Flowers solitary, open all day in full sunshine, subsessile, the peduncle enclosed in the sheath of the basal leaf-like bracts. Sepals 5 or rarely 6-7, about equal in length or the two outer ones longer and more acute than the membranously margined inner ones, remaining erect in the fully open flower and so causing the petals to form a cup in which the lower part of the stamens is enclosed. Petals 3-4-seriate, tapering from near the middle downwards, obtuse or emarginate, yellow, golden or orange; staminodes absent. Stamens erect, papillate. Ovary more or less concave above, or slightly raised in the very middle. Glands 6, 8, 9 or 10 distant ones. Placentas parietal. Stigmas 7-14, slender, papillate. Capsule subglobose below, flat above, valves winged, the wings reaching a little more than halfway up; expanding keels erect, dark brown, contiguous in the lower half, then widely divergent, shortly aristate, almost reaching the apex of the valve, loculi covered, tubercle absent. Seeds very dark brown subobovate, minutely muricate.

Species: 7, Cape Province: Karroo, Free State. (Type: *R. albinota* (Haw.) N. E. Br.)

x3
nat
x2
tip
mid
base
old leaf
B.O.Carter.

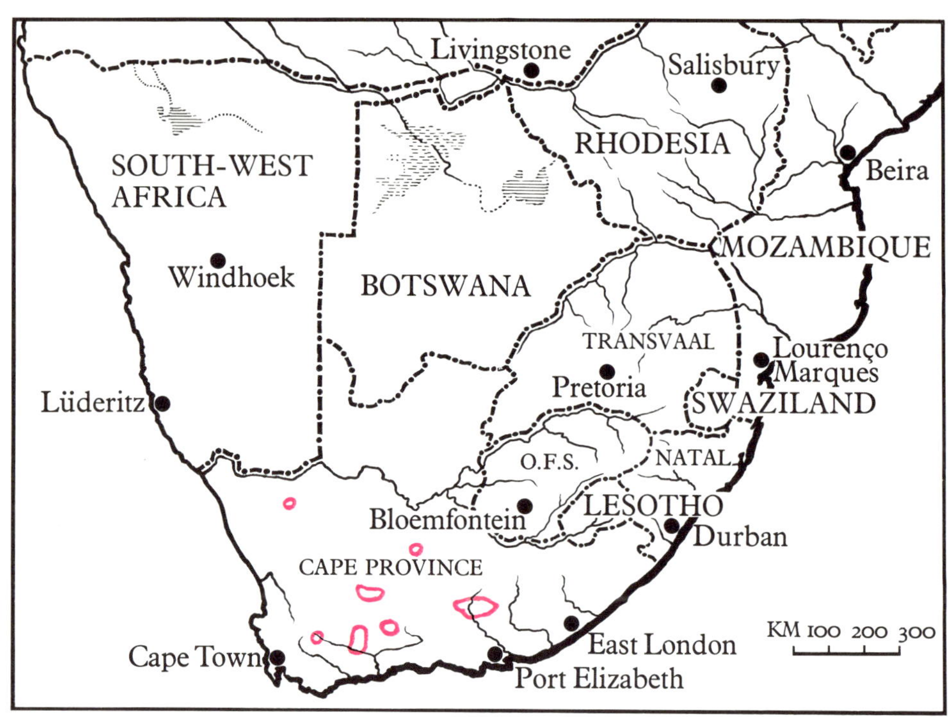

Rhinephyllum

Rhinephyllum N. E. Brown
From the Greek *rhine* = file
and *phyllon* = leaf.

N. E. Brown, Gard. Chron. 81: 92; fig. 39, 40. 1927. - Labarre (red.), Mesembr. 286; fig. 159. 1931. - v. Pöllnitz, Aufteilung, 63. 1933. - Jacobsen, Sukk. 172; fig. 177. 1933. - Pax, Natürl. Pfl. 218. 1934. - Jacobsen, Succ. Pl. 233; fig. 224. 1935. - Bolus, Notes, 40 (with key of species); pl. 8. 1937. - Jacobsen, Verzeichnis, 169. 1938. - Goossens, Blomplante, 144. 1940. - Jacobsen, Herre, Volk, Mesembr. 71 (in key), 122. 1950. - *Phillips, Genera, 315. 1951.* - Jacobsen, Handbuch, 1613; fig. 1307-1309. 1955. - Schwantes, Fl. Stones, 138, 339; pl. 44b. 1957. - Jacobsen, Handbook, 953 (in system of Schwantes), 957 (in key of Bolus), 972 (in key of Herre & Volk), 1361 (with key of species); fig. 1573-1576. 1960; Lexikon, 485, t.193/2. 1970.

Neorhine Schwantes in Monatsschr. Kakt. Ges. 2: 22. 1930. - v. Pöllnitz, Aufteilung, 57. 1933. - Pax, Natürl. Pfl. 221. 1934. - Jacobsen, Succ. Pl. 224, 238. 1935; Verzeichnis, 155. 1938. - Jacobsen, Herre, Volk, Mesembr. 71 (in key), 119. 1950. - Phillips, Genera, 321. 1951. - Jacobsen, Handbuch, 1569; fig. 1270. 1955. - Schwantes, Fl. Stones, 138, 339. 1957. - Jacobsen, Handbook, 953 (in system of Schwantes), 957 (in key of Bolus), 972 (in key of Herre & Volk), 1309; fig. 1526. 1960.

Peersia L. Bolus, Ann. S. Afr. Mus. 9: 143 & Fl. Pl. S. Afr. 7: pl. 264 (coloured). 1927. - v. Pöllnitz, Aufteilung, 59, 1933. - Pax. Natürl. Pfl. 221. 1934. - Jacobsen, Verzeichnis, 158. 1938. - Bolus, Notes, 135. 1938. - Phillips, Genera, 321. 1951. - Schwantes, Fl. Stones, 338. 1957.

Short-stemmed perennial succulents, with woody or thickened and fleshy rootstock and abbreviated, or sometimes slender woody and more elongated branches with mostly 2-4 pairs of leaves. Leaves opposite, slightly united at the base, thickened upwards or clavate, upper side flat, back surface round or slightly keeled, rough from being covered at the upper part with small hard white tubercles, edges and keel sometimes with irregularly distant teeth. Flowers solitary, terminal, pedicellate with or without bracts, white to yellow, opening during the afternoon, sometimes also at night, scented, to about 3 cm in diam. Sepals 5, down to its union with the ovary, with *R. pillansii* (*Neo-rhine*) tubularly connate above the ovary. Petals in 1-5 series, free, scarcely longer than the sepals. Stamens numerous, erect, posteriors bearded at the base. Ovary inferior, shallow, slightly elevated at the central part on the top and with the margin glands (disc) raised, rarely distant and separated as with *R. pillansii*, *R. frithii* and *R. macradenium*, the latter 2 with 5 large, at the apex truncate glands. Placentas on the floor of the shallow chambers. Stigmas 5, ascending-spreading, subulate, somewhat plumose. Capsule 5-locular, small, shortly obconical, flat on the top, with the sutures slightly raised and gaping; valves deltoid, reflexed when expanded, expanding keels as long as the valves, contiguous into a central keel throughout, with broad membranous marginal wings that are not united in pairs between the valves; loculi-roofs none or reduced to a limb (*R. macradenium*), tubercle none. Seeds numerous in each loculus, globose-ovoid, slightly compressed, with a nipple on one end, smooth or with minute tubercles along the margins (*R. macradenium*).

Species: 12, Cape Province: Karroo and Little Karroo. (Type: *R. muirii* N. E. Br.)

Rhinephyllum frithii   265

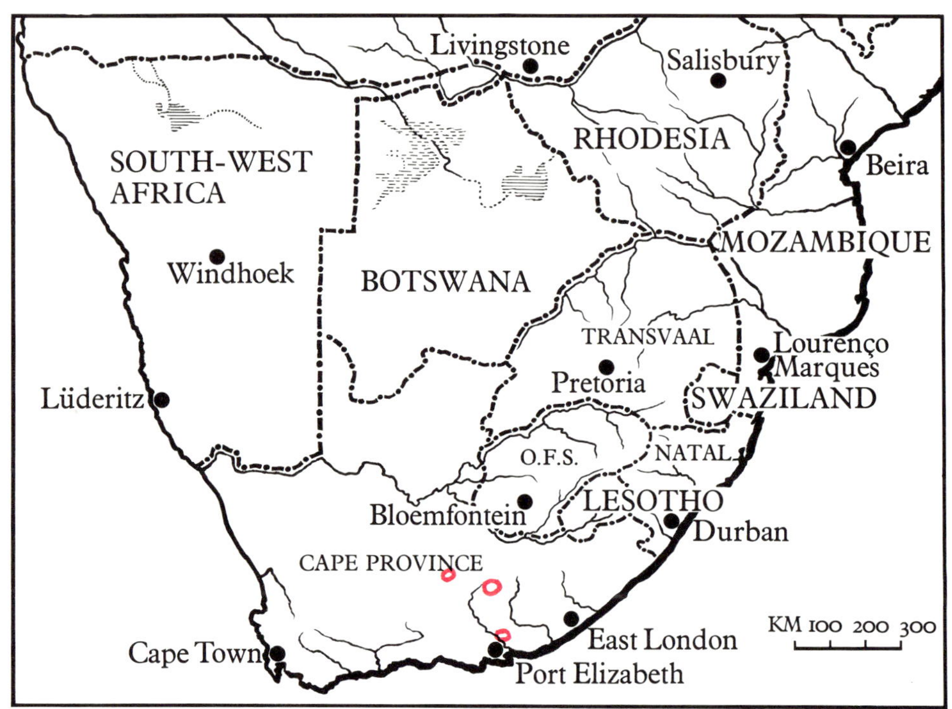

Rhombophyllum

Rhombophyllum Schwantes
From the Greek *rhombos* = lozenge
and *phyllon* = leaf.

Schwantes, Z. Sukk. 2: 180. 1926; 3: 23. 1927. - v. Pöllnitz, Aufteilung, 64. 1933. - Jacobsen, Sukk. 173; fig. 178. 1933. - Pax, Natürl. Pfl. 221. 1934. - Jacobsen, Succ. Pl. 238; fig. 225, 226. 1935; Verzeichnis, 169. 1938. - Jacobsen, Herre, Volk, Mesembr. 71, 73 (in key), 122; fig. 8. 1950. - Phillips, Genera, 321. 1951. - Jacobsen, Handbuch, 1616; fig. 1310, 1311. 1955. - Schwantes, Fl. Stones, 91, 338; fig. 23, 24; pl. 24a. 1957. - Jacobsen, Handbook, 952 (in system of Schwantes), 962 (in key of Bolus), 972 & 973 (in key of Herre & Volk), 1364; fig. 1577, 1578. 1960; Lexikon, 486, t.193/3 & 4. 1970.

Small shrubs, or caespitose, very succulent plants with fleshy, often turnip-like roots, with short branchlets. Leaves densely crowded, opposite, somewhat connate at the base, decussate, semi-cylindrical, keeled towards the apex, lower surface drawn forward like a chin, upper surface more or less linear, expanded towards the apex or almost obliquely rhomboidal, entire or armed with 1-2 short teeth, or keel towards the apex hatchet-like, flat, margins often whitish, smooth, glossy, intense green, with whitish dots which may be visible against light, about 5 cm long and 2 cm broad and thick. Flowers dichotomous, one to three, with 2,5 cm long pedicels, often with distinct bracts, yellow, to about 4 cm in diam. Sepals 5, subequal, lanceolate. Petals about 3-series, narrow linear, yellow, apices outside reddish. Stamens many, erect, glabrous or sometimes bearded at the base. Ovary flat. Glands distant, separated. Placentas parietal. Stigmas 5, filiform. Capsule 5-locular, loculi-roofs stiff, tubercle flat, minute or very large, white with 2 spherical, hyaline warts. Seeds light-brown, pear-shaped with nipple.

Species: 3, Eastern Cape Province & Graaff-Reinet. (Type: *R. rhomboideum* (SD) Schw.)

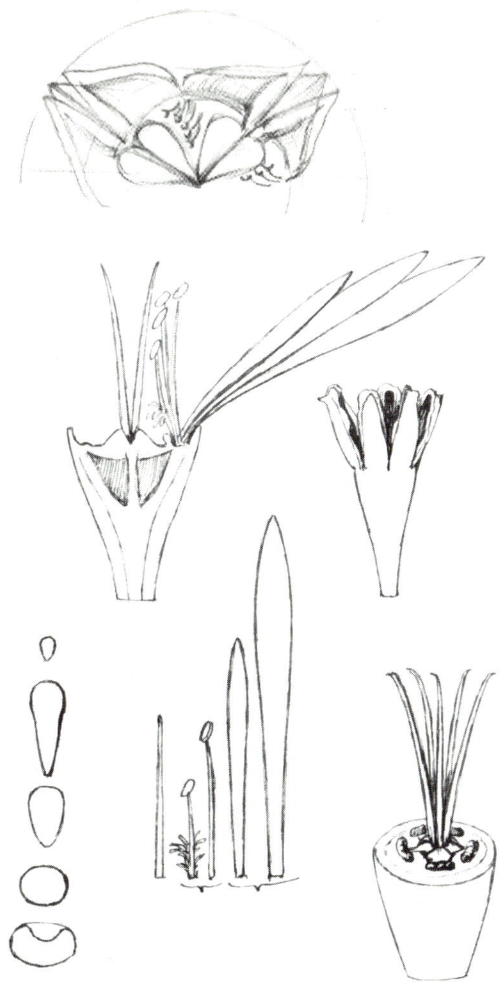

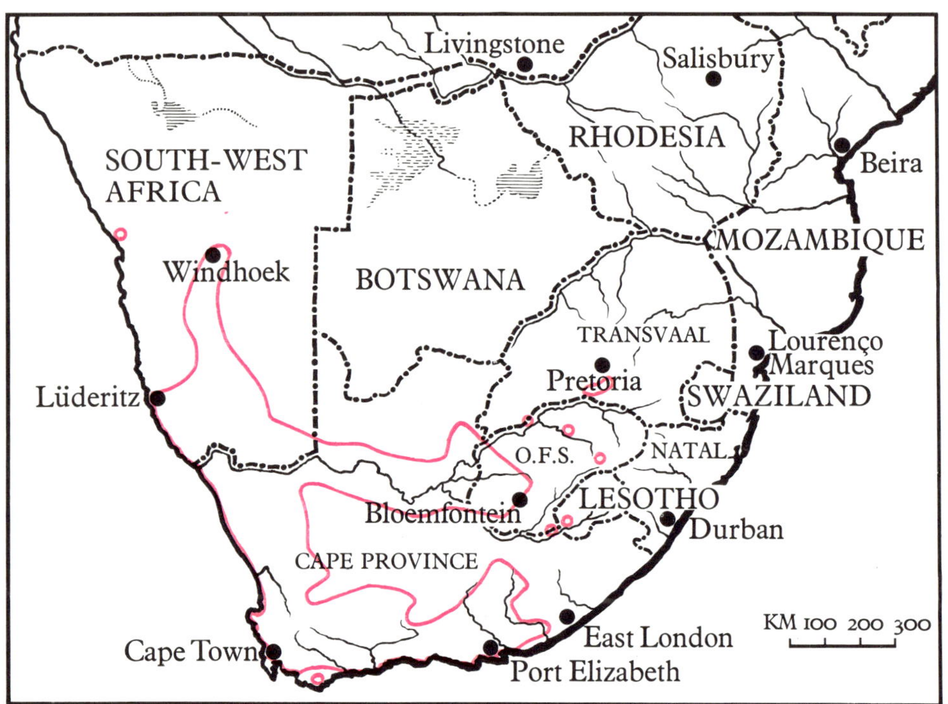

Ruschia

Ruschia Schwantes
Named in honour of the late Mr
Ernst Rusch, Sen., Farm Lichtenstein,
S.W.A.

Schwantes, Z. Sukk. 2: 186. 1926; 3: 18, 105. 1927. - Bolus, Notes, 13 with pl. 3E; 23 with fig.; 27 with fig. 1; 65 with pl. 9; 72 with fig. 4. 1928. - N. E. Brown, Gard. Chron. 87: 14, 32. 1930. - v. Pöllnitz, Aufteilung, 65. 1933. - Jacobsen, Sukk. 174; fig. 180-182. 1933; Succ. Pl. 241; fig. 123. 1950. - Adamson, Salter, Fl. Cape Pen. 387. 1950. - *Phillips, Genera, 316. 1951.* - Jacobsen, Handbuch, 1618; fig. 1312, 1313, 1314. 1955. - Schwantes, Fl. Stones, 76 (with 12 subgroups), 338; fig. 20a; pl. 19a & b; pl. 93a & b (capsule). 1957. - Jacobsen, Handbook, 952 (in system of Schwantes), 962 (in key of Bolus), 968, 969, 973, 975 (in key of Herre & Volk), 1367 (with 12 subgroups acc. to Prof. Schwantes); fig. 1579, 1580, 1581, 1582. 1960; Lexikon, 486, t.194/1-4, 195/2, 3. 1970. - Friedrich in Merxmüller, Prodromus, 109. 1970.

Large or small shrubs, erect or decumbent, or prostrate, with woody, rarely tuberous roots, or tufted with short branchlets, these often covered with old leaves, persistent sheaths. Leaves opposite, amplexicaul, the leaves often only shortly connate, the free blades 3-angled, long connate, thus forming bodies, these oblong-spherical, with or without windows at the apex, rarely included in white, membranous sheaths (rests of former pairs of leaves); mostly mucronate, the keel entire or armed with blunt teeth, stiff, texture firm, bluish-green, nearly always set with darker transparent dots, glabrous or ciliate with hairs, rarely with glittering papillae, very variable as regards their size. Flowers axillary or terminal, solitary, or 2-nate, or in a many-flowered inflorescence, often with distinct bracts which may be very like ordinary leaves, stalked or also sessile, pink, red, violet or white, diurnal or in some species nocturnal, mostly medium-sized. Sepals 4-5 free down to the junction with the ovary. Petals 1-several series, linear or linear-spathulate, often flaccid-recurved. Stamens often conically collected, often hairy at the base, surrounded by many or only a few staminodes, which are also hairy at the base. Ovary inferior, slightly or strongly convex or cone-shaped. Glands often very distinct, not separated, crenulated, dark-green. Placentas parietal. Stigmas 4-5, free, broom-like to almost filiform free almost to the base, without style. Capsule 4-often 5-locular, very hard, with firm central column; valve-wings connate with the wall of the valves, thus looking as if not present; expanding keels much diverging; loculi-roofs present, stiff, often with spur-like projects at the mouth, just above the tubercle; tubercle large, at the back often cap-like or wrinkled. Seeds middle-sized, yellowish or brownish, more or less flat spherical, with rounded, much angled, chess-board-like marks, rough.

Species: About 350, South West Africa, Cape Province, Free State, Transvaal, Lesotho. (Type: *R. rupicola* (Engl.) Schwantes.)

M.M.Page. XII. 1922.

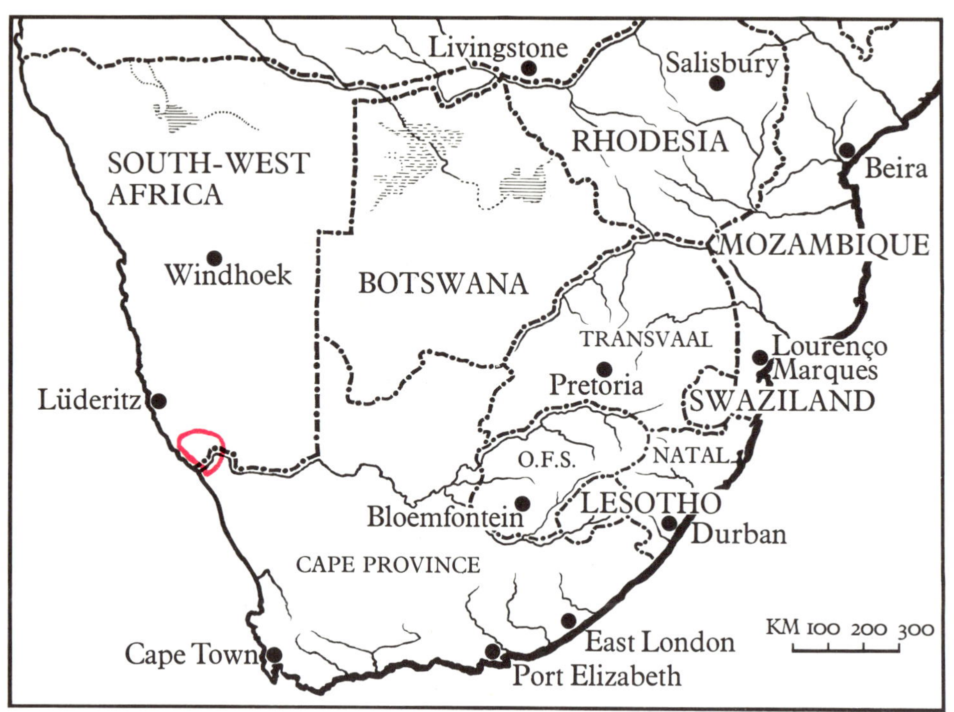

Ruschianthemum

Ruschianthemum Friedrich
Named in honour of the late
Mr Ernst Rusch, Jun., Windhoek, and
the Greek *anthemis* = flower.

Friedrich, Mitt. Bot. Staatssammlung München, 3: 563; fig. 2 no. 3. 1960; in Merxmüller, Prodromus, 121. 1970. - Jacobsen, Lexikon, 506. 1970.

Ruschia micropetala L. Bolus, Notes, 76. 1929.

Stoeberia gigas (Dinter), Dinter et Schwantes, Z. Sukk. 3: 17. 1927.

Perennial shrubs, woody, near the genus *Stoeberia*. Leaves succulent, decussate, connate at the base, ascending or spreading, nearly level above with the margins somewhat angulate, keeled towards the top. Flowers terminal and much branched cymes with many flowers, flowers small, with pedicels and 2 bracts, receptacle broadly obconical. Sepals 5, triangular, fleshy, 2 outer ones slightly larger. Petals indistinctly 1-seriate, filiform, shorter or a bit longer than the sepals. Staminodes filiform with the stamens conically collected. Ovary above nearly convex and obtuse 5-lobed. Stigmas 5, short, free. Capsule, matured, turbinate, receptacle soon perishes, its woody nerves persistent, superior parts of the carpels connate, operculum soon rejected; inferior parts of the carpels segregated, mericarps in the superior parts with the expanding keels and the rudimentary wings ornate; tubercle large, filling the greatest part of the carpels and the seeds are formed between it and the wall of the carpel. Seeds only one or two in each carpel.

Species: 1, Southern part of South West Africa and in the Richtersveld. (Type: *R. gigas* (Dint.) Friedr.)

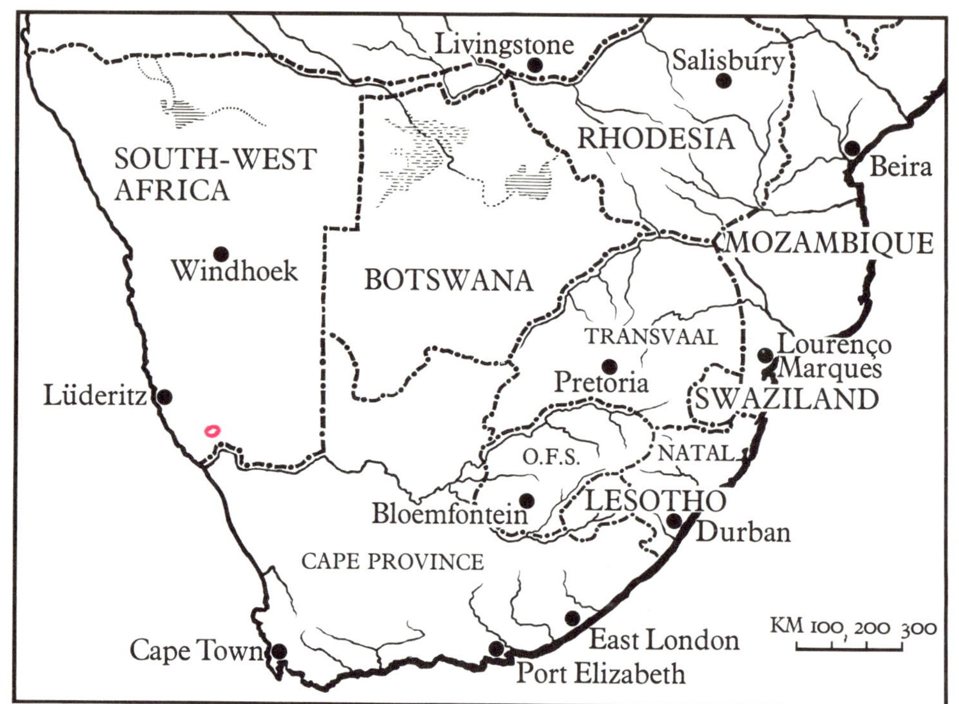

Ruschianthus

Ruschianthus L. Bolus
Named in honour of the late
Mr Ernst Rusch, Jun., Windhoek,
S.W.A., and the Greek *anthos* = flower.

Bolus, J. S. Afr. Bot. 27: 62. 1961. -
Jacobsen, Lexikon, 506, t.195/1. 1970.
- Friedrich in Merxmüller, Prodromus,
121. 1970.

Tufted small succulent perennials with
woody or nearly tuberous roots, old
branches at the base closely covered with
the remains of old leaves, flowering
branches with 4 leaves, green parts finely
rough and pale grey-green. Leaves
erect, falcate, surface flat, acuminate, at
the base convex, compressed towards
the apex, keel at the lower side obscure,
pinkish. Flowers 1-3-nate, pedicel short,
at the base with bracts resembling the
leaves. Receptacle globose-obovate.
Sepals 5, keeled, acuminate, the inner
ones with narrow membranous margins.
Petals about 5-seriate, connate at the
base, rounded at the points, in the inside passing to staminodes, pale lemon-
yellow. Stamens and staminodes produced at the base of the corolla, 8-
seriate, anteriors without papillae, posteriors with papillae. Ovary concave at
the outer wall, in the centre suddenly
raised and forming a truncate cone.
Glands slightly distant from each other,
semi-circular, obscurely fine crenulate.
Placentas parietal. Stigmas 5, subulate, incurved. Capsule open as in *Delosperma*,
without loculi-roofs and tubercle.
Seeds light-brown, oval, with nipple.

Species: 1, Southern South West
Africa, Numeis. (Type: *R. falcatus*
L. Bol.)

Ruschianthus falcatus   273

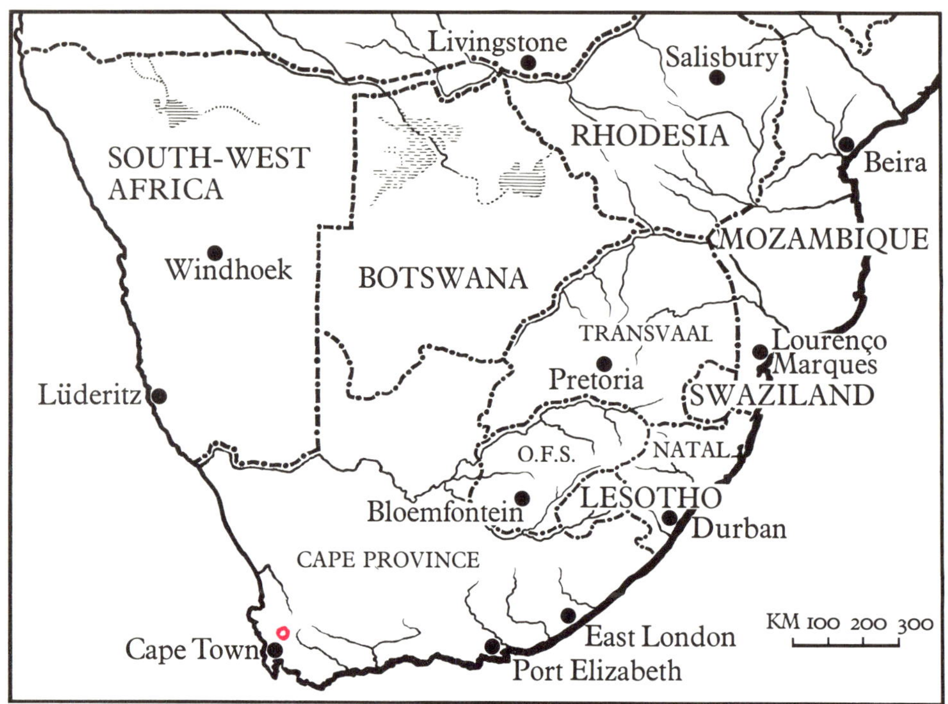

Saphesia

Saphesia N. E. Brown
From the Greek *saphes* = different,
since the genus differs from all the others.

N. E. Brown, Gard. Chron. 205. 1932. -
Pax, Natürl. Pfl. 218. 1934. - Jacobsen,
Verzeichnis, 189. 1938. - Goossens,
Blomplante, 149 (in key) 1940. - Jacobsen, Herre, Volk, Mesembr. 61 (in key).
126. 1950. - *Phillips, Genera, 316. 1951.* -
Herre, Sukkulentenkunde 5: 51 with
5 fig. 1954. - Jacobsen, Handbuch, 1676;
fig. 1315. 1955. - Schwantes, Fl. Stones,
52; fig. 12 (capsule); pl. 11a. 1957. -
Jacobsen, Handbook, 954 (in system of
Schwantes), 957 (in key of Bolus), 968
(in key of Herre & Volk), 1410; fig.
1583. 1960; Lexikon, 506, t.195/4.
1970.

Perennial succulent with a long caudex
(about 1 m), stems branching with distinct internodes, woody. Leaves alternate, sessile, not connate, flat, flaccid,
about 5 cm long and to 8 mm broad,
drying off during the resting period
in summer. Flower terminal, long stalked,
yellow and white, about 4 cm in diam.
Sepals, unequal, 5. Petals 3-4-seriate,
numerous, very acute, free, shorter
than sepals. Stamens numerous, erect
in a mass, surrounded by staminodes.
Ovary inferior. Placentas on the floor of
the very shallow chambers. Stigmas 5,
erect, filiform. Capsule 5-locular, differs
in structure from all the known capsules in Mesembr. Opens by drying out,
like *Dianthus*, and scatters the seeds.
The 5 cell-walls reach almost to the end
of each valve. When ripe, the ends of
the valves split apart for about half
the height of the capsule. From this point
a bridge-like structure stretches to the
base of the capsule where it is fastened
at its middle point to the placenta, that is,
the point of attachment of the seed.

The lowest part of the capsule bulges
outwards at the rim so that a sort of
curved circular passage is formed within,
which is interrupted in five places by
the bridges previously mentioned. Some
of the seeds seem to be grown into this
marginal, arched space; I could particularly see seeds closely pressed to the
bridges; they were the only ones that
had not fallen out of the capsule. One
gets the impression that perhaps the
object of the "circular passage" below
the rim of the capsule is to prevent some
of the seeds from being shot out.
(Schwantes, l.c.p. 54). Seeds flat, bean-
or horseshoe-shaped, with distinct
navel, verrucose, dark-brown.

Species: 1, South-western Cape,
Kalbas Kraal. (Type: *S. flaccida* (Jacq.)
N. E. Br.)

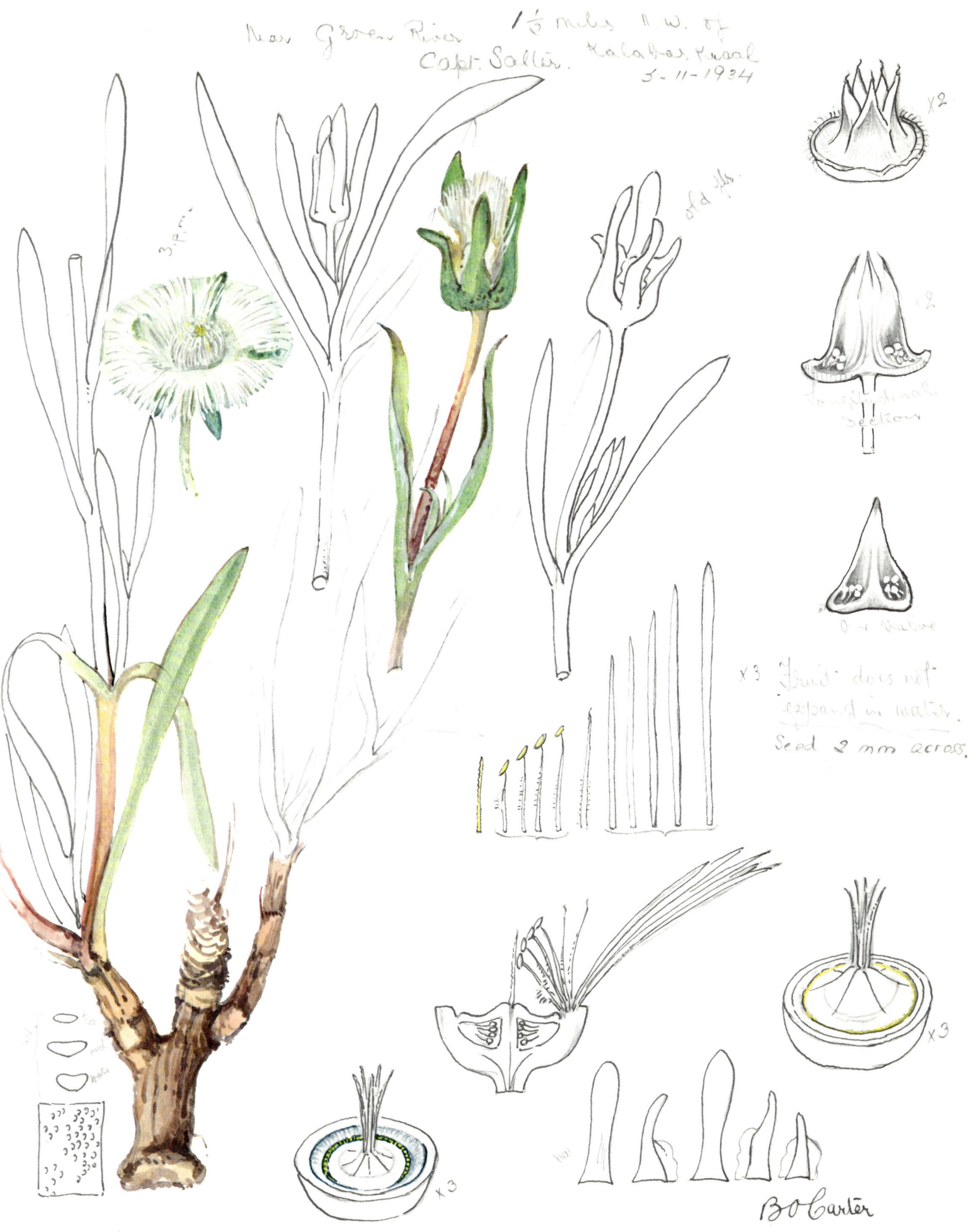

Saphesia flaccida 275

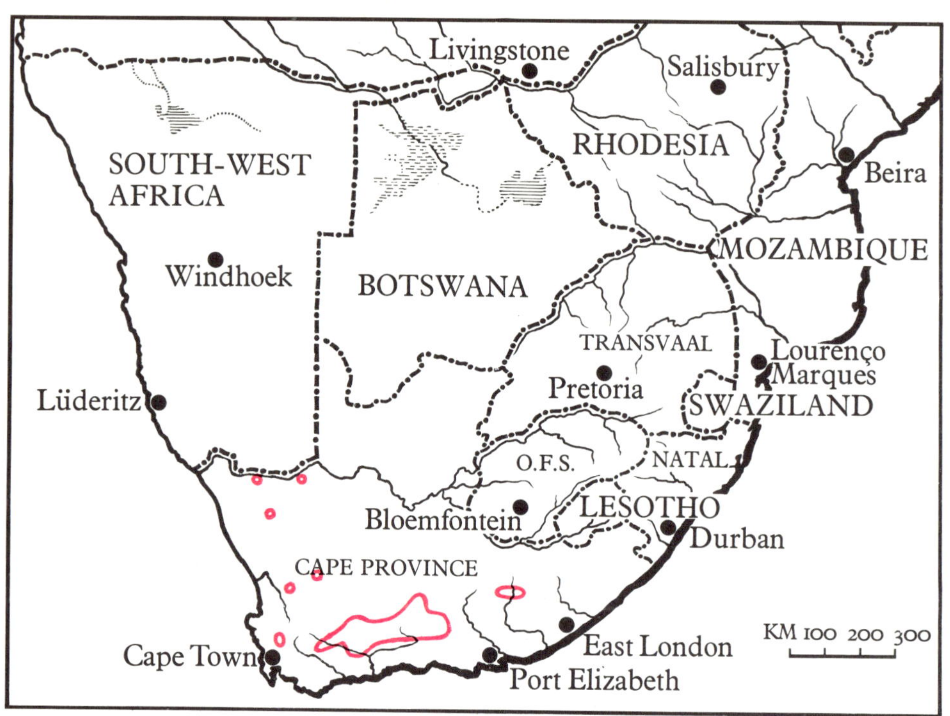

Sceletium

Sceletium N. E. Brown
From the Latin *sceletium* = skeleton, because the leaves become skeletonised.

N. E. Brown, Gard. Chron. 78: 412 (in key). 1925. - Phillips, Genera, 245. 1926. - Bolus, S. Afr. Gardening, 399. 1927; Notes, 91; fig. 19 & 20. 1928. - v. Pöllnitz, Aufteilung, 66. 1933. - Jacobsen, Sukk. 177; fig. 183. 1933. - Pax, Natürl. Pfl. 218. 1934. - Jacobsen, Succ. Pl. 244; fig. 231. 1935; Verzeichnis, 189. 1938. - Goossens, Blomplante, 148 (in key), 1940. - Jacobsen, Herre, Volk, Mesembr. 55 (in key), 126. 1950. - *Phillips, Genera, 316. 1961.* - Jacobsen, Handbuch, 1677; fig. 1316, 1317. 1955. - Schwantes, Fl. Stones, 28, 35; pl. 6b. 1957. - Jacobsen, Handbook, 951 (in system of Schwantes), 956 (in key of Bolus), 966 (in key of Herre & Volk), 1410; fig. 1584, 1585, 1586. 1960. Lexikon, 506, t.196/1. 1970.

Perennial undershrubs with a short stem and long or short spreading or prostrate branches, very rarely somewhat shrubby, papillose when young. Leaves opposite, sessile, slightly united at the base, ovate-lanceolate, soft, not dotted, minutely papillate, withering to a skeleton and persisting, to about 5 cm long and 1,5 cm broad. Flowers terminal, solitary or a few in a cyme, pedicels may have 1-2 pairs of bracts, whitish to yellowish, to about 4 cm in diam. Sepals 5, with a short tube above its union with the ovary, 2 lobes often leafy and larger than the rest. Petals numerous, several series, changing to staminodes, all united at the base into a short tube, arising where the calyx-tube is united with the ovary. Stamens numerous, connivent-erect, more or less concealed by the much longer staminodes. Ovary partly superior. Placentas axile, style 0. Stigmas 4-5, sometimes but not always, concealed under the stamens and staminodes. Capsule 4-5-locular, $\frac{1}{2}$ or more than $\frac{1}{2}$-superior; valves with 1 stout acute expanding-keel down its centre, loculi without roofs or tubercles. Seeds compressed, orbicular D-shaped, with appressed pouch forming valvewings. See L. Bolus, Notes 1, 91, fig. 19. 1928., also occuring with other species of this genus.

Species: 22, Namaqualand, Karroo so far as Cradock. (Type: *S. tortuosum* (L.) N. E. Br.)

All the species contain the poisonous principle "mesembrine" a relative of cocaine and other substances. After fermentation, leaves are dried again and chewed. It is the "kougoed" or "channa" of the Coloured people. (Jacobsen, l.c., p. 40.)

If taken in certain quantities it causes drunkeness. Perhaps it may give a valuable medicine. Other members of the *Mesembryanthemaceae* contain mesembrine too but not in quantities as large as found in *Sceletium*. In this genus it is not formed in Europe and northern countries, but it is in North Carolina (U.S.A.). Next to *Sceletium* follows *Mesembryanthemum* (*M. crystallinum* and relatives), but it contains salt in large quantities which is very troublesome to those who take it.

Sceletium sp.   277

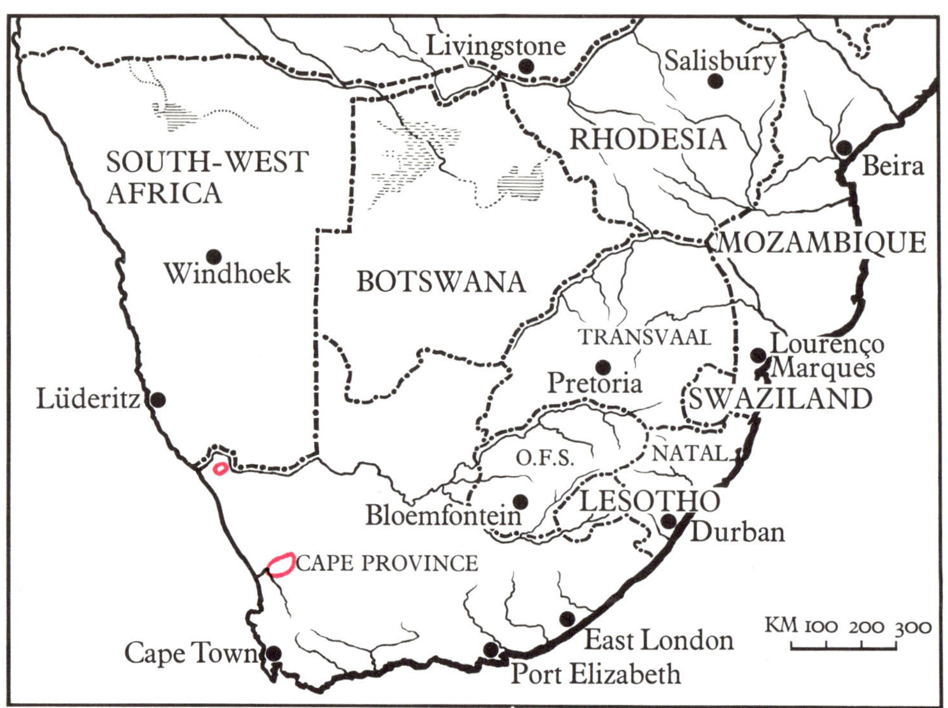

Schlechteranthus

Schlechteranthus Schwantes
Named in honour of the late
Mr Max Schlechter, Springbok
and Port Nolloth.

Schwantes, Monatsschr. Kakt. Ges. 16. 1929. - v. Pöllnitz, Aufteilung, 72. 1933. - Pax, Natürl. Pfl. 221. 1934. - Jacobsen, Verzeichnis, 190. 1938. - Jacobsen, Herre, Volk, Mesembr. 68 (in key), 126. 1950. - Phillips, Genera, 321. 1951. - Jacobsen, Handbuch, 1681. 1955. - Schwantes, Fl. Stones, 338. 1957. - Jacobsen, Handbook, 952 (in system of Schwantes), 962 (in key of Bolus), 971 (in key of Herre & Volk), 1414. 1960; Lexikon, 507. 1970.

Dwarf shrub with woody roots, branchlets very crowded with leaves, internodes not visible. Leaves opposite, united up to half their lengths into bodies 5 mm long and 5 mm wide, upper surface flat, back surface keeled, usually mucronate at the end, armed on the end of the keel with 1-2 short teeth, smooth, glossy, bluish, often very much reddened. Flowers solitary, nearly sessile, pedicel bracted in the middle, bracts clasping the pedicel, purple-rose. Sepals 6, anteriors acute, with keel, posteriors with membranous margins. Petals 2-3-seriate, lax, inner ones erect, later on recurved, angustate, obtuse. Stamens 4-seriate, papillate like the staminodes. Ovary on top convex. Disc annular, rather inconspicuous, denticulate. Placentas parietal. Stigmas 11-12, very narrowly subulate. Capsule 11-12-locular, half-globose, small; expanding keels at first parallel, later on diverging, ending with a short point; loculi-roofs marginal recurved as with *Pleiospilos*, with spur-like processes, which nearly close the mouth. Seeds pear-shaped, somewhat flat and rough, reddish-yellow.

Species: 2, Richtersveld: Kubus and Alexander Bay. (Type: *S. maximilianii* Schw.)

Schlechteranthus maximilianii   279

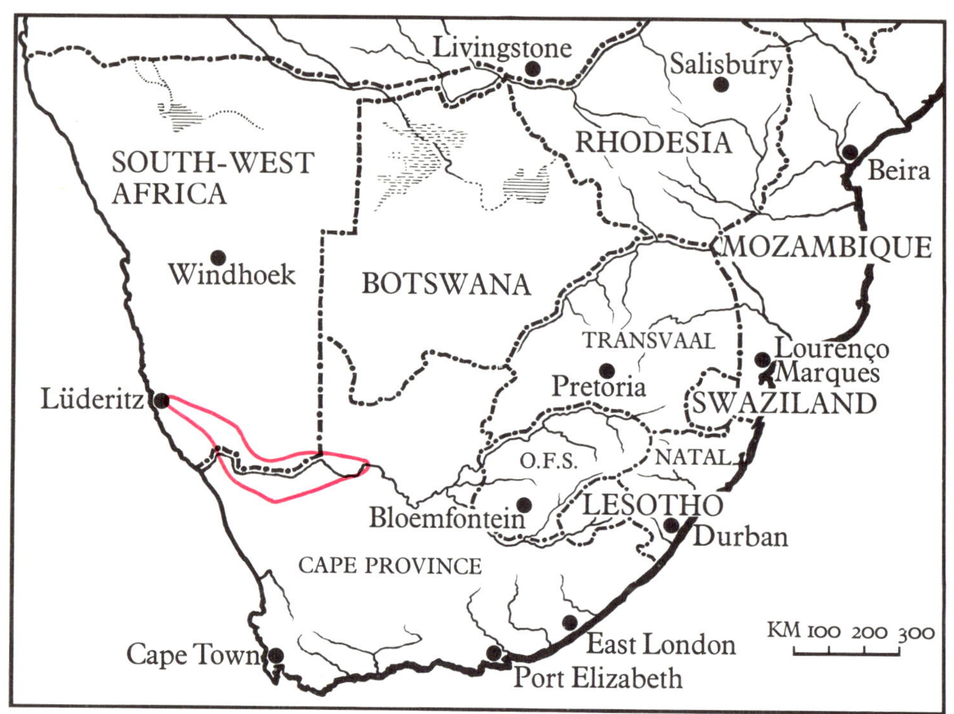

Schwantesia

Schwantesia Dinter
Named in honour of the late Prof.
Dr G. Schwantes, Kiel, Germany.

Dinter, Möllers Gärtnerztg. 42: 234; fig. 202. 1927. - v. Pöllnitz, Aufteilung, 65. 1933. - Jacobsen, Sukk. 178; fig. 184. 1933. - Pax, Natürl. Pfl. 221. 1934 - Jacobsen, Succ. Pl. 245; fig. 232. 1935; Verzeichnis, 190. 1938. - Jacobsen, Herre, Volk, Mesembr. 74 (in key), 126. 1950. - Phillips, Genera, 322. 1951. - Jacobsen, Handbuch, 1681; fig. 1318, 1319. 1955. - Schwantes, Fl. Stones, 175, 340; fig. 28 (seedlings); pl. 52b. 1957. - Bolus, Notes, 346 (with key of species); pl. 98-102. 1958. - Jacobsen, Handbook, 953 (in system of Schwantes), 959 (in key of Bolus), 973 (in key of Herre & Volk), 1414 (with key of species); fig. 1587-1589. 1960; Lexikon, 507, t.196/1, 197/4. 1970. - Friedrich in Merxmüller, Prodromus, 122. 1970.

Dwarf compact plants with the internodes enclosed in the leaf-sheaths, stem up to 1 cm in diam., primary branches with age densely enclosed with hardened remains of previous years' leaves, flowering branchlets 2-leaved. Leaves opposite, erect or ascending, those of one pair unequal, narrow boat-shaped or rhomboid to spathulate, keeled, surface flat or somewhat convex above, edges rounded, lower side more or less semi-spherical, hard, thick, bluish-green, whitish dotted, upper surface finely granulate, apices more or less broadened, obtuse with 3-7 blue-green teeth, which differ very much as regards their size, rather thick, broad, distant and brownish pointed, keel projected like a chin and often toothed, about to 5 cm long. Flowers solitary, terminating the axillary shoots of the previous year, pedicels to 5 cm long, without bracts, light to golden-yellow, about 6 cm in diam. Sepals 5, 2 of them large, sharply keeled, with a few small teeth, 3 shorter, rounded, with a membranous margin, cap-like contracted, with a stout erect tooth. Petals numerous, 1-series, linear-lanceolate, rounded at the apices, to about 2,5 cm long. Stamens 4-series, only in the middle with short hairs. Ovary convex. Disc annular, crenulated, rather low, yellowish-green. Placentas parietal. Stigmas 5, greenish, apices reflexed. Capsule 5-locular, central column weak; expanding keels contiguous, with broad wings; loculi-roofs reduced to a limb; tubercle none. Seeds glittering, rough.

Species: 11, Southern South West Africa, Richtersveld, Bushmanland. (Type: *S. rüdebuschii* Dinter.)

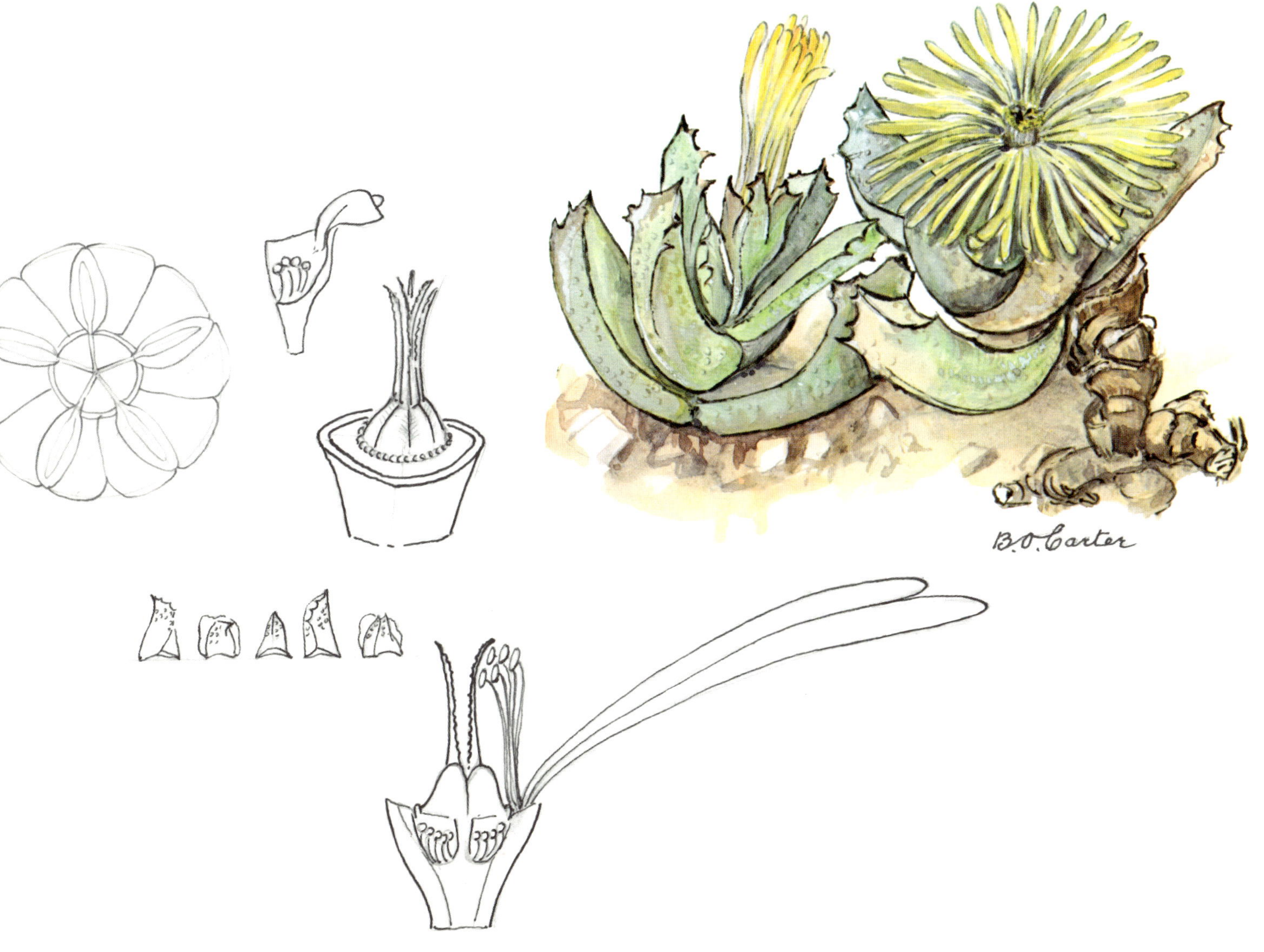

B.O.Carter

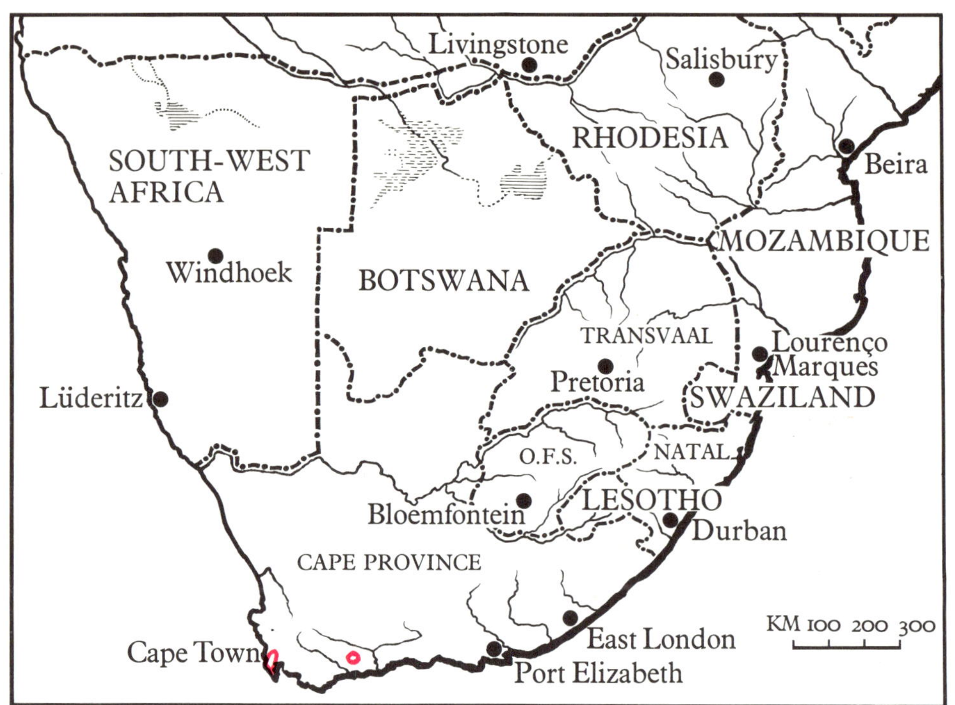

Scopelogena

Scopelogena L. Bolus
From the Greek *scopelos* = stone, rock
and *genos* = child, descendant.

Bolus, J. S. Afr. Bot. 28: 1962. - Jacobsen, Lexikon, 508, t.197/1. 1970.

Lampranthus verruculatus (L.) L. Bolus in Jacobsen, Handbuch, 1548. 1955. - Jacobsen, Handbook, 1212; fig. 1423a. 1960.

Woody shrublets forming cushions when old; green parts smooth, very soft, bluish-green. Leaves crowded, connate, incurved, erect, 3-angled to cylindrical or falcate, bluntish, shortly mucronate, mealy-grey pruinose, sometimes somewhat reddened, to about 3,5 cm long. Flowers in cymes, these often 3-8 cm high, more or less branched with 7-17 flowers, pedicels 4-15 mm long, on the base up to the middle with bracts, resembling the leaves, flowers yellow, about 1,5 cm in diam. Sepals 5, acute or pointed, subequal, the two inner ones with broad membranous margins. Petals 2-series, no staminodes. Stamens erect, posteriors up to the middle papillate. Ovary divided, convex, compressed. Disc annular, crenulate. Stigmas filiform, longer than the stamens, green. Placentas parietal. Capsule 5-locular, on the back nearly clavate and acute or obconic, top convex, sutures almost compressed, when once open, they never close again completely; valves without wings; expanding keels from the beginning diverging, not pointed, up to the middle connate with the valves; loculi-roofs broad but short, nearly covering the loculi; without tubercle. Seeds numerous at the base of the loculi, dark-brown, obovate, obscurely tuberculate.

Closely related to *Lampranthus*, but the valves of the open capsule are without wings, the loculi-wings cover the loculi only partly, the capsule once opened does not quite close again. This is the main difference from *Lampranthus*.

Species: 2, South-Western Cape (Table Mt., Lion's Head, Swellendam). (Type: *S. verruculata* (L.) L. Bolus.)

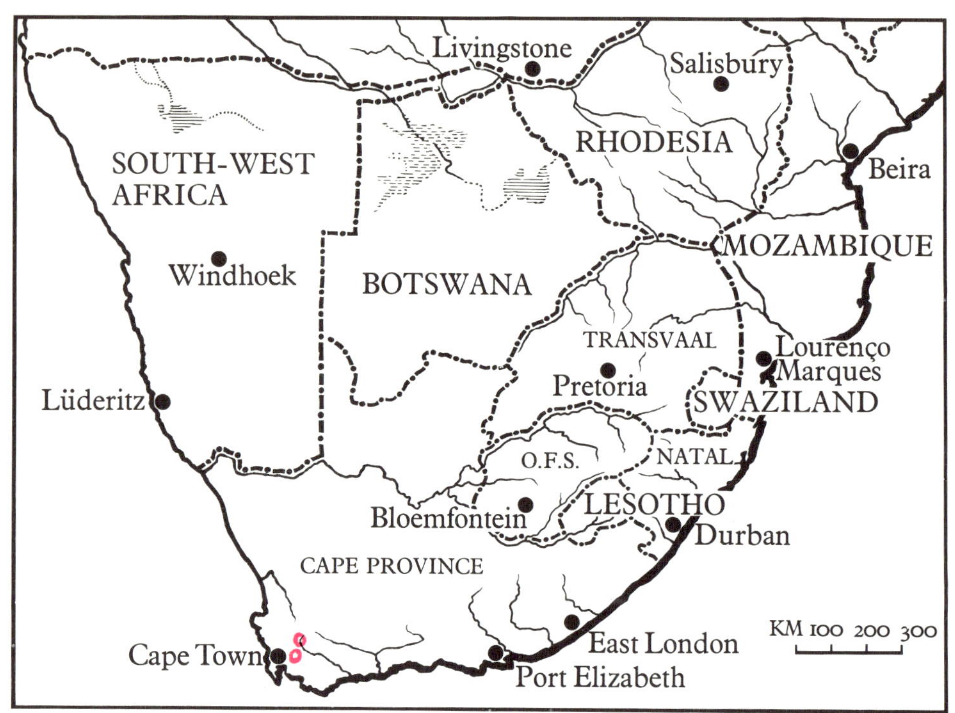

Semnanthe

Semnanthe N. E. Brown
From the Greek *semnos* = outstanding,
distinguished and *anthos* = flower.

N. E. Brown, Gard. Chron. 81: 12 (in key). 1927. - v. Pöllnitz, Aufteilung, 66. 1933. - Jacobsen, Sukk. 179; fig. 185. 1933. - Pax, Natürl. Pfl. 218. 1934. - Jacobsen, Succ. Pl. 246; fig. 233. 1935; Verzeichnis, 191. 1938. - Goossens, Blomplante, 147 (in key). 1940. - Jacobsen, Herre, Volk, Mesembr. 66 (in key), 126; fig. 11. 1950. - Phillips, Genera, 322. 1951. - Jacobsen, Handbuch, 1685; fig. 1320. 1955. - Schwantes, Fl. Stones, 70, 339. 1957. - Jacobsen, Handbook, 953 (in system of Schwantes), 960 (in key of Bolus), 970 (in key of Herre & Volk), 1417; fig. 1590. 1960; Lexikon, 509, t.197/2. 1970.

Best description: Berger, Mesembr. & Portulac. 193; fig. 38 (1–4). 1908.

Erect shrub, up to 1 m high, with spreading, stout, 2-angled branches, which are green in the beginning and later on grey with reddish margins. Leaves opposite, connate at the base, incurved, 3-angled, almost sabre-shaped, acute, upper surface flat or slightly grooved, very much laterally compressed, mucronate, margins cartilagenous, finely dentate, keel edge lacerated or cartilaginously dentate, smooth, light green, with transparent dots, to about 5 cm long and more than 1 cm broad. Flowers solitary or 2 together, short-stalked, violet-red, about 5 cm in diam. Sepals: 5-6, unequal, lanceolate-trigonous, acuminate with toothed margins. Petals numerous, free, narrow linear, posteriors smaller, concealing the stamens, anteriors longer than the sepals, recurved. Stamens short, concealed by many incurved petals and staminodes. Ovary dish-like with raised margin. Placentas parietal. Stigmas 10,

very small. Capsule 10-locular, after having been wetted, valves erect-spreading and after first opening, do not close again; expanding keels diverging, ending in awn-like points; loculi-roofs stiff; tubercle none. Seeds pear-shaped with tubercles.

Species: 1, South-Western Cape, Paarl and Stellenbosch mountains. (Type: *S. lacera* (Haw.) N. E. Brown.)

Semnanthe lacera   285

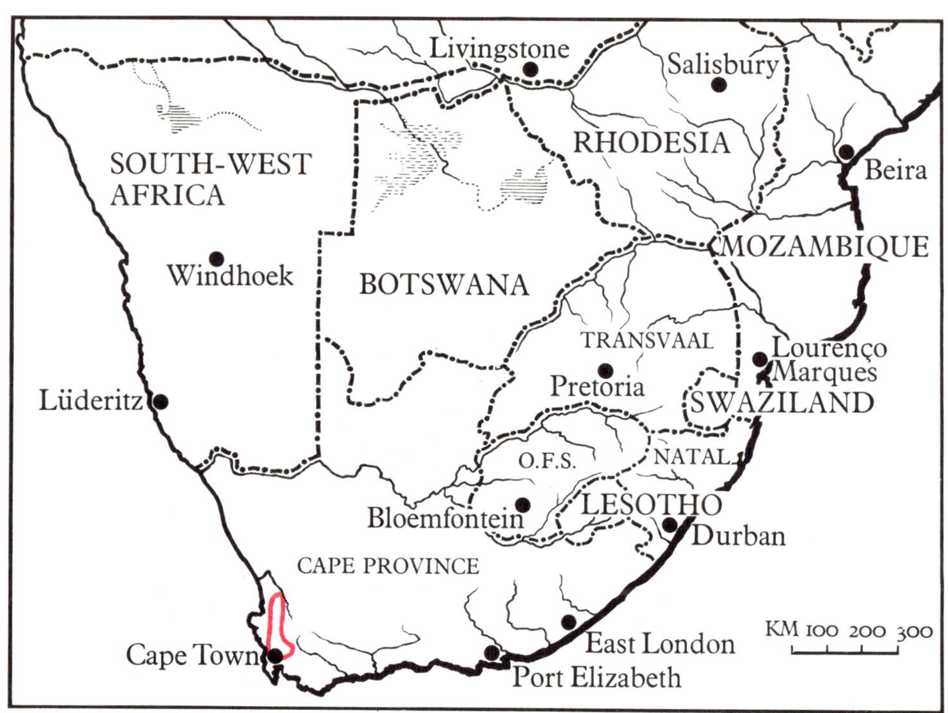

Skiatophytum

Skiatophytum L. Bolus
From the Greek *skia* = shade and
phyton = plant.

Bolus, S. Afr. Gard. 435. 1927; Notes, 15; fig. IV B (capsule). 1928. - Schwantes, Mitt. Inst. Allg. Bot. Hamburg, 8: 162 (capsule). 1929. - v. Pöllnitz, Aufteilung, 67, 1933. - Jacobsen, Verzeichnis, 191. 1938. - Jacobsen, Herre, Volk, Mesembr. 58 (in key), 126. 1950. - Adamson, Salter, Fl. Cape Pen. 375. 1950. - *Phillips, Genera, 316. 1951.* - Jacobsen, Handbuch, 1685; fig. 1321. 1955. - Schwantes, Fl. Stones, 56, 341; fig. 13, 14 (capsule); pl. 12. 1957. - *Leistner, J. S. Afr. Bot. 24: 89–102. 1958.* - Jacobsen, Handbook, 954 (in system of Schwantes), 958 (in key of Bolus), 967 (in key of Herre & Volk), 1417; fig. 1591. 1960.

Gymnopoma N. E. Brown, Gard. Chron. 83: 194. 1928. - v. Pöllnitz, Aufteilung, 44. 1933. - Pax, Natürl. Pfl. 215. 1934. - Goossens, Blomplante, 148 (in key). 1940; Lexikon, 509, t.197/3. 1970.

Biennual or annual, root stout, fleshy, branching. Leaves opposite or alternate, not united at the base, flat, petiolate, or spathulate-lanceolate, smooth, shining, margins, entire, waved in the earlier stages, crowded at the base of the stem, more distant on flowering branches, a little fleshy, with a thick central nerve and distinct lateral nerves, about 8 cm long and 2 cm wide. Flowers 2-5 in a terminal cyme, sometimes solitary, pedicellate, pedicel about to 6 cm long, without bracts, thickened towards the point, opening at noon, snow-white, about 3 cm in diam. Sepals 5, unequal, free. Petals numerous, in several series, linear-lanceolate, reflexed, free. Stamens numerous, erect, filaments slender. Ovary inferior, shortly conical and 5-angled at the top. Placentas on the outer walls of the chambers; stigmas 5, subulate, overtopping the stamens. Capsule large, obconic, 5-locular; valves without expanding keels and marginal wings, flattish within, do not expand on being wetted, but separate and spread when the capsule is ripe and dry and remain so without closing again. Seeds compressed, somewhat kidney-shaped, tuberculate.

Species: 1, South-Western Cape. (Type: *S. tripolium* (L.) L. Bolus) In shade!

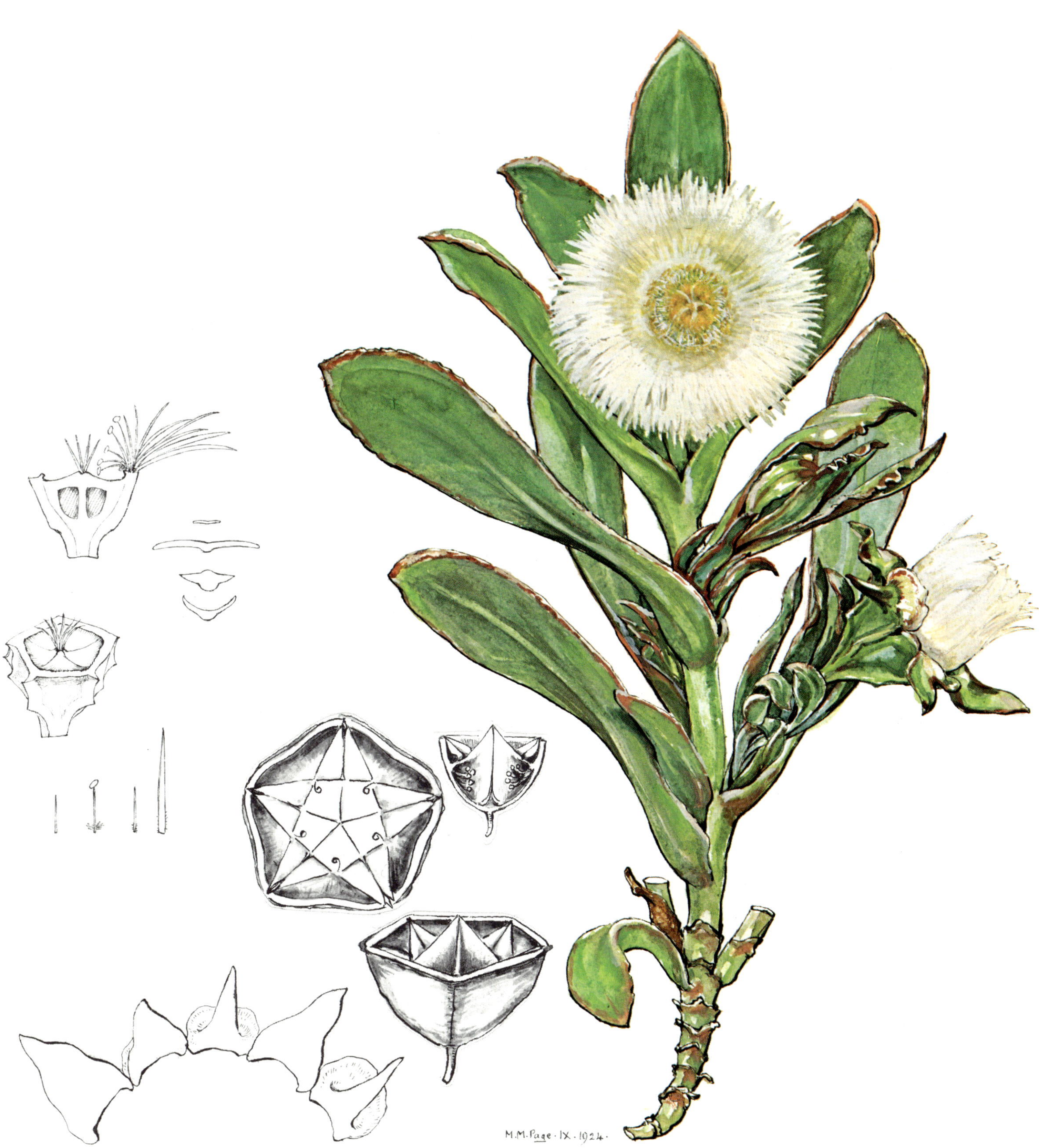

Skiatophytum tripolium   287

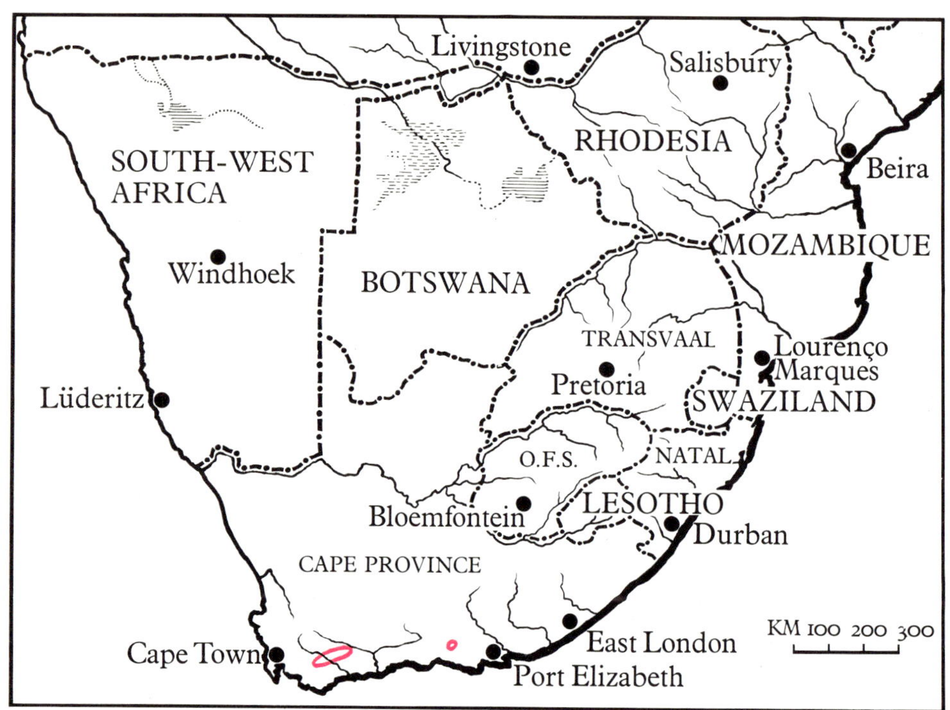

Smicrostigma

Smicrostigma N. E. Brown
From the Greek *smicro* = small and
stigma = stigma.

N. E. Brown, Gard. Chron. 87: (in key), 186. 1930. - v. Pöllnitz, Aufteilung, 67. 1933. - Pax, Natürl. Pfl. 219. 1934. - Jacobsen, Verzeichnis, 191. 1938. - Goossens, Blomplante, 146 (in key). 1940 - Jacobsen, Herre, Volk, Mesembr. 65 (in key), 126. 1950. - *Phillips, Genera, 317. 1951.* - Jacobsen, Handbuch, 1686. 1955, - Schwantes, Fl. Stones, 342. 1957. Jacobsen, Handbook, 953 (in system of Schwantes), 960 (in key of Bolus), 970 (in key of Herre & Volk), 1418; fig. 1592. 1960; Lexikon, 509, t.198/1. 1970.

Low shrubby succulent perennial, branches succulent, becoming woody with age, with distinct internodes, entirely glabrous. Leaves opposite, united at the base and continuous with the stem, small, deltoid-trigonous with the tips recurved, with a short reddish, terminal spine, set with fine transparent dots, shorter than 1,25 cm. Flowers solitary, terminal, subsessile or very shortly pedicellate, after the first opening remain open, pink, about 3 cm in diam. Sepals 5, unequal, acuminate, 2 keeled, 3 with a membranous margin. Petals numerous, about 3-series, free, cuneately linear, horizontally spreading, obtuse, sometimes posteriors papillate at the base. Stamens in several series superposed, short, all inflexed, often filaments bearded. Staminodes very numerous, in 3-4 series, filiform, all inflexed and partly concealing the stamens,

bearded below, recurved at the tips. Ovary inferior, flattish on the top. Glands scarcely visible. Placentas on the outer walls of the chambers. Stigmas 7-10, very short or minute. Capsule 7-10-locular, obconic, slightly convex, with 7-10 prominent sutures, valves horizontal, spreading, without or with very narrow wings; expanding keels with rounded edges, contiguous at the basal part, with diverging acute or awned tips without marginal wings; loculi roofed with flexible loculus-wings, without tubercle. Seed compressed-ovoid, with a short point, minutely tuberculate, blackish.

Species: 1, Cape Province: from Montagu and Robertson. (Type *S. viride* (Haw.) N. E. Br.)

Smicrostigma viride   289

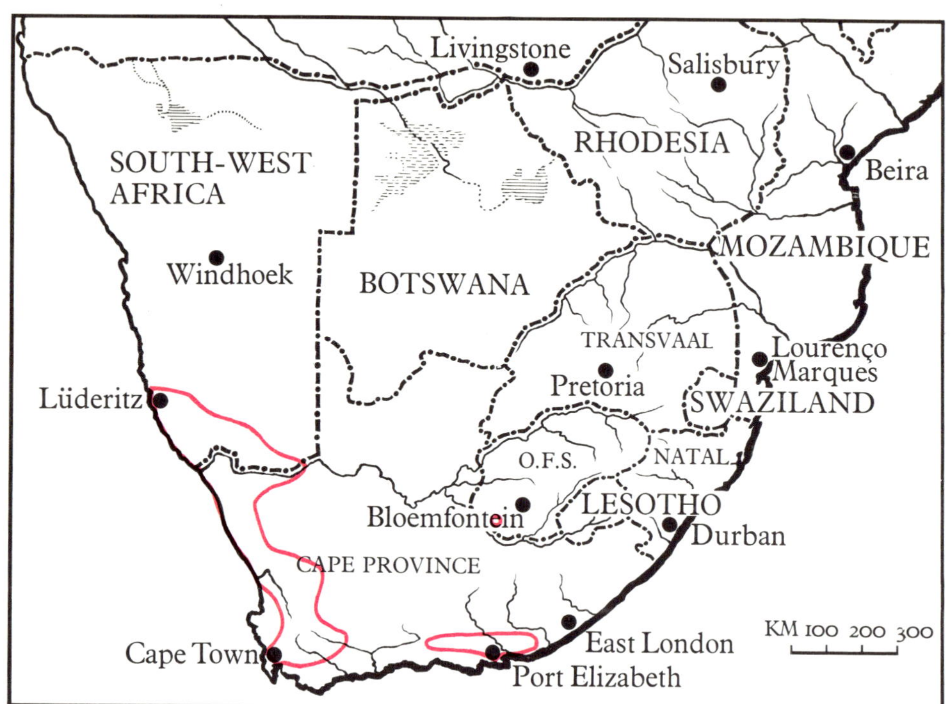

Sphalmanthus

Sphalmanthus N. E. Brown
From the Greek *sphalme* = stumble
and *anthos* = flower (on account of the
rooting runners).

N. E. Brown, Gard. Chron. 78: 433 (in key). 1925; 84: 13; fig. 34, 35. 1928. - Bolus, Notes, 54, 94, 102, 124; fig 21 A.1-4. 1928. - v. Pöllnitz, Aufteilung, 67. 1933. - Pax, Natürl. Pfl. 219. 1934. - Jacobsen, Verzeichnis, 191. 1938. - Goossens, Blomplante, 142 (in key). 1940. - Jacobsen, Herre, Volk, Mesembr. 57 (in key), 126. 1950. - *Phillips, Genera, 317. 1951.* - Jacobsen, Handbuch, 1686. 1955. - Schwantes, Fl. Stones, 27, 35. 1957. - Jacobsen, Handbook, 957 (in key of Bolus), 967 (in key of Herre & Volk), 1311, 1419. 1960; Lexikon, 509, t.198/2, 4, 1970. - Friedrich in Merxmüller, Prodromus, 126. 1970.

Succulent perennials papulose on all the green parts, rootstock tuberous or fleshy; stems prostrate, herbaceous or becoming woody at the base, the creeping ones often with very long branches, rooting at the nodes and these roots are often rape-shaped, sometimes the branches are woody, often also withering and marcescent, most of them grow in sandy soil. Leaves alternate or opposite or those of the flowering part alternate and the remainder opposite, sessile, semiterete, erect or spreading, often marcescent and then spinescent, to about 4 cm long. Flowers solitary and terminal or, by the growth of an axillary branch or branches, 2-6 to a stem and becoming one by one lateral and opposite the leaves, or in lax terminal leafy or bracteate cymes, pedicellate, yellowish, green or flesh-coloured, to about 4 cm in diam. Sepals 5, subequal, leafy, forming a short tube above the ovary, or without it. Petals numerous, 3 and more series, united into a short tube at the base, passing into staminodes which are adnated at the corolla-tube and fall off with it. Stamens numerous, arising from the tube of the corolla, erect in several series. Ovary partly superior or inferior. Placentas axile. Stigmas 4-5, subulate or short and pointed; style 0. Capsule superior, very convex or dome-like in the upper part, with much raised sutures, smooth and transparent when young, so that one will see the seeds lying in the chambers (type-species!), usually 5- (sometimes 4-) locular, valves widely spreading or recurved; expanding keels contiguous, forming a central keel, united at the base to the septa and so forming the locular partitions, with broad erect and flap-like marginal wings unfolded upon them; loculi open without loculi-wings or tubercles. Seed compressed, rounded or horseshoe-shaped in outline, minutely tuberculate.

Species: 35, southern part of South West Africa, Namaqualand, Vanrhynsdorp, South-Western Cape, Worcester. Free State: Fauresmith. (Type: *S. canaliculatus* (Haw.) N. E. Br.)

M. M. Page. 1. 1921.

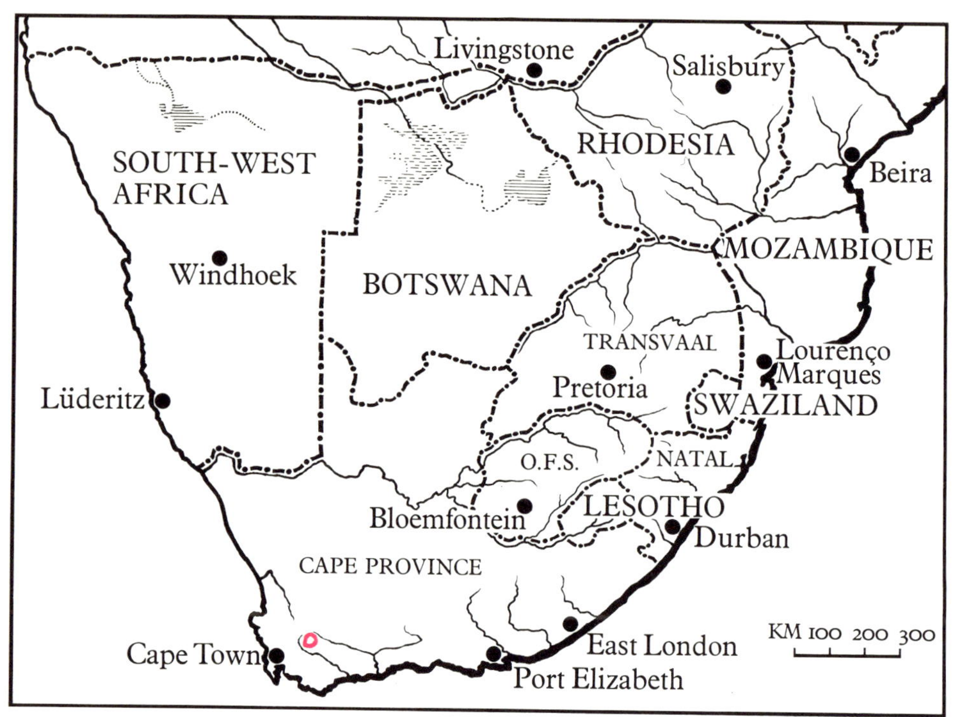

Stayneria

Stayneria L. Bolus
Named in honour of Mr W. F. Stayner,
Curator, Karroo Garden at Worcester.

Bolus, J. S. Afr. Bot. 27: 47. 1961. -
Jacobsen, Lexikon, 516. 1970.

Erect, stout, large, loosely branched,
smooth shrub, with woody, long branches, the internodes often 3-4 cm long,
branches full of the hardened remains
of former leaves. Leaves ascending,
sometimes subfalcate, clasping the stem
like a sheath, sheath impressed
lined, above linear - lanceolate with an
acute keel, slightly convex, or rounded,
light green, about to 7 cm long and
0,8 cm in diam. Flowers 3-nate, main
pedicel of the cymes about 1 cm long,
with bracts, bracts 2 cm long, the outer
pedicels up to 5 mm long, with or without bracts, those 6-7 mm long. Sepals
5-6 compressed, keeled above, acute,
all more or less with membranous margins. Petals 1-2-seriate, sometimes
round at the top, rarely emarginate,
snow white. Staminodes numerous,
matured red-brown. Stamens conically
collected, filaments about 5-seriate,
posteriors papillate up to the middle.
Ovary slightly convex. Glands obscurely
annular. Stigmas 6-9, often 7-8, narrow subulate. Capsule 6-9 or 7-8-locular,
woody, rigid, sometimes viscid, glo-
bose-obconical, obscurely ribbed, on
top about 2 mm raised, sutures, lightly
compressed; valves widely spreading;
expanding keels continuous at the base
then widely diverging, without wings;
loculi-roofs well developed; tubercle
none. Once open it will not quite close
again. Seeds obovate, brown, 1 mm long.
This genus is nearly related to *Ruschia*
but differs in having 7-9, very rarely 6
stigmas; the capsule opens repeatedly
but never closes again completely but
only about $\frac{3}{4}$ of the way.

Species: 1, Cape Province: Nuy,
district Worcester. (Type: *S. littlewoodii* L. Bol.)

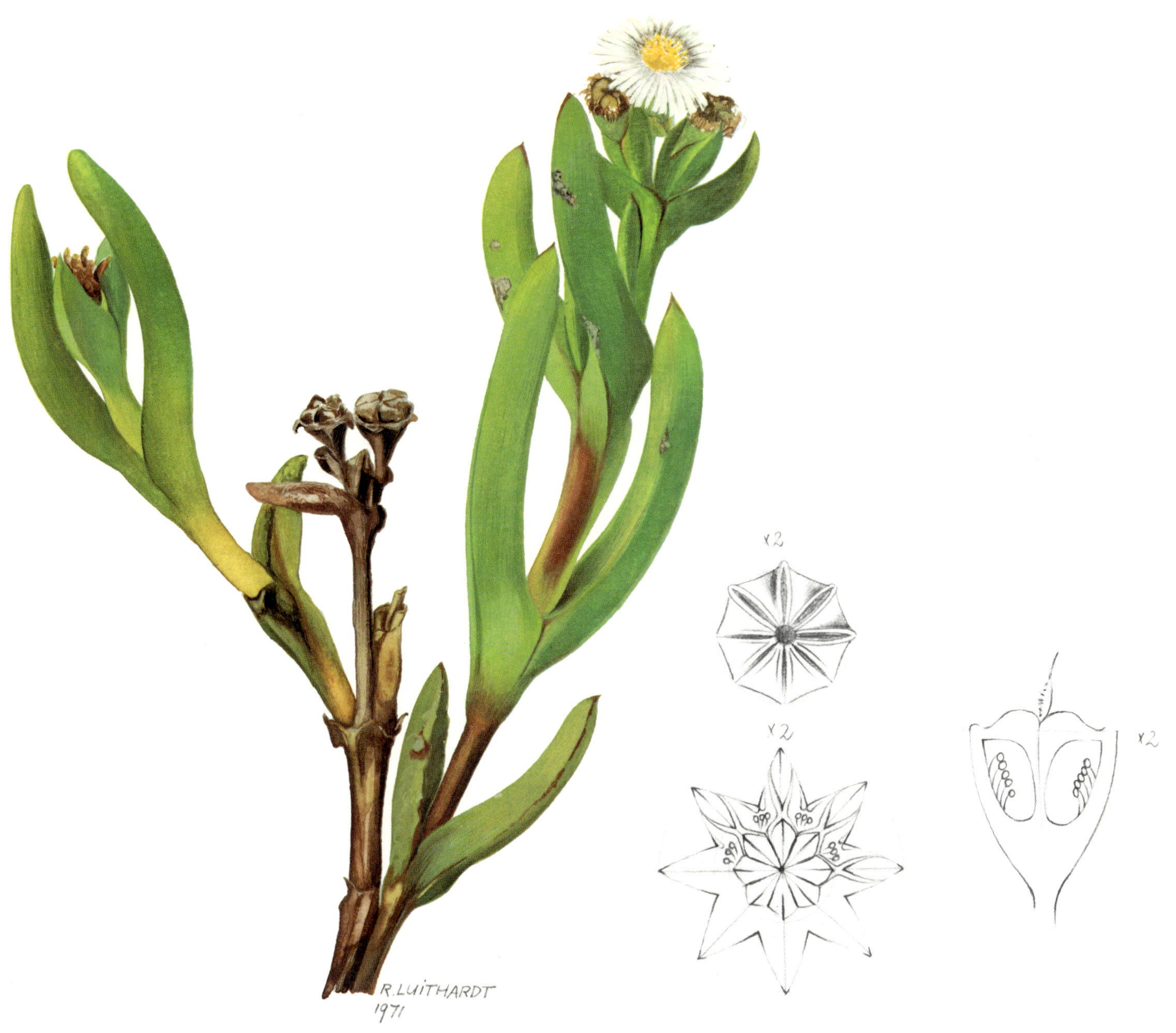

Stayneria littlewoodii   293

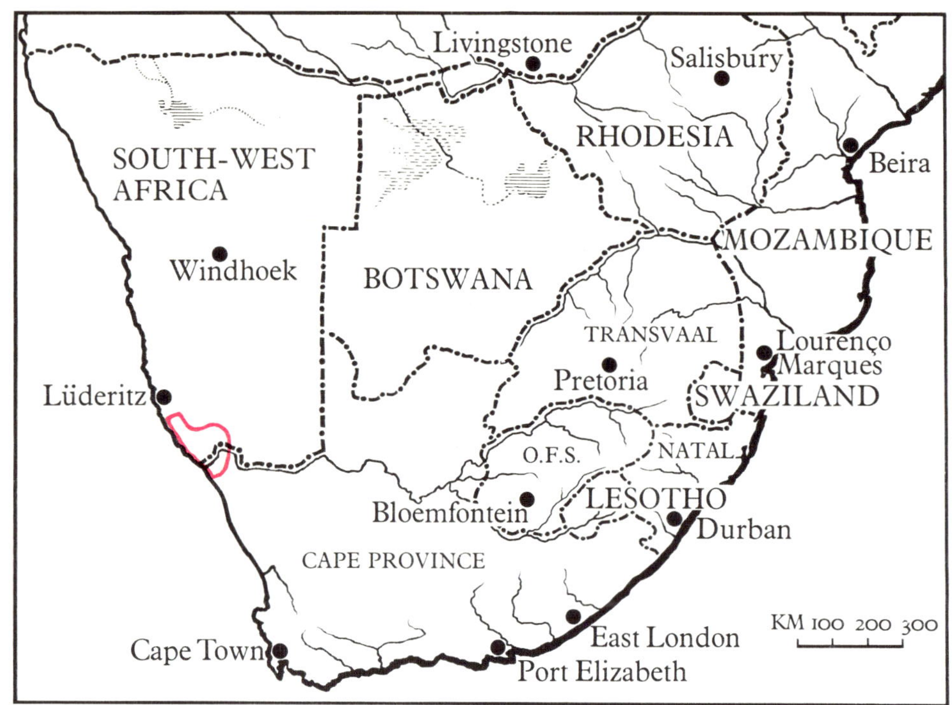

Stoeberia

Stoeberia (Dint. et Schwant.) emend. Friedr.
Named in honour of the late
Mr E. Stoeber, Lüderitz, S.W.A.

Dinter et Schwantes, Z. Sukk. 3: 14 (in key), 17. 1927. - v. Pöllnitz, Aufteilung, 68. 1933. - Pax, Natürl. Pfl. 221. 1934. - Jacobsen, Verzeichnis, 193. 1938. - Jacobsen, Herre, Volk, Mesembr. 63 (in key), 127. 1950. - Phillips, Genera, 322. 1951. - Jacobsen, Handbuch, 1688; fig. 1322. 1955. - Schwantes, Fl. Stones, 82, 338. 1957. - Jacobsen, Handbook, 952 (in system of Schwantes), 961 (in key of Bolus), 969 (in key of Herre & Volk), 1419; fig. 1593. 1960. - *Friedrich, Mitt. Bot. Staatssammlung, München, part 3: 554; fig. p. 557; fig. p. 561; fig. p. 562. 1960.* - Jacobsen, Lexikon, 516. 1970. - Friedrich in Merxmüller, Prodromus, 128. 1970.

Shrubs, woody, much branched. Leaves decussate, ascending or spreading, slightly connate at the base, succulent, boat-shaped or nearly clavate, above nearly flat with slightly angulate margins, keeled towards the apex. Flowers terminal in a much branched inflorescense, flowers pedicelled with 2 bracts. Receptacle obconical. Sepals 5-6, triangular, light flesh-coloured. Petals obscurely 1-seriate. Stamens conically collected; staminodes filiform. Stigmas 5-(6-) short, free, slightly pinnate. Capsule matured turbinate or subglobose, more or less deep 5-(6)-locular, valves triangulate; expanding keels arcuate; loculi-wings present; tubercle large. Seeds pear-shaped, very rough.

Species: 2, southern part of South West Africa. (Type: *S. beetzii* (Dint.) Dint. et Schwant.)

Stoeberia bicolorata   295

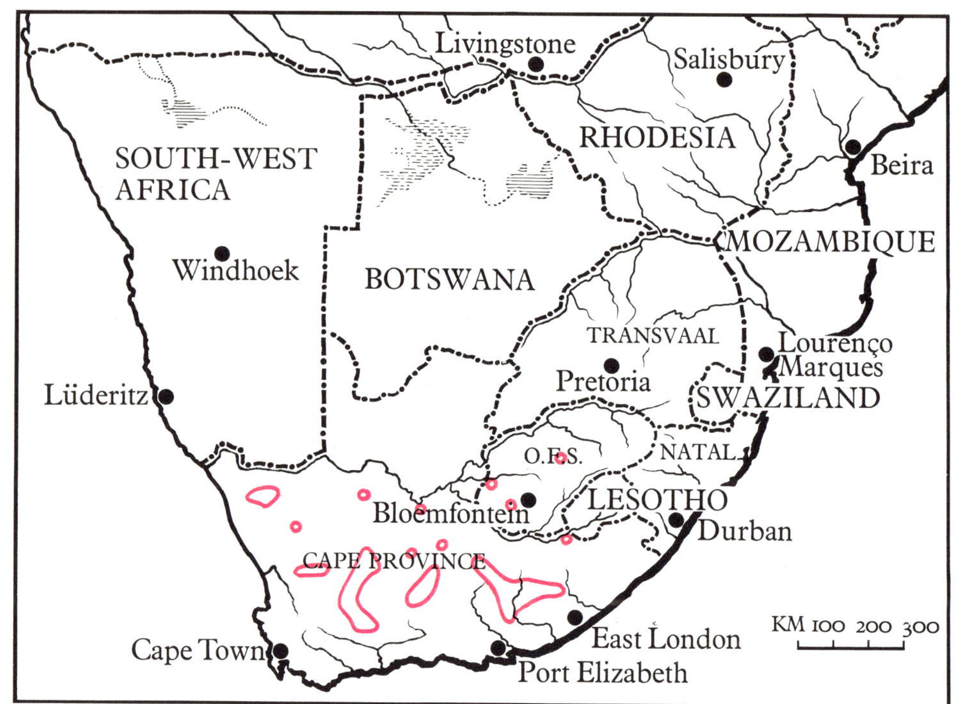

Stomatium

Stomatium Schwantes
From the Greek *stomation* = open
mouth.

Schwantes, Z. Sukk. 2: 175. 1926; *3: 30. 1927.* - v. Pöllnitz, Aufteilung, 68. 1933. - Jacobsen, Sukk. 179; fig. 186. 1933. - Pax, Natürl. Pfl. 219. 1934. - Jacobsen, Succ. Pl. 246; fig. 234-236. 1935; Verzeichnis, 193. 1938. - Jacobsen, Herre, Volk, Mesembr. 70 (in key), 127; fig. 16. 1950. - Phillips, Genera, 322. 1951. - Jacobsen, Handbuch, 1689; fig. 1323-1330. 1955. - Schwantes, Fl. Stones, 139, 339; pl. 45a. 1957. - Jacobsen, Handbook, 953 (in system of Schwantes), 957 (in key of Bolus), 972 (in key of Herre & Volk), 1420; fig. 1594-1601. 1960. - Jacobsen, Lexikon, 516, t.198/5, 6, 1970.

Agnirictus Schwantes, Monatsschrift Kakt. Ges. 2: 21. 1930. - v. Pöllnitz, Aufteilung, 24. 1933. - Pax, Natürl. Pfl. 220. 1934. - Jacobsen, Verzeichnis, 11. 1938.

Perennial, tufted or clumped succulents, small, very short-stemmed, branchlets with 4-6 pairs of leaves. The two leaves of a pair often unequal in length, decussate, crowded, connate and more or less vesicularly inflated below, thickfleshy, shortly triangular or broadly spathulate or elongated lanceolate, semicylindrical in transverse section at the base, more or less keeled towards the apex, the lower surface sometimes drawn forward like a chin, the margins and sometimes the keel more or less armed with a few short, broad teeth or rarely not toothed, epidermis dull, soft, more or less transparently tuberculate, green or grey-green, generally mouth-like, 2-5 times longer than broad, about 4 cm long. Fowers solitary, sessile or short pedicelled, without bracts, yellow or white, opening during afternoon or at night, scented (hyacinth or musklike), often very finely radiate, about 3 cm in diam. Sepals 4-6, vase- or rarely conical to short-cupular to tubularly connate. Petals numerous, in several series, narrow-linear, usually also tubularly connate. Staminas numerous, not papillate. Ovary conical above. Glands distinct, emarginate. Placentas parietal. Stigmas 5-6, short to very short, free at the base or connate and forming a short style, often broadened or thickened (club or button-shaped). Capsule 5-6-locular with deep loculi; expanding keels contiguous at the base, later somewhat diverging, with narrow or broad wings; loculi-roofs reduced to a limb, or well developed; tubercle none. Seeds short pear-shaped, light-brown to brown, with round or irregular shaped warts or tubercles, about 1 mm long.

Species: 40, Cape Province: Namaqualand, Bushmanland, Vanrhynsdorp, Clanwilliam, Karroo, Free State (Jagersfontein, Theunissen) (Type: *S. suaveolens* Schwant.)

B.O.Carter
M. M. Page

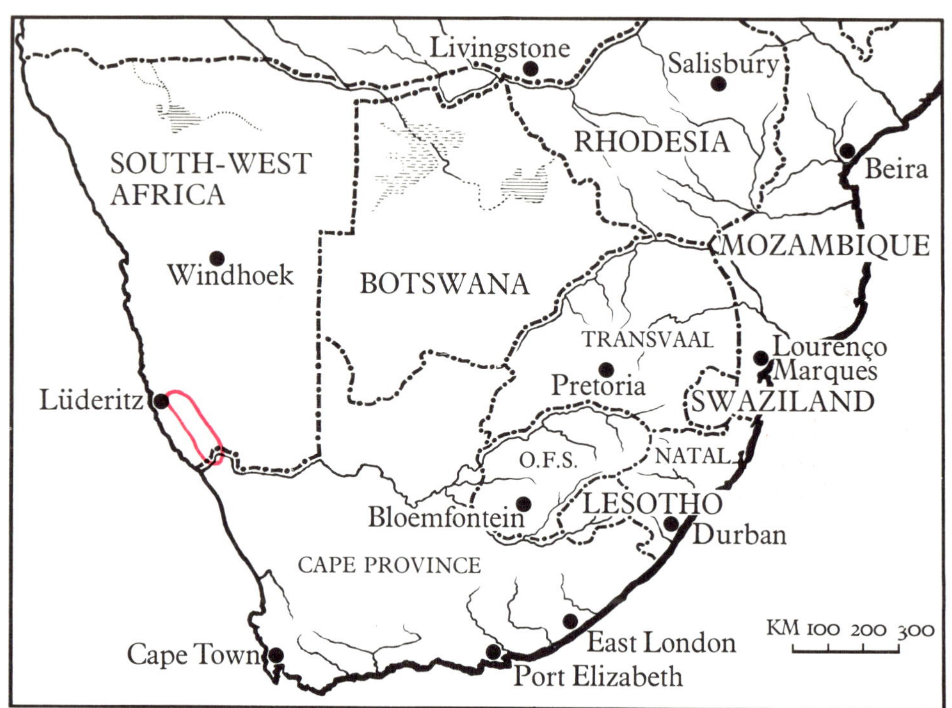

Synaptophyllum

Synaptophyllum N. E. Brown
From the Greek *synaptos* = gnarled
and *phyllos* = leaf.

N. E. Brown, Gard. Chron. 78: 412
(in key). 1925; 83: 254. 1928. - Bolus,
S. Afr. Gard. 19: 49. 1929. - v. Pöllnitz,
Aufteilung, 69. 1933. - Pax, Natürl.
Pfl. 219. 1934. - Jacobsen, Verzeichnis,
195. 1938. - Goossens, Blomplante,
148 (in key). 1940. - Jacobsen, Herre,
Volk, Mesembr. 53 (in key), 127. 1950. -
Phillips, Genera, 322. 1951. - Jacobsen,
Handbuch, 1698. 1955. - Schwantes,
Fl. Stones, 35. 1957. - Jacobsen, Hand-
book, 951 (in system of Schwantes),
956 (in key of Bolus), 965 (in key of
Herre & Volk), 1426; fig. 1602. 1960;
Lexikon, 518, t.199/2. 1970. - Friedrich
in Merxmüller, Prodromus, 129. 1970.

Most complete description: Fedde's,
Rep. 19: 190. 1924.

Annual plant, branching from the base,
glabrous, glossy, finely papillose, the
whole plant fresh purple or salmon red,
stem forked 2 or 3 times, leaves opposite,
sessile, amplexicaul connate at the
base, much broader than thick, flat,
more or less semi-orbicular in outline,
upper leaves more narrowed, fleshy,
about 1,5 cm long and broad. Flowers in
small inflorescences, short-stalked,
white, about 7 mm in diam. Sepals 5,
often 2 leafy and larger than the others.
Petals 3-4-series, posteriors shorter.
Stamens connate at the base with the
petals to about 6 mm. Placentas axile.
Stigmas 4-5, stout, subulate, papillate.
Capsule 4-5-locular, valves with large,
erect, incurved wings; expanding keels
contiguous; without loculi-roofs and
tubercles. Seeds small, either flat or
winged.

Species: 1, Lüderitz, S.W.A. (Type:
S. juttae (Dint. et Berger) N. E. Br.)

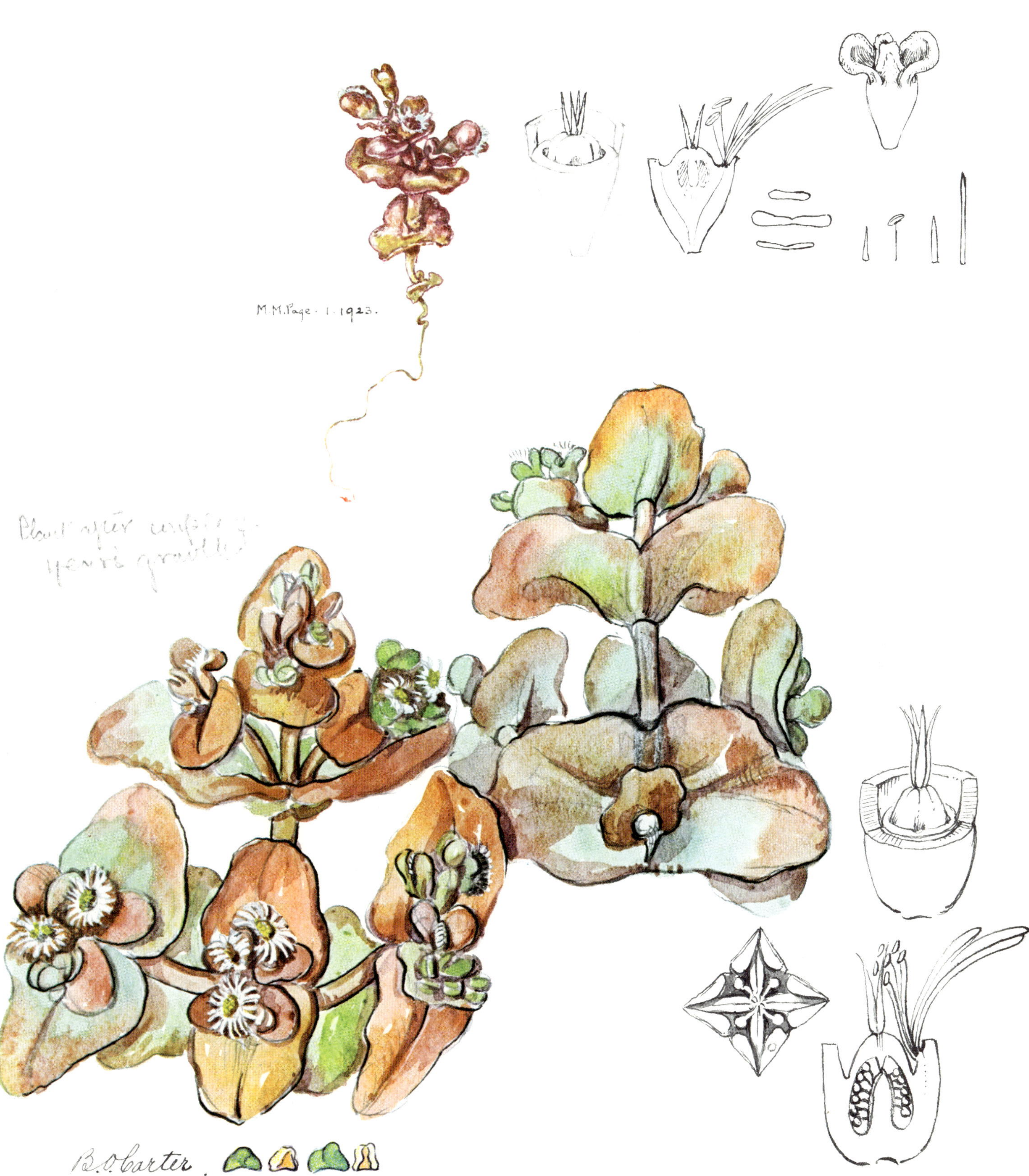

M. M. Page · 1 · 1923.
B. O. Carter.

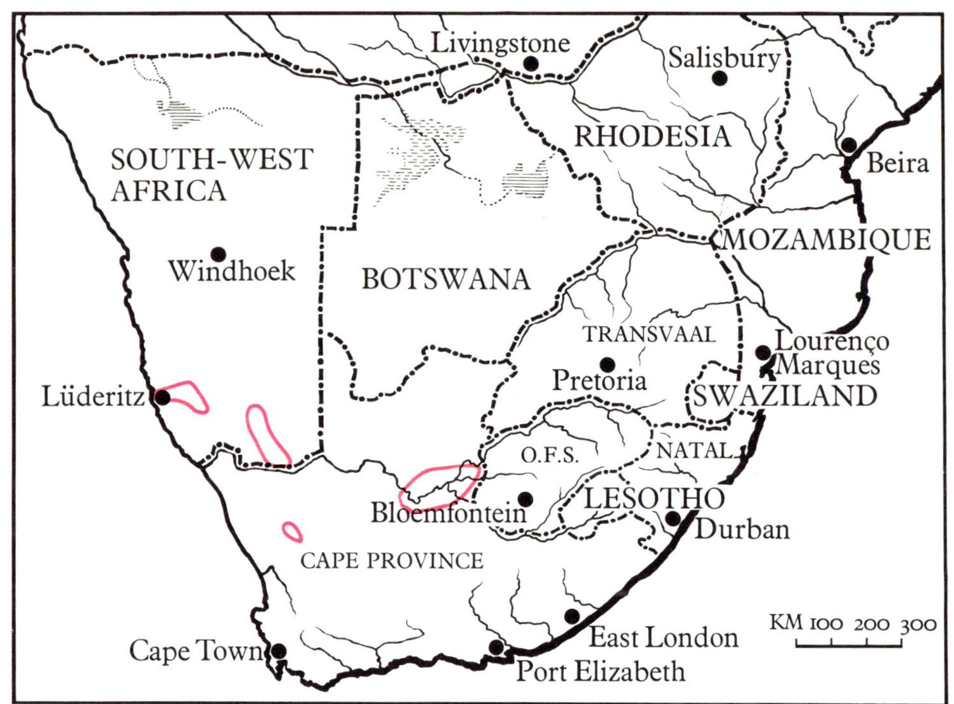

Titanopsis

Titanopsis Schwantes
From the Greek *titanos* = chalk and
opsis = appearance.

Schwantes, Z. Sukk. 2: 178. 1926. -
N. E. Brown, Gard. Chron. 88: 279 (in
key). 1930; 89: 13; fig. 8. 1931. - Labarre (red.), Mesembr. 295; fig. 165.
1931. - v. Pöllnitz, Aufteilung, 69. 1933 -
Jacobsen, Sukk. 180; fig. 187. 1933. -
Pax, Natürl. Pfl. 219. 1934. - Jacobsen,
Succ. Pl. 249; fig. 237, 238. 1935; Verzeichnis, 195. 1938. - Jacobsen, Herre,
Volk, Mesembr. 70 (in key), 127; fig. 14.
1950. - *Phillips, Genera, 318. 1951.* -
Jacobsen, Handbuch, 1699; fig. 1332-
1336. 1965. - Schwantes, Fl. Stones,
158, 339; pl. 48b. 1957. - Jacobsen,
Handbook, 953 (in system of Schwantes),
959 (in key of Bolus), 972 (in key of
Herre & Volk), 1427; fig. 1604-1608.
1960. Lexikon, 518, t.199/3, 4. - 1970. -
Friedrich in Merxmüller, Prodromus,
130. 1970.

Verrucifera N. E. Brown, Gard. Chron.
88: 278 (in key), 513; fig. 213-215. 1930. -
Labarre (red.), fig. 93. 1931. - v. Pöll-
nitz, Aufteilung, 71. 1933. - Jacobsen,
Verzeichnis, 198. 1938. - Goossens,
Blomplante, 144 (in key). 1940;

Short-stemmed, very succulent perennials, with firm, fleshy, sometimes rape-
shaped rootstock, each branchlet with
about 6-8 leaves. Leaves opposite,
crowded in dense rosettes, spathulate,
semi-teretely or trigonously clavate,
covered with raised whitish or tinted
pustules, pustules formed of cells filled
with calcium. Flowers solitary, sessile, or
pedicellate, without bracts, yellow, to
about 2 cm in diam. Sepals 5-6, subequal,
with pustules like the leaves, 3-4 with
membranous margin. Petals, numerous,
1-2-series, free, linear, spreading in
one plane, reflexed. Stamens numerous,
loosely erect or collected into a cone,
filaments sometimes bearded at the base.
Ovary inferior or 1½- or more than ½-
superior, flattish or conical at the top.
Glands annular, sometimes crenulated.
Placentas on the outer walls or on the
floor of the ovary-chambers. Stigmas 5-6,
or 10 in one species, filiform or subulate.
Capsule 5-6, in one species 10-locular,
broadly and shortly obconic, flat or
slightly convex on the top with prominent sutures; expanding keels rather
stout and subcontiguous at their basal
half, then widely diverging, minutely

toothed on the inner edge, excurrent at
the margin about half-way up the valve
and ending in an awn reaching to the
tip of the valve, with a membranous margin to the part below the awn; loculi
roofed with membranous semitrans
parent loculi-wings; tubercle absent.
Seeds many in a loculus, subglobose,
with a small point, very minutely tuberculate.

Species: 6, southern part of South
West Africa, Bushmanland, Prieska,
Hopetown and Kimberley. (Type:
T. calcarea (Marl.) Schwantes.)

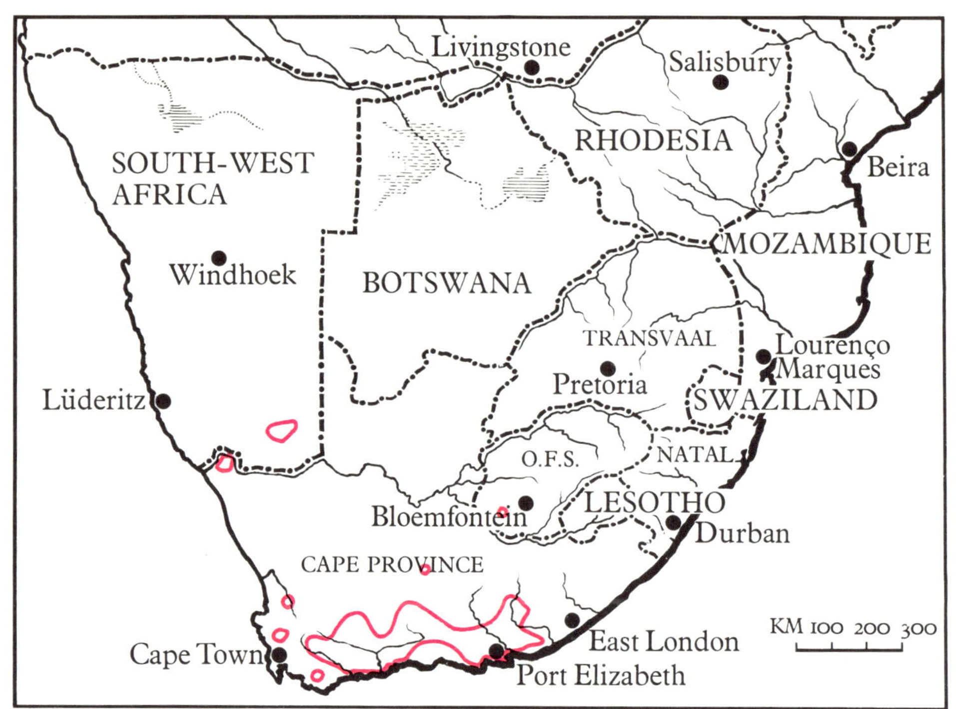

Trichodiadema

Trichodiadema Schwantes
From the Greek *trix* = hair and
diadema = crown.

Schwantes, Z. Sukk. 2: 187. 1926. -
Bolus, Notes, 57, 61, 63; fig. pl. 1 (p. 57).
1928. - N. E. Brown, Gard. Chron. 87:
72 (in key). 1930. - v. Pöllnitz, Aufteilung, 70. 1933. - Jacobsen, Sukk. 182,
fig. 189. 1933. - Pax, Natürl. Pfl. 220.
1934. - Jacobsen, Succ. Pl. 252; fig. 240.
1935; Verzeichnis, 196. 1938. - Goossens, Blomplante, 144 (in key), 1940. -
Jacobsen, Herre, Volk, Mesembr. 62
(in key), 127. 1950. - Phillips, Genera,
322. 1951. - Jacobsen, Handbuch, 1704;
fig. 1337-1341. 1955. - Schwantes, Fl.
Stones, 86, 338; pl. 22a & b. 1957. -
Jacobsen, Handbook, 952 (in system of
Schwantes), 961 (in key of Bolus), 969
(in key of Herre & Volk), 1430; fig. 1609-
1612. 1960. - Bolus, J. S. Afr. Bot. 29:
17 (with key of 6 subgroups). 1963;
32: 234 (improved key of groups and
subgroups). 1966; Jacobsen, Lexikon,
519, t.199/5, 6. 200/2. 1970 - Friedrich in
Merxmüller, Prodromus, 133. 1970.

Perennial shrubs with long, slender,
arched branches or very short-stemmed
and of tufted habit; roots woody or
tuberous. Leaves opposite, slightly connate, semi-cylindrical - cylindrical, glistening with papillae on the surface, the
tip usually with a cluster of more or
less spreading bristles (the "diadem") to
about 3 cm long. Flowers solitary,
short-stalked, white, light to dark-
violet-red, to about 5 cm in diam. Sepals
5-8, papillate and also with a fringe of
bristles at the tips. Petals 1-several
series, linear to lanceolate. Stamens
column-like or conically collected, sometimes hairy at the base; staminodes
present. Ovary flat or convex above.
Placentas parietal. Stigmas 5-8, lanceolate, shorter than the stamens. Capsule
5-8-locular; expanding keels contiguous, towards the points somewhat diverging, wings broad and roundish;
loculi-roofs present or reduced to a limb;
tubercle none. Seeds pear-shaped,
brown or yellowish, with minute warts
and small grooves.

Species: 30, South West Africa,
Namaqualand, Karroo, Little Karroo,
Eastern Cape Province. (Type:
T. barbatum (L.) N. E. Brown.) Abys-
sinia (1 species).

Trichodiadema marlothii   303

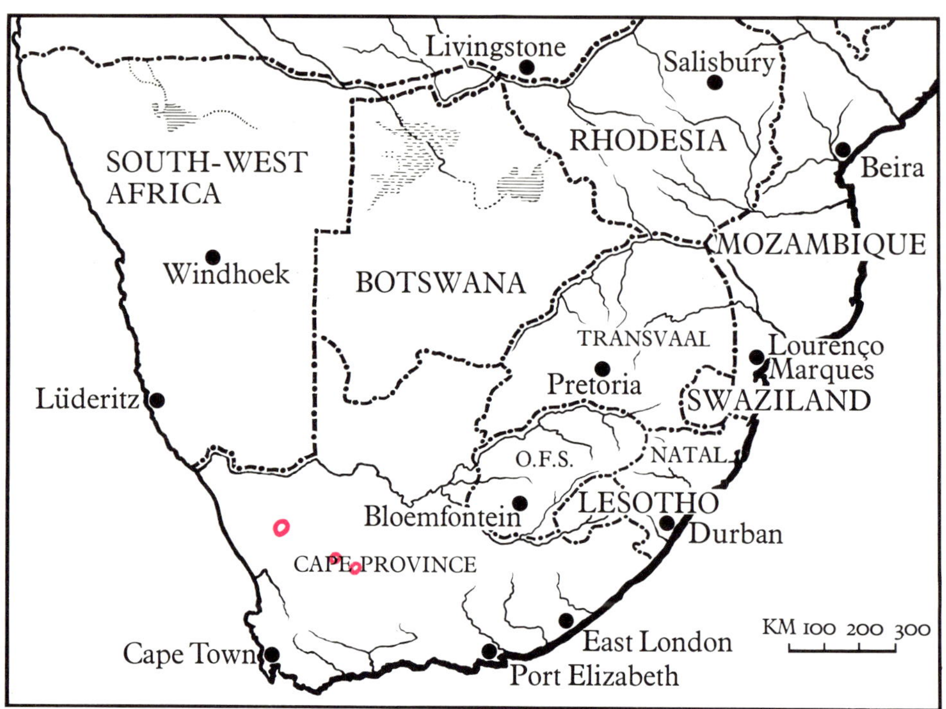

Vanheerdea

Vanheerdea L. Bolus
Named in honour of Mr P. van Heerde, Springbok.

Bolus, Notes, 136. 1938. - Jacobsen, Herre, Volk, Mesembr. 78 (in key), 128. 1950. - *Phillips, Genera, 318, 1951.* - Jacobsen, Handbuch, 1710; fig. 1341. 1955. - Schwantes, Fl. Stones, 111, 338; pl. 31b. 1957. - Jacobsen, Handbook, 952 (in system of Schwantes), 959 (in key of Bolus), 975 (in key of Herre & Volk), 1435 (with key of species); fig. 1613-1615. 1960; Lexikon, 521, t.200/3, 4, 1970.

Rimaria N. E. Brown in L. Bolus, Notes, 67, 68 (with key of species); pl. 10, 13. 1937.

Stemless succulent perennials with the growths usually crowded on the top of a deeply descending rootstock, rarely solitary. Growths subglobose or compressed-ovoid formed of two (or, when forming a new growth, four) hemispherical or thick ovate equal or unequal leaves united for half or two-thirds of their length, and their free parts pressed closely together, entire or with minute teeth on their edges and sometimes along a slight keel on the top, in one species with a distinct window on top, smooth, glabrous or puberulous with microscopic soft points, green or whitish or dove-grey, often brownish, without dots, about to 7 cm long and 5 cm in diam. Flowers solitary or up to 3 together, or at least the ovary exserted and bearing a pair of bracts included in the body of the growth, pedicels with smaller bracteoles, bright yellow to orange-yellow, to about 5 cm in diam. Sepals 5-9, free down to their union with the ovary, very broad at the base, unequal with rounded apex, some with narrow membranous margin. Petals, in about 3 series, numerous, free, linear, anteriors with rounded apex, interiors emarginate. Stamens numerous, erect, either conically or column-like collected, filaments of the posteriors papillate at the base, staminodes none. Ovary slightly convex, inferior. Glands rather obscurely crenulated. Placentas parietal, on the floor of the chambers. Stigmas 7-15, erect, filiform, papillate to the apex; style none. Capsule 7-15-locular, shallow, flattish on top; valves when wetted widely spreading or more or less reflexed; expanding keels more or less diverging, fringed, with broad membranous wings; loculi roofed with flat flexible membranous, semitransparent loculus-wings, only in *V. primosii* reduced to a limb and narrow; tubercle none. Seeds many in each loculus, very small, subglobose, smooth, with a nipple at one end, deep brown.

Species: 4, Bushmanland, Vanrhynsdorp district. (Type: *V. roodiae* (N. E. Br.) L. Bol.)

magnified.
Inside of leaf magnified.
seed wings
B. Carter. 10—1931.
11—1931.

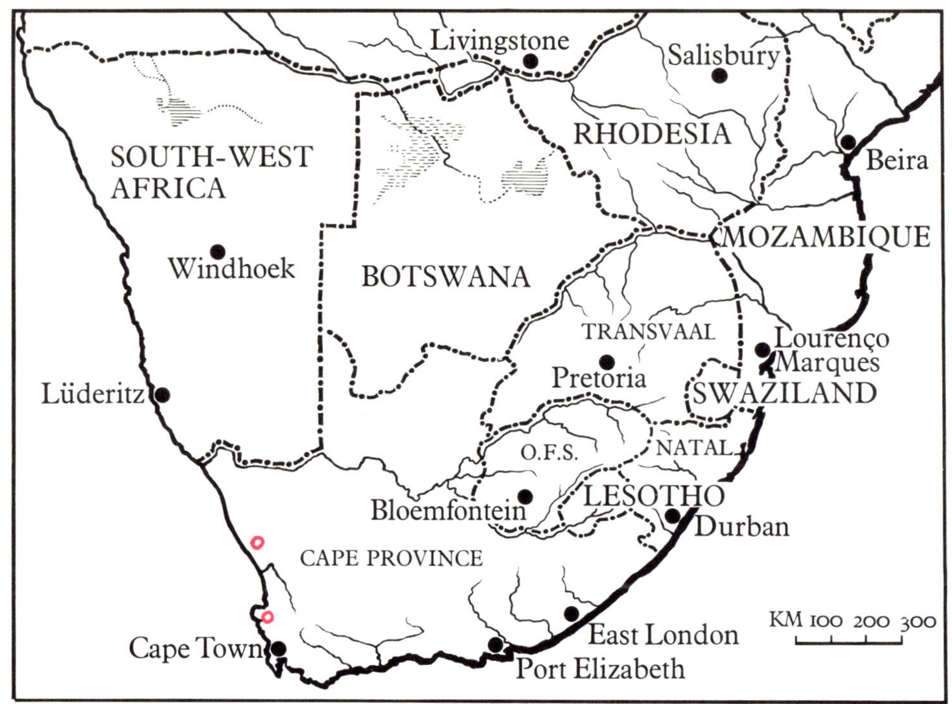

Vanzijlia

Vanzijlia L. Bolus
Named in honour of the late Mrs
Dorothy van Zijl, wife of Judge-President
H. S. van Zijl.

Bolus, Fl. Pl. S. Afr., 7: pl. 262 (coloured). 1927. - Phillips, Genera, 247. 1926. - N. E. Brown, Kew Bull. 61. 1929; Gard. Chron. 87: 71 (in key). 1930. - v. Pöllnitz, Aufteilung, 71. 1933. - Pax, Natürl. Pfl. 220. 1934. - Jacobsen, Verzeichnis, 198. 1938. - Goossens, Blomplante, 145 (in key). 1940. - Jacobsen, Herre, Volk, Mesembr. 68 (in key), 128. 1950. - Phillips, Genera, 322. 1951. - Jacobsen, Handbuch, 1711; fig. 1342. 1955. - Schwantes, Fl. Stones, 338. 1957. - Jacobsen, Handbook, 952 (in system of Schwantes), 963 (in key of Bolus), 971 (in key of Herre & Volk), 1437; fig. 1616. 1960; Lexikon, 522, t.200/1. 1970.

Low glabrous succulent shrubs, branches wiry, prostrate or creeping. Leaves opposite, leaf pairs dimorphic: (1) pair of leaves small, oblong or spherical, nearly completely connate, with or without small free points of the leaves. (2) pair of leaves longer, formed of more or less semi-cylindrical to cylindrical leaves, only connate at the base and here often inflated, smooth, dotted or not dotted, about 14 mm long and 4 mm in diam. Flowers solitary, pedicelled or sessile, white or pink, up to 6 cm in diam. Sepals 5, green, with membranous margin. Petals 2-series, linear. Stamens conically collected, concealing the stigmas, surrounded by 2-3 series of staminodes. Ovary flat above, placenta parietal. Stigmas 10, subulate, erect. Capsule 10-locular; expanding keels diverging, with wings; loculi-roofs present; tubercle present. Seeds: smooth with nipple.

Species: 3, Cape Province: from Lamberts Bay to Elandsbay and Hoedjies Bay as also in the Karroo Mts. of the district Vanrhynsdorp. (Type: *V. annulata* (Berger) L. Bol.)

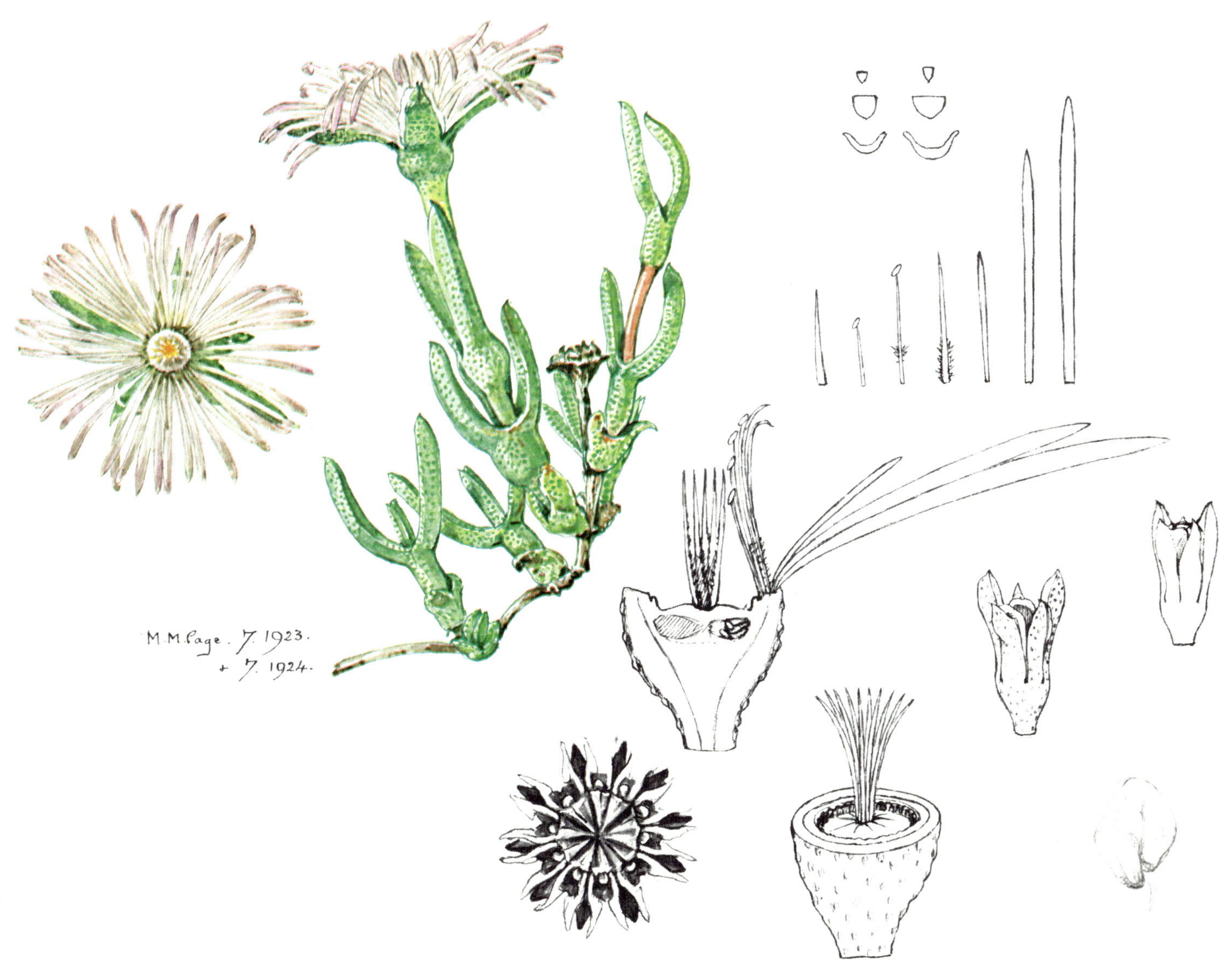

Vanzijlia annulata   307

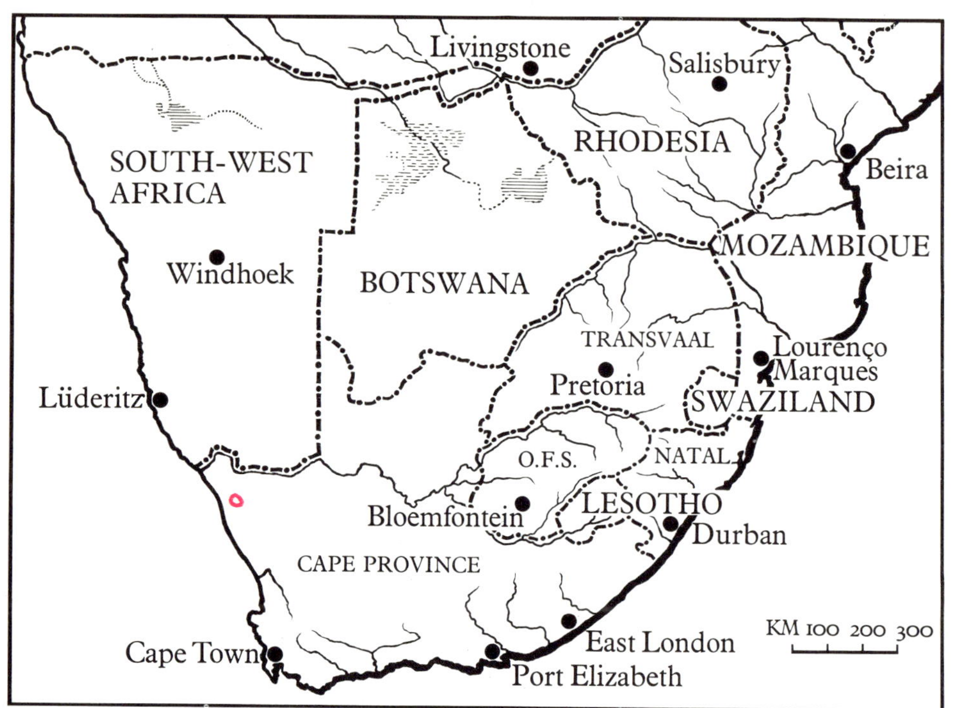

Wooleya

Wooleya L. Bolus
Named in honour of
Major C. H. F. Wooley.

Bolus, J. S. Afr. Bot. 27: 48. 1961. -
Jacobsen, Lexikon, 522. 1970.

Namaquanthus farinosus L. Bolus, in
Notes, 334. 1958.

Erect, compact, stout shrub, vegetative
parts ash-grey, pruinose. Leaves erect or
spreading, slightly convex above, roundish below, keel reduced to a line,
laterally convex, blunt, up to 3,5 cm
long and 9 mm in diam. Flowers solitary, pedicel obscurely 4-angulate,
short, with bracts at the base, white.
Sepals 4, the outer ones broadly egg-
shaped, the inner ones without membranous margin. Petals 2-series. Stamens erect, numerous, outer filaments
from the middle to the base ciliate,
the inner ones bearded nearly up to
the middle; staminodes only a few.
Ovary concave at the margin, slowly
rising to the centre, where it is slightly
convex. Disc annular, finely crenulate.
Placentas parietal. Stigmas 11-12, subulate. Capsule 11-12-locular, obovate,
top convex with compressed sutures.
Valves when wetted spreading, with
wings, wings above slightly connate
with the expanding keels, touching each
other at the points; expanding keels
parallel, towards the apex often slightly
diverging; loculi-roofs well developed;
tubercle none. Seeds smooth.

Species: 1, Namaqualand: Hondeklip
Bay and Kleinzee. (Type: *W. farinosa*
(L. Bolus) L. Bolus.)

Wooleya sp.   309

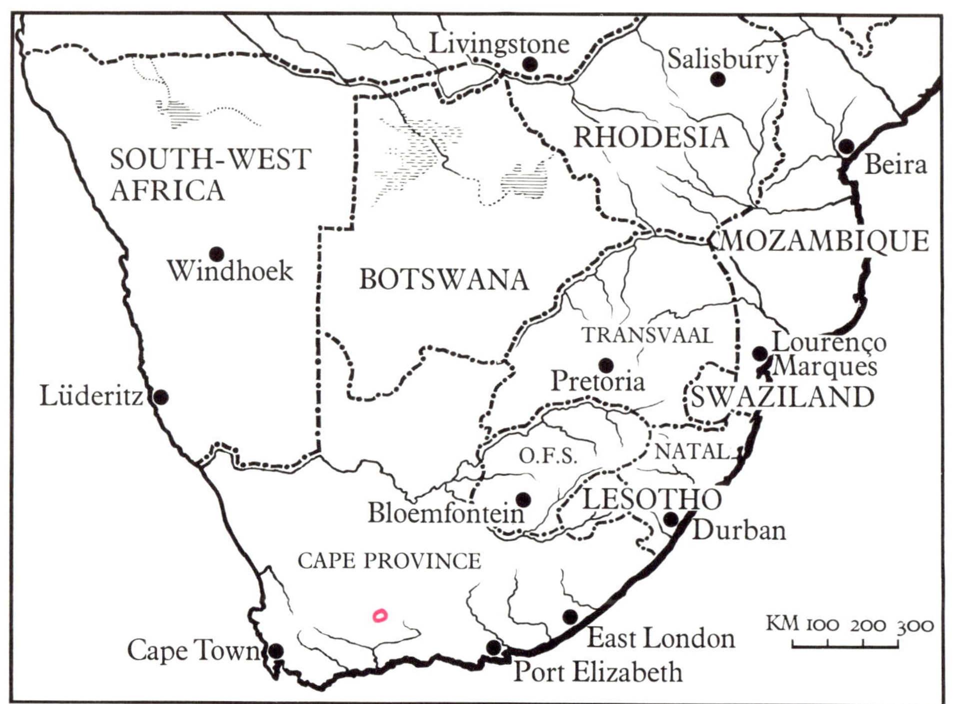

Zeuktophyllum

Zeuktophyllum N. E. Brown
From the Greek *zeuktos* = joint, connect
and *phyllon* = leaf.

N. E. Brown, Gard. Chron. 81: 12 (in key). 1927. - v, Pöllnitz, Aufteilung, 71. 1933. - Pax, Natürl. Pfl. 220. 1934. - Jacobsen, Verzeichnis, 198. 1938. - Jacobsen, Herre, Volk, Mesembr. 68 (in key), 128. 1950. - Phillips, Genera, 222. 1951. - Jacobsen, Handbuch, 1712. 1955. - Schwantes, Fl. Stones, 342. 1957. - Jacobsen, Handbook, 954 (in system of Schwantes), 958 (in key of Bolus), 971 (in key of Herre & Volk), 1437; fig. 1617. 1960; Lexikon, 522, t.200/5. 1970.

Mesembr. suppositum L. Bolus *in Ann. Bol. Herb. 4: 1. 1925.*

Robust, glabrous, small shrublet with woody stem, up to 10 cm high. Branches ascending, densely 4-6-leaved, old leaves persistent. Leaves ascending, for 2-5 mm united at the base, swollen, 3-angled, somewhat compressed laterally, obtusely keeled, not dotted, without teeth, glabrous, 2-3 cm long, 5-10 mm wide, 12 mm thick, bluish purple. Flower solitary, ca. 3 cm in diam., with a short stalk nearly enclosed into the two bracts. Calyx 5-lobed with a turbinate tube, lobes unequal, oblong - ovate, 2 with membranous margins, up to 1,5 cm long. Petala numerous, linear at the base, attenuate, yellowish-pink, 1,5 cm long, nearly 1 mm broad, free, changing to staminodes. Staminas numerous, in many series, erect, distinct, filaments of the inner ones shorter, hairy, hairs exceedingly long; anthers and pollen aureus to nearly albidus. Disc annular, crenulated. Ovarium top convex. Stigmas 10, linear to filiform,

acuminate, 25 mm long. Capsule 10-locular, very flat, not thin; expanding keels very broad, contiguous, covering the basal parts of the valves, convex above, with no sharp edges, ending with a small, stiff, obtuse wing; without loculi-roofs and tubercles. Seeds smooth, compressed, 3-angled in outline.

Species: 1, from the Little Karroo, Riversdale, div. Phisantefontein and Bakoond. (Type: *Z. suppositum* (L. Bolus) N. E. Brown.)

R. DARROLL '71.

Bibliography

Adamson, R. S. and Salter, T. M.: *Flora of the Cape Peninsula.* 1950.

Berger, A.: *Mesembrianthemen und Portulacaceen.* Eugen Ulmer, Stuttgart. 1908.

Boer, H. W. de and Boom, B. K.: Lithops, An analytical key for the Genus. *Nat. Cact. & Succ. Journal*, vol. 19/3, p. 34, Sept. 1964; as also Mededeling no. 239, Instituut voor de veredeling van tuinbougewassen. Wageningen, Holland. April 1965.

Bolus, H. M. L.: Novitates africanae, in *Annals of the Bolus Herbarium*, Cambridge University Press, Fetter Lane, London. Vol. 1, 1914-15; vol. 2, 1916–18; vol. 3, 1920–24; vol. 4, 1925–28.

Notes on Mesembryanthemum and allied Genera, University of Cape Town, Bolus Herbarium. Vol. 1, 1928; vol. 2, 1935; vol. 3, 1958.

Notes on Mesembryanthemum and allied Genera in *South African Gardening and Country Life*, The Specialty Press, Wynberg, Cape. Vol. XVII, 1927 et seq.

Brown, N. E.: New and old species of Mesembryanthemum with critical Notes. *Journal of the Linnean Society: Botany*, vol. 45, July 1920. London.

Burtt-Davy, J.: *A Manual of the Flowering Plants and Ferns of the Transvaal with Swaziland.* Parts 1 & 2. London, 1926.

Cactussen en Vetplanten, Maandblad van de Nederlandsche Vereeniging van Liefhebbers van Cactussen en Vetplanten. Amsterdam, vol. 1, 1935 et seq.

Deutsche Kakteen-Gesellschaft, Berlin.
Monatsschrift für Kakteenkunde, vol. 1, 1891 et seq.
Zeitschrift für Sukkulentenkunde. Bd. 1, 1923/24; Bd. 2, 1925/26; Bd. 3, 1927/28.
Monatsschrift der Deutschen Kakteen-Gesellschaft. Bd. 1, 1929; Bd. 2, 1930; Bd. 3, 1931; Bd. 4, 1932.
Kakteenkunde. 1933, 1934, 1935, 1936.
Jahrbuch. Bd. 1. Juli 1935 – Dez. 1936.
Kakteen und Andere Sukkulenten. 1937–1966 et seq.

Dillenius, J.: *Hortus Elthamensis* seu plantarum rariorum, quas in horto suo Elthami in Cantio coluit J. Sherard. Delineationes et descriptiones. 1732.

Dinter, K.: *Neue und wenig bekannte Pflanzen in Südwest-Afrika.* Okahandja, 1914.
Botanische Reisen in Südwest-Afrika. Posen, 1918.
Botanische Reisen in Südwest-Afrika in Fedde's Repertorium, Beihefte, vol. 3, 1921.
Sukkulentenforschungen, I. in l.c. vol. 23, 1923; II. Herrnhut (Saxonia) 1928.
Beiträge zur Flora von Südwest-Afrika, I & II in l.c. 19, 1923.

Südwestafrikanische Lithops-Arten, in Index, R. Graessner, Horticulture, Perleberg, 1928.
Diagnosen südwestafrikanischer Pflanzen, Fedde's Repertorium, 29, 1931.

Fedde's *Repertorium specierum novarum regni vegetabilis.* Berlin, vol. 1, 1905 et seq.

Flowering Plants of South Africa. Edited by J. B. Pole-Evans. London & Cape Town, vol. 2, 1922; vol. 3, 1923; vol. 7, 1927.

Gardeners' Chronicle, London, 1920 et seq.

Gartenwelt, Verlag Paul Parey. Berlin–Hamburg, vol. 41, 1926 et seq.

Goossens, A. P.: *Suid-Afrikaanse Blomplante, Sleutel tot die families en geslagte.* Johannesburg, 1940.

Harvey, Sonder, Thiselton, Dyer: *Flora Capensis.* Dublin, 1859–1920.

Haworth, A.: *Observations on the genus Mesembryanthemum*, 2 vols. London, 1795.
Miscellania naturalia. London, 1803.
Synopsis plantarum succulentarum. London, 1812.
Supplementum. London, 1819.
Revisiones plantarum succulentarum. London, 1821.
Andr. Bot. Rep. XI, 1797–1811.

Hooker, W. J.: *Icones plantarum.*

Hortus Academicus Lugduno-Batavensis. 1587–1937. Haarlem, 1938.

Ihlenfeldt, H.-D.: Entwicklungsgeschichtliche, morphologische und systematische Untersuchungen an Mesembryanthemen. Dissertation. Kiel, 1958.
Gedruckt: Fedde's Repertorium specierum novarum regni vegetabilis. Bd. 63. 1960.

Ihlenfeldt, H.-D. und Straka, H.: Aufteilung der alten Aizoaceae, Berichte der Deutschen Botanischen Gesellschaft, Bd. 74: 485–492. 1962.

Jacobsen, H.: *Die Sukkulenten.* 1933.
Succulent Plants. Williams & Norgate. 1935.
Index specierum generis Mesembryanthemi L. cum separatis generibus, in Fedde's Repertorium specierum novarum regni vegetabilis, Beihefte. Vol. 106, 1938, with 3 supplements.
Handbuch der Sukkulenten Pflanzen. VEB. Gustav Fischer, Verlag, Jena, 1955. 3 Bände; Band no. 3: Mesembryanthemum.
A Handbook of Succulent Plants, 3 vols.; vol. 3: Mesembryanthemums (Ficoidaceae). Blandford Press. London, 1960.
Das Sukkulentenlexikon. VEB Gustav Fischer Verlag. Jena, 1970.

Jacobsen, H., Herre, H. and Volk, O. H.: *Mesembryanthemaceae.* Ludwigsburg. 1950.

Jacquin, N. J.: *Plantarum rariorum Horti Schoenbrunnensis, descriptiones et icones.* Vienna. 4 vols. 1797–1804.

Journal of South African Botany, Kirstenbosch, vol. 18/19, 1952/53 et seq.

Krainz, H.: Jahrbücher der Schweizerischen Kakteen-Gesellschaft: "Sukkulentenkunde", No. 1. 1947 et seq.

Krainz, H. and Roshardt, Pia: Sukkulenten. Silva Verlag, Zürich.

Labarre, E. J., N. E. Brown, Dr A. Tischer, M. C. Karsten: *Mesembryanthema*, L. Reeve & Co. Kent, England. 1931.

Linné, C.: *Species plantarum.* Uppsala, 1753.

Marloth, R.: *Das Kapland, insonderheit das Reich der Kapflora, das Waldgebiet und die Karroo.* Jena, 1908.
The Flora of South Africa. Vol. 1–4. 1915.

Merxmüller, H.: *Prodromus einer Flora von Südwestafrika*, 27. Aizoaceae, bearbeitet von H. Chr. Friedrich. Verlag von J. Cramer, 3301 Lehre. (August) 1970.

Mitteilungen aus dem Institut für Allgemeine Botanik in Hamburg. 8 Band. Heft 1. 1929

Mitteilungen der Botanischen Staatssammlung München, No. 3, 1951 et seq.

Möller's Deutsche Gärtnerzeitung, vol. 41, 1926 et seq.

Nel, G. C.: *The Gibbaeum Handbook.* Edited by P. G. Jordaan and E. W. Shurly. Blandford Press. London, 1953.
Lithops. 1947.

Pax, F. and Hoffmann, K.: *Aizoaceae-Mesembryanthemum* in A. Engler: *Die Natürlichen Pflanzenfamilien.* Second Edition, vol. 16c, 1934.

Phillips, E. P.: *The Genera of South African Flowering Plants.* Cape Town. 1926. Second Edition. 1951.

Pöllnitz, K. v.: *Die Aufteilung der Gattung Mesembryanthemum.* L. Fedde's Repertorium, vol. 32, 1933.

Rauh, W.: *Die grossartige Welt der Sukkulenten.* Verlag Paul Parey, Hamburg and Berlin. 1966.

Rawe, R.: *Succulents in the veld.* H. Timmins. Cape Town, 1968.

Rendle: *Journal of Botany, British and Foreign.* Vol. 65. London, 1927 et seq.

Ryutanji, Y.: *A Color Photo Album of Cacti & Succulents.* 3 vols. Tokyo.

Salm, J. M.: *Monographia generum Aloes et Mesembryanthemi.* Düsseldorf. 1836–1849.

Schwantes, G.: Die systematische Gliederung der Familie Mesembryanthemaceae in *Sukkulentenkunde*, no. 1. of Schweiz. Kakt. Ges. 1947.
Die Früchte der Mesembryanthemaceae in Mitteilungen aus dem Botanischen Museum der Universität Zürich, no. 193, 1952.
Flowering Stones and Midday Flowers. Ernest Benn Ltd. London, 1957.
His many publications about Mesembryanthema are listed in *The Cultivation of the Mesembryanthemaceae*, Blandford Press, 1953, where 107 works are cited between 1916 and 1952.

Straka, H.: Anatomie und entwicklungsgeschichtliche Untersuchungen an Früchten paraspermer Mesembryanthemen in Nova Acta Leopoldina, vol. 17. Leipzig, 1955.

Succulenta, Maandblad van de Nederlands–Belgische Vereeniging van Liefhebbers van Cactussen en Andere Vetplanten, tot 46ste Jaargang, 1967.

Thunberg, C. P: *Flora capensis, plantas promontori Bonae Spei Afric.* Edited Schultes, 1823.
Prodromus plantarum capensium in Promontori Bonae Spei Afric. annis 1772–75 collectarum. Uppsala, 1794–1800.
Nov. Act. Nat. Cur., VIII, 1791.

Transactions of the South African Philosophical Society. Cape Town. Vol. 18, 1907.

Transactions of the Royal Society of South Africa. Cape Town. Vol. 1, 1909/10; vol. 2, 1910/12; vol. 4, 1914/15.

Wulff, H. H. D.: Die Polysomatie des Wurzelperiblems der Aizoaceae in *Berichte der Deutschen Botanischen Gesellschaft*, Berlin, vol. 57, part 6, 1940.
Untersuchungen zur Zytologie und Systematik der Aizoaceae-Subtribus Gibbaeinae, Schwantes, in *Botanisches Archiv.* Leipzig, vol. 45, 1944.

Glossary *compiled by Mary Lavis*

Accrescent: enlarged and persistent

Acerose: needle-shaped

Achene: a one-seeded small dry indehiscent fruit

Acies: the edge (of stem)

Aculeate: prickly

Acuminate: tapering to a point in hollow curves

Acute: tapering to a point in straight lines

Adnate: united with another body or series

Alternate: applied to leaves inserted individually at different levels along the branch

Annual: a plant which completes its full cycle in a year

Annular: arranged in a ring

Anther: the part of the stamen which contains the pollen

Awn: a bristle-like appendage

Axil: the angle between a leaf and a branch

Axile: used of the attachment of ovules to the axis or to the inner angle of the chambers of the chambered ovary

Axillary: arising from the axil

Appressed: flattened down

Argenteus: silver

Aril: an extra coat to the seed

Barbatus: bearded

Basal: used of the attachment of ovules to the base of the ovary

Biennial: a plant which completes its full life-cycle in two years

Bifid: divided into two parts

Bisexual: having both stamens and pistil in the same flower

Bract: a modified leaf found below an inflorescence or lower down on the peduncle

Bracteole: a modified leaf or scale immediately below a flower

Calyx: the outer envelope of the flower, consisting of a number of sepals which may be free or joined together, and fleshy, leaf-like or partly membranous

Canaliculate: channelled (used of leaves)

Capsule: a dry fruit consisting of 2 or more carpels that, at maturity, let the seeds fall out

Carina: a keel

Carpel: a modified leaf bearing the ovules; one more, either free or united, forming the female organ of the flower.

Connate: joined

Cyme: an inflorescence in which each successive branch ends in a flower, the branching being continued from below

Decumbent: reclining but with the summit ascending

Dehiscent: opening spontaneously when ripe; most capsules in Mesembryanthemaceae open when wet

Dimorphic: of two forms

Dissepiment: septum

Dissected: deeply divided

Emarginate: with a notch at the apex

Entire: with an even margin

Embryo: the rudimentary plant still enclosed in the seed

Exserted: projecting beyond the tube of the calyx or corolla

Filament: the stalk of the stamen

Fimbriate: fringed

Foliaceous: leaf-like, leaf-bearing

Free: not united (sepals, petals, etc.)

Free central: used of the attachment of ovules to an outgrowth arising from the floor of the ovary

Glabrous: devoid of hairs (smooth)

Gland: an organ secreting fluid; nectary

Glaucous: sea-green, covered with a bloom like a plum, giving a greyish appearance

Imbricate: overlapping

Incanus: hoary white

Incised: notched at the margin

Included: not projecting

Indehiscent: not opening naturally

Innate: joined to the filaments by its base

Internode: the portion of a stem between 2 nodes

Lacerate: deeply and irregularly divided

Linear: long and narrow with parallel edges

Mericarp: the one-seeded portion of a fruit (schizocarp) which splits away as a perfect fruit

Multiflora: many-flowered

Mutabilis: changeable

Muticus: blunt

Niveus: snow-white

Node: the point on the stem at which one or more leaves are borne

Nut: a one-seeded fruit with a hard shell, does not split open when ripe

Nutans: nodding

Opposite: applied to leaves arising in pairs opposite one another at the nodes

Ovary: the part of the pistil which contains the ovules

Ovule: a small body contained in the ovary or on the carpel, it ultimately becomes a seed

Ovate: about twice as long as broad, tapering to the tip

Ovoid: oval

Parietal: used of the attachment of ovules to the outer wall in an ovary

Pedicle: stalk of an individual flower

Peduncle: stalk of an inflorescence

Perennial: a plant which lasts several years

Petiole: the stalk of a leaf

Placenta: the cushion-like outgrowth to which the ovules are attached

Plumose: feathered

Procumbent: lying along the ground

Puberulous: minutely pubescent

Pubescent: covered with fine soft hairs, downy

Pulvinate: cushion-shaped

Pulvinus: a swollen joint

Punctate: dotted

Receptacle: the expanded top of the flower-stalk on which the parts of the flower or the florets are inserted

Saccate: pouched

Scabrous: rough to the touch

Schizocarp: a fruit which splits into a number of one-seeded carpels

Seed: the product of the ovule after fertilization

Sessile: not stalked

Shoot: a young branch or twig

Staminode: an aborted or vestigial stamen (may be petaloid or forming a nectary)

Stamen: the male part of the flower consisting of anther (bearing pollen) and filament

Stigma: the part of pistil which receives the pollen

Stipules: leaf-like or scale-like appendages of the leaf

Style: the part of the pistil lying between the stigmas and the ovary

Terete: circular in transverse section

Trimorphic: of three forms

Turbinate: cone-shaped

Type specimen: that from which the original description of a species was drawn

Vaginate: sheathed

Valve: one of the parts produced by the splitting of a capsule when ripe

Vulgaris: common

Xerophyte: a plant capable of existing with a small water supply

BIBLIOGRAPHY

J. C. Willis: *Flowering Plants & Ferns*
M. R. Levyns: *Guide to Flora of the Cape Peninsula*

Index to Descriptions of the Genera of the Mesembryanthemaceae